Hunting the Largest Animals

Hunting the Largest Animals

Native Whaling in the Western Arctic and Subarctic

Allen P. McCartney

Editor

Studies in Whaling No. 3
Occasional Publication No. 36
© 1995 The Canadian Circumpolar Institute
University of Alberta

Canadian Cataloguing in Publication Data

Main entry under title:

Hunting the largest animals

(Studies in whaling ; no. 3) (Occasional publication ; no. 36)
Includes bibliographical references.
ISBN 0-919058-95-7
ISSN 0838-133X-36

1. Whaling—Arctic Regions.   2. Eskimos—Hunting.
3. Eskimos—Social life and customs.   I. McCartney, Allen P.
II. Canadian Circumpolar Institute.   III. Series in whaling ; no. 3.
IV. Series: Occasional publication series
(Canadian Circumpolar Institute) ; no. 36.
E99.E7H86 1995   639.2'8'091644   C95-910612-X

© 1995 The Canadian Circumpolar Institute
Cover Design by Art Design Printing Inc.
Printed by Art Design Printing Inc., Edmonton, Alberta, Canada

(This book was made possible with funds received from the Japan Small-Type Whaling Association)

Table of Contents

Editor's Note vii

Introduction ix
    Roger K. Harritt, Carol Zane Jolles, and Allen P. McCartney

The Archaeological Imagination, Zooarchaeological Data, the
Origins of Whaling in the Western Arctic, and "Old Whaling"
and Choris Cultures 1
    Owen K. Mason and S. Craig Gerlach

The Development and Spread of the Whale Hunting Complex in
Bering Strait: Retrospective and Prospects 33
    Roger K. Harritt

Whale Traps on the North Pacific? 51
    Don E. Dumond

Prehistoric Use of Cetacean Species in the Northern Gulf of Alaska 63
    Linda Finn Yarborough

Whale Size Selection by Precontact Hunters of the North American
Western Arctic and Subarctic 83
    Allen P. McCartney

Prehistoric Beluga Whale Hunting at Gupuk, Mackenzie Delta,
Northwest Territories, Canada 109
    T. Max Friesen and Charles D. Arnold

An Ethnoarchaeological Investigation of Inuit Beluga Whale
and Narwhal Harvesting 127
    James M. Savelle

Whales, Mammoths, and Other Big Beasts: Assessing Their
Roles in Prehistoric Economies 149
    David R. Yesner

Siberian Eskimos as Whalers and Warriors *Hans-Georg Bandi*	165
Whaling Surplus, Trade, War, and the Integration of Prehistoric Northern and Northwestern Alaskan Economies, A.D. 1200-1826 *Glenn W. Sheehan*	185
And Then There Were None: The "Disappearance" of the *Qargi* in Northern Alaska *Mary Ann Larson*	207
Paul Silook's Legacy: The Ethnohistory of Whaling on St. Lawrence Island *Carol Zane Jolles*	221
Contemporary Alaska Eskimo Bowhead Whaling Villages *Stephen R. Braund and Elisabeth L. Moorehead*	253
Sex and Size Composition of Bowhead Whales Landed by Alaskan Eskimo Whalers *Howard W. Braham*	281
Speaking of Whaling: A Transcript of the Alaska Eskimo Whaling Commission Panel Presentation on Native Whaling *Edited by Carol Zane Jolles*	315
Whaling: A Ritual of Life *Herbert O. Anungazuk*	339

Editor's Note

In 1992, Roger Harritt, Carol Jolles, and I organized a symposium on native whaling, to be held at the Alaska Anthropological Association meeting in April, 1993, at Anchorage. We encouraged many colleagues to join us to report on some aspect of their archaeological, ethnohistoric, ethnological, or biological research about whaling. Further, Roger contacted Maggie Ahmaogak, Executive Director of the Alaska Eskimo Whaling Commission, to invite her and several Commissioners to participate on behalf of that organization. We were delighted to know that by the time of the meeting, some 15 persons had agreed to present papers and that all the members of the AEWC planned to attend. Two half-day sessions were held for the presentation of papers and discussions.

Roger, Carol, and I decided at the close of the meetings to attempt to publish the presented papers and others from persons who could not participate in the symposium (Bandi, Braham, Sheehan, and Yesner). I volunteered to edit the papers, and subsequently contacted Dr. Milton Freeman (University of Alberta) to inquire if the Canadian Circumpolar Institute might be interested in publishing the papers in their *Studies in Whaling* series. Upon approval of our request to the CCI Publications Committee, the authors prepared revised papers, which were then reviewed and edited, that make up this volume. One paper given at the meeting that is not included in this volume is "Whale Hunting in the Northwestern Bering Sea (Lakhtin Culture)," by Alexander Orekhov (International Pedagogical University, Magadan, Russia; translated by Katerina Solovjova and Richard Bland).

I wish to acknowledge with thanks the continuous advice and assistance of Roger and Carol in this project; they have provided me with reliable direction throughout. I also wish to thank the anonymous reviewers who read these papers, Mary Herrington, my department secretary, who has provided unending computer advice, and Cindy Reekie, CCI's Publications Manager, for her guidance. Special credit goes to Jim Savelle, who gave generously of his time and experience during the final production stages of this volume.

We hope readers of this volume will discover some new dimensions about the rich whale hunting traditions of the Western Arctic and Subarctic.

Allen McCartney

Introduction

Roger K. Harritt
Carol Zane Jolles
Allen P. McCartney

Whales are outstanding among living animals, not only because they are the largest mammals, but also because humans have focused upon them for such disparate economic, cultural, and ecological reasons. During the 1800s, European and American commercial whalers viewed whales as a limitless resource and exploited them almost to the point of extinction. Current wildlife conservationists, deploring their loss, have emphasized their preservation, and cetacean biologists now study behavior and reproduction among remaining whale populations with great interest. Whales also figure prominently in contemporary popular culture, as both adults and children seem equally fascinated by these intriguing and intelligent behemoths.

This volume offers yet another perspective on whales and whale hunting: that of northern native societies that have depended upon small and large whales for centuries. Residents of Alaskan and Western Canadian Arctic coastal regions have pursued these colossal animals as sources of food and fuel, but whale hunting also serves as a center for native cultural tradition and spiritual sustenance. These papers highlight past and present whale practices, making reference to the belief systems which support them.

The diversity of papers represented in this volume — a veritable "smorgasbord" of overlapping issues and treatments — developed less by design than by happenstance. Yet, the diversity itself serves as a powerful reminder that there are myriad ways to study precontact, historical, and contemporary whale hunting traditions. Whaling, particularly the technological and subsistence elements of the hunt, has long been a focus for northern researchers. Studies conducted by Margaret Lantis, Froelich Rainey, Robert Spencer, Charles Hughes, James VanStone, Don Foote, Ernest Burch, and John Bockstoce, to name a few of the more prominent researchers, added greatly to our understanding of the practical and spiritual strategies employed by native men, whose responsibility it was to confront whales directly. More recent studies by Richard Nelson and Rosita Worl, among others, have helped to record contemporary whale and other marine mammal hunting patterns. Still, many technological and ideological aspects of whale hunting await exploration. Attention to women's roles in whaling, for example, has only recently been a focus, with the work of Barbara Bodenhorn and Edith Turner. The present collection of papers expands upon these previous studies, and adds dimension and scope to our understanding of whaling traditions and processes from the perspective of both men and women as complementary expert practitioners.

Several themes run through these papers. One of the most obvious, articulated especially by symposium participants, is that native whale hunting is a living tradition in northwestern Alaska. Unlike most areas of the world where whale hunting has entered the realm of remembered tradition and history, in the Western Arctic subsistence whaling thrives in an unbroken continuity with an ancient adaptive pattern. That pattern, not surprisingly, is a cooperative endeavor in which both men and women perform critical tasks and shoulder great responsibilities. It also includes an on-going process of adjustment in order to achieve subsistence goals, now limited and defined by international mandate.

Another theme is reevaluation. Many authors ask new questions of old data or pose questions that formerly were not addressed. Northern anthropologists continually try to see the cultural world from different perspectives. Thus, several papers ask that we see this world differently, particularly where native peoples are concerned, with regard to the use and significance of whales.

Some papers illustrate that human predation of large to small whales is based on species-specific behaviors. Similar capture and butchery techniques are found wherever a whale species is found in the North. Where belugas are hunted, for example, we are likely to find that hunters drive animals into shallow waters to facilitate harpooning them. Large whales require hunters in boats to approach surfacing animals from behind, in order to cast harpoons or darts into them. Such ubiquitous human responses to species, while not peculiar to sea mammals, imply that procurement patterns will tend to converge on a relatively narrow range of procedures for successfully taking animals and using their products. Thus, human adaptations to whales permit comparisons to be made across large northern regions.

Papers that focus on social, political, and economic dimensions of whaling, such as community structure, contributions of women, spiritual or religious elements of whaling, trading networks, and warfare, call our attention to the pervasive impacts that whale hunting has had and continues to have on a society. While the techniques of whaling, from sighting to harpooning to butchering, often receive our attention as the "doing" of whaling, enormous quantities of meat and blubber greatly influence a community's structure and define its ethos. Such "gifts" result from the whale "offering itself" to humans within a complex web of spiritual connection. These social and spiritual structures, created by and surrounding whale hunting, have as much reality as the material weapons and products of the hunt proper.

Finally, these papers stress a collaborative theme. While we may approach the study of native whale hunting from a variety of directions, the greatest gain in understanding past and present use of whales seems likely to come from merging zoological, historical, and cultural data into a holistic view of whale predation and its sociocultural impacts upon those who engage in it.

*****

The papers of this volume are arranged in rough chronological order of their focus. There is no sharp division between the precontact period and that following the mid-18th century, when Russian colonists and other Euroamericans began to affect native whaling societies. Although anthropologists attempt to impose a chronology on the cultural record, we also recognize the timeless quality of whale hunting which permits us to slip back and forth between whaling of some millennia ago and that of today.

The Archaeological Imagination, Zooarchaeological Data, the Origins of Whaling in the Western Arctic, and "Old Whaling" and Choris Cultures

Owen K. Mason
Alaska Quaternary Center
University of Alaska Museum
P.O. Box 756960
Fairbanks, AK 99775-6960

S. Craig Gerlach
Department of Anthropology
P.O. Box 757720
University of Alaska Fairbanks
Fairbanks, AK 99775-7720

Abstract. The working consensus among arctic archaeologists is that the first efforts at whaling in the Western Arctic started with the Old Whaling culture at Cape Krusenstern, on the oldest whale bone ridge, No. 53, dated to 2900-2800 C-14 yrs B.P. Taphonomic and faunal data do not exist to support whale hunting over the scavenging of whale bone and/or tissue. We advise abandoning the biased taxon "Old Whaling" in favor of the Chukchi Archaic, which reports notched bifaces over a long period. The younger, Choris residents of whale bone ridges 44 to 52 were linked with reindeer "herding" in an imaginative analysis by Giddings. Choris sites date between 2900-2100 C-14 yr B.P., concurrent with a warmer interval during the Neoglacial expansion. Beach ridge sites (n=182) at Cape Espenberg and Cape Krusenstern indicate that the extremely mobile Choris people camped in pole structures for short-term sealing and caribou hunting and may also have cached and stored goods. Any role of whale in the diet had few consequences for increased sedentism and social stratification. Choris represents a brief period of cultural interaction, perhaps even immigration, among Alaskan and Chukotkan peoples, including antecedent (3100 C-14 yrs B.P.) and contemporary examples. Reanalysis of faunal data supports evidence of a subsistence economy specialized on ringed seal and caribou.

INTRODUCTION

The assumption that Old Whalers actually hunted baleen whales is a rather persistent belief in the archaeological imagination in the Western Arctic, based on the publications of J. Louis Giddings (1960, 1961b, 1962, 1967). The actual data for whaling is another matter. No similar case is offered for the other inhabitants of the Cape Krusenstern "whale bone ridges" (Nos. 44-53),[1] the Choris people, who may have hunted beluga whales (Giddings and Anderson 1986:229). Most archaeologists assume that Old Whaling represents a "temporary stay of a year or little more by some small group of coastal outlanders" who had no "impact at all upon Alaskan cultural developments of the time" (Dumond 1977:101-103). Whaling in the Western Arctic was developed by either Old Bering Sea, Birnirk, Punuk, or Thule peoples, depending on the researcher (Collins 1937a; Larsen and Rainey 1948; Rudenko 1961; Taylor 1963; Bockstoce 1972, 1979; Stanford 1973; McGhee 1969/70, 1972, 1981; Dekin 1984; McCartney 1984; Sheehan 1985; Minc and Smith 1989; Harritt 1991, 1992; Krupnik 1993).

The casual use of faunal percentage data continues to fuel speculative reconstructions of subsistence change in northwestern Alaska (Bockstoce 1979; Anderson 1983; Ackerman n.d.). To avoid circular reasoning, the analysis of data must precede any theoretical constructs about the cultural feedbacks that derived from harvesting the bounty of the bowhead. Quantitative faunal analyses (Gerlach n.d.) from three Choris-age sites indicate a reliance on winter sealing and caribou hunting. The relationship of Choris people to whale bone accumulations on Cape Krusenstern is unclear and may have involved scavenging.

Old Whaling and Choris sites date to several centuries between 3000 to 2200 C-14 yrs B.P., within a critical juncture in climatic terms. The Old Whaling culture co-occurs with a stormy interval within the Chukchi Sea, while the Choris people lived during a brief amelioration of climate during the overall Neoglacial expansion. In general, Choris people witnessed a lessening of storms and, perhaps, warmer temperatures (Mason and Gerlach n.d.). Terminologically, Choris "culture" is an ill-advised construct that disguises its nature as a horizon (*sensu* Willey and Phillips 1958). The Choris horizon is important as the first period during which ceramics were used in the North American Arctic, despite the presence of ceramics in adjacent northeastern Siberia for several thousand years (Ackerman 1982). This pottery represents the initiation of a larger complex of economic and/or social relationships between Siberia and Alaska. Presumably, pottery appeared as a result of direct trade contacts, migration, or stimulus diffusion and not independent invention. Despite its taxonomic importance, Choris pottery is found at only a few sites and its importance to internal cultural dynamics is unclear. Choris people were apparently only the third group to live adjacent to the northwestern Alaskan coast following sea level stabilization at c. 4000 C-14 yrs B.P. (Mason and Jordan 1993) and the first to rely heavily on sea mammals (Dumond 1977:125, 1984:76). The appearance of labrets in northwestern Alaska is an indication of this symbolic identity between humans and sea mammals; Choris labretifery is among the oldest in the world. The greatest paradox involving Choris involves the co-occurrence of Choris sites with the "whale bone" ridges of Cape Krusenstern and the confusing relationship of the Choris horizon with the slightly older Old Whaling and Devil's Gorge materials (see below).

---

1. Cape Krusenstern beach ridges are numbered between 1 and 114; No. 1 is the youngest and No. 114 is the oldest (Giddings 1962). The series is not sequential but has a disjunctive character due to erosional disconformities (Mason and Ludwig 1990). Whale bone may also be found on ridges as young as No. 41, according to a statement by Giddings (1962:36) that "the large jawbones, vertebrae, and ribs of baleen whales" were "buried under a few inches or a foot or more of peat and moss."

THE INTERREGNUM OF OLD WHALING/CHORIS IN NORTHWESTERN ALASKA

The cultural affinities of Choris and Old Whaling assemblages continue to trouble both archaeological taxonomists and historians of northern material culture. Both contain anomalous artifacts in the broad context of culture history throughout northwestern Alaska. Old Whaling, however, with its few preserved organic tools, closely resembles the Northern Archaic notched point traditions of the interior that extend over 3000 yrs of time, to as recently as 2000 C-14 yrs B.P. Choris, by contrast, entered northwestern Alaska as a sudden eruption of new ideas and practices. In this sense, Choris is an archaeological horizon, in the sense of Willey and Phillips (1958:32): "highly specialized artifact types, new technologies, peculiar ritual assemblages ... that indicate a rapid spread over a wide geographic space." Although a number of critical lithic elements are lacking, some archaeologists (e.g., Giddings and Anderson 1986) see clear links between Arctic Small Tool tradition (ASTt) and Choris, as "a curious combination of artifact types" (Anderson 1988:103) related to the presumably older bifacial tool complexes rooted in or somehow related to the Northern Archaic tradition. Other archaeologists, most notably, Dumond (1977, 1984), place a clear break between ASTt and Choris. Much confusion stems from the use of Onion Portage and Cape Krusenstern as informal type localities for the Choris and Old Whaling phenomena.

Choris discoveries made since 1986 by the National Park Service (Schaaf 1988) and others (Mason, unpublished data) have substantially increased the number of Choris localities known along Kotzebue Sound. Although many archaeologists doubt the validity of "inland Choris" on the basis of a noticeable decline in trait elements, it appears that the inland component is possibly as well defined as the coastal component. However, in examining the evidence for whaling, we discuss only the coastal facies of Choris: the context of the reported sites, their geographic distribution, occupational densities, radiocarbon chronology, and technological and symbolic characteristics.

"Old Whaling" tools include a wide range of side and end scrapers and distinctive well-flaked large bifaces, some with side notching. The most noteworthy tools are the several "lamp" fragments in House 21 at Cape Krusenstern. Otherwise, many of the tools could be classified as a facies or a distant descendent of the Northern Archaic tradition, as Gillispie (1985) suggested and as Willey (1966:431) noticed much earlier. Ackerman (1984, n.d.) expanded Old Whaling to include the Wrangel Island Devil's Gorge site. We recommend the further inclusion of the Palisades assemblages and the southern Alaskan Security Cove materials as a co-tradition within a broader cultural tradition termed the Chukchi Archaic. Old Whaling, coined by Giddings (1961a:164), is a designation that is prejudicial to any discussion of faunal data and should be abandoned, if for no other reason than its exclusive occurrence on Cape Krusenstern represents a phase or component rather than a regional or temporal unit. It is time to phase Old Whaling out of the literature.

Tool Function vs. Faunal Data as Evidence of Subsistence Economies

Recently, Ackerman (n.d.) offered several assertions about the existence of whaling in the Chukchi Archaic "Paleo-Eskimo" cultures, based on the observation that "the side notched and lanceolate point forms as well as the large bifacial knives found in the Old Whaling site are tool types *undoubtedly* utilized in the hunting and butchery of whales (supported by the whale bone found on [Cape Krusenstern] beach ridge 53"; our italics). As recounted below, these assertions are the same proposed by Giddings (1961a, 1967) over 30 years ago. Despite a certain intuitive attraction, arguments based upon apparent tool function are rather circular. First, the case for whale hunting is unfounded in the absence of harpoon foreshafts or other items for inset of the purported weaponry. Quite possibly, physical limitations prevented the butchery of whales using the small tools (e.g., the microblades of Denbigh times) and required technical innovations. However, lacking experimental data or chemical analyses, it is impossible to establish that this is the case. A preliminary start in this direction would be research on blood residues. As a note, the occurrence of whale bones on Choris ridges is not used as evidence of whale hunting or consumption by Ackerman or others.

GEOGRAPHIC AND CLIMATIC CONTROLS ON THE CULTURES OF NORTHWEST ALASKA

Alaska physiography exerts a strong influence on site distribution by its controlling factors on resource availability and access corridors. Northern Archaic cultures are found throughout northern and central Alaska, but Chukchi Archaic or Old Whaling sites are limited to the margins of the Chukchi Sea (Fig. 1). For the most part, Choris sites are limited to the tundra biome in the north and west of Alaska. The Alaska coast alternates between low bluffs of unconsolidated silts and bedrock cliffs over much of the coast from

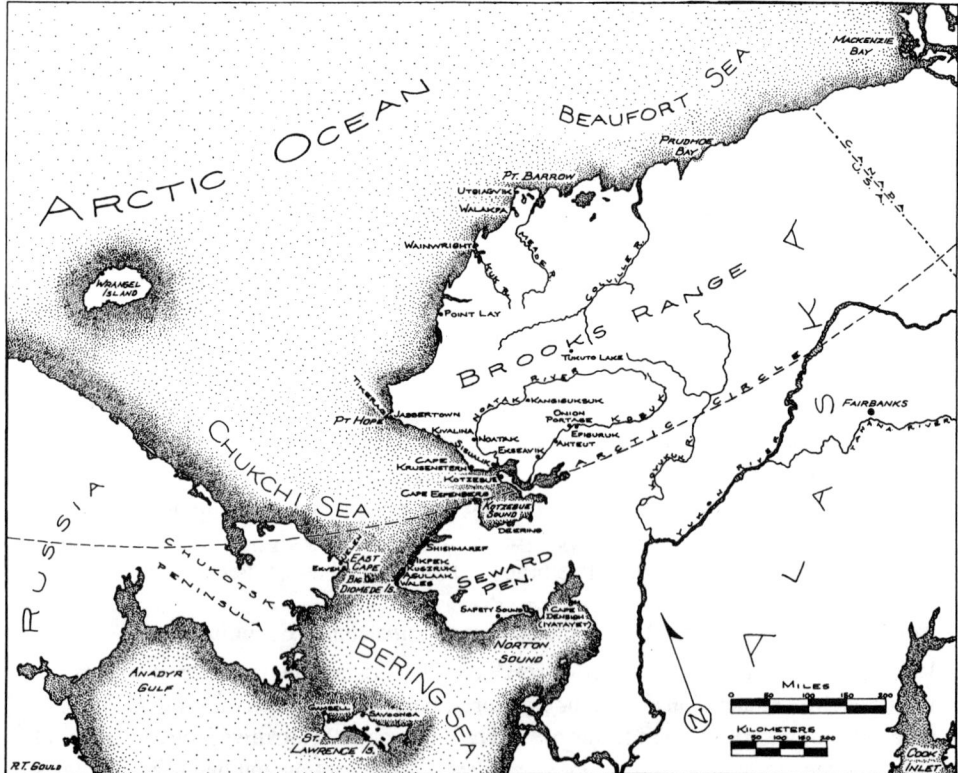

Figure 1. Map showing northern and western Alaskan archaeological sites.

Cape Lisburne to Cape Krusenstern. Sand and gravel barrier island chains and spits are comparatively common; spits or beach ridge complexes occur at most capes. The capes and promontories constrict ice flow and favor lead formation, which provides access for maritime sea mammal hunters. Spits are comparatively recent geologic features, forming with the stabilization of sea levels after 4000 C-14 yrs B.P. (Mason 1990; Mason and Jordan 1993; Mason et al. n.d.). Evidence of coastal occupations have inevitably been lost as rising sea levels transgressed, eroded, and reassembled coastal sediments at a variable rate during the last several thousand years (Mason and Jordan 1994). Ice is a perennial feature of Alaskan seas and rivers that eases travel, but requires technological ingenuity and modifications to hunting strategies (Lucier and VanStone 1991b).

Ecological and climatic controls over the late Holocene Alaskan archaeological record are discussed more fully in Mason and Gerlach (n.d.). Several points are important for discussing the Old Whaling and Choris phenomena. Bering Strait serves as a conduit for nutrients and water drawn across a gradient from the higher Bering Sea into the lower elevation (-0.5 m) Chukchi Sea of the Arctic Ocean. The east and west coasts and shelves of the two seas differ considerably, in terms of nutrients, water temperature, and ice cover (Walsh et al. 1989; Coachman and Hansell 1993). The west shore, Chukotka, is deep, rocky, and has a narrow shelf, while the east shore, Alaska, has shallow bays and a wide shelf. These factors translate into ecological differences that define discrete archaeological cultures.

Two distinct streams of water flow side by side through Bering Strait (Walsh et al. 1989; Coachman 1993). To the west, cold upwelling Anadyr Water, the major contributor, yields rich nutrients (95% of nitrates added to the system; Walsh et al. 1989:301), while few nutrients are added from the east, the nutrient-poor, less saline, and comparatively warm Alaska Coastal Water. The richness of nutrients (i.e., nitrate) in the Anadyr Water translates into high surface and benthic productivity within several "hot spots" in the southern Chukchi Sea (Springer and McRoy 1993). The principal hot spot is in the southern Chukchi Sea, 50 km southwest of Point Hope, and extends toward the northern Chukotsk coast. High nutrient concentrations also lead to heightened primary production off Cape Dezhneva (East Cape) and at the mouth of Anadyr Gulf. The influence of Anadyr Water continues to be felt west to Wrangel Island, throughout the East Siberian Sea to the margin of the Laptev Sea (Walsh et al. 1989:345). The amount of ice cover influences the number of days available for photosynthesis and, hence, productivity. The effects of high nutrient fluxes translate into biomass production, benthic nutrients, and eventually in high numbers of walrus and whales. Significantly, whale populations follow the nutrient "hot spots" in their annual progress to the ice edge (Fraker 1989).

What are the consequences for humans? First, clearly, in terms of walrus and whales, Kotzebue Sound and Seward Peninsula should be comparatively impoverished compared to regions near the hot spots. Three regions are best positioned to benefit from the productivity of Anadyr Water: Point Hope, the greater East Cape region (including the Diomedes), and northern St. Lawrence Island. These are, of course, the hot spots of western arctic prehistory. Eastern St. Lawrence Island and the Wales region are close to hot spots. Significantly, although Point Hope is closest, the region nearest Point Hope is comparatively poor. People could either journey to the hot spot or wait until ice leads funnelled walrus and whales closer to shore after feeding.

Climate and Its Role During the Chukchi Archaic and the Choris Horizon

Based on regional climatic data (see detailed discussion in Mason and Gerlach n.d. and Mason and Jordan 1993), northwestern Alaska witnessed several profound climatic transitions. A brief cooling, the onset of the Neoglaciation, occurred at 3300 C-14 yrs B.P. and continued for several centuries, apparently broken by brief spells of warmer conditions in the following millennium of 3000-2000 C-14 yrs B.P. The Neoglacial brought intense fall storms to northwestern Alaska; it is this increased precipitation that accounts for glacial expansion and for colluviation at Onion Portage. The timing of Onion Portage colluviation, at 3600-2800 C-14 yrs B.P. and 2200-1600 C-14 yrs B.P. (Anderson 1988), are correlative with a major erosional cycle on Chukchi Sea beach ridges, at 3300-3100 C-14 yrs B.P. (Mason and Jordan 1993). This erosional period correlates with increased storminess and expanded glaciation in several Alaskan mountain ranges (Mason and Jordan 1993). The intervening period, 2800-2200 C-14 yrs B.P., of stable surfaces at Onion Portage was also a time of surface stability at Cape Espenberg.

In paleoclimatic terms, the Choris culture period falls within a 100-300 year-long warm oscillation within the Neoglacial, 3300 to 2000 C-14 yrs B.P. Unfortunately, lacking the record from settlements situated on pack ice, a serious skew enters into the record: terrestrial settlements are favored during warm periods but may not reflect population density. Contrary to temperate zone preconceptions, warmer, stable weather and early breakup could bring hardship to arctic peoples as is seen in the tendency toward coastal migration during warming episodes and interior congregation and population increase during cold periods (Mason and Gerlach n.d.). Early sea ice breakup can decrease the availability of a staple resource, ringed seal, an ice dependent species (Stanford 1973). In any event, the record is biased in that stable pack ice sites (Lucier and VanStone 1991b) are irretrievably missing from the archaeological record.

Three principal animal groups provide the base of the Choris economy: ringed seal (*Phoca hispida*), pink and dog salmon, and caribou (*Rangifer tarandus*). Walrus and whale, two animal groups important in ethnohistoric times, were apparently rarely used by Choris people, based on extant faunal data (Gerlach et al. n.d.). Traditionally, walrus served a variety of purposes, as cordage, boat covers, and dog food. In the absence of dog traction, the latter use would be less important and the former presumably could be replaced by the ringed seal. Choris people's use of whale is explored below. Two of the most economically viable species are associated with inland access points: salmonids and caribou. Although both species frequent the coast, technological limitations or density factors favored hunting on rivers or within mountain passes, a pattern that persists in large measure in northwestern Alaska to the present. Thus, seasonal movements between mountains and the shore were probably the favored strategy during Choris times, and, perhaps, during the Chukchi Archaic period as well. In the following, we discuss both the Chukchi Archaic tradition and Choris in technological terms and assess the role of whaling in their subsistence pursuits. This approach is justified because most supporting arguments concerning whaling origins in the Western Arctic are based on technological attributes, in lieu of faunal evidence for whaling.

TECHNOLOGY AND CHRONOLOGY OF THE CHUKCHI ARCHAIC

The most distinctive characteristic of the Chukchi Archaic tradition is its absence of ceramics and microblade technology. Otherwise, the phenomenon is primarily defined by lithic technological attributes. Several types of bifaces occur at Cape Krusenstern; the wider pieces (>4 cm) are termed whaling harpoon or lance blades (Giddings and Anderson 1986:265), despite evidence "of a slight dullness" that indicates cutting wear on some edges (Giddings and Anderson 1986:235). Side-notching, often marked by grinding, is an important technical attribute and includes points and end scrapers. The notched points fall within two general size classes, either narrow (<2 cm width) and small (<5 cm length) or broad (> 3 cm in width) and long (> 4 cm in length). The flaking technique of the notched pieces is variable, at times collateral and precise, while other notched points are simply edge retouched flakes. Technological ability seems to have varied by household, and it is difficult to characterize the variability in all the Cape Krusenstern houses. Flake tools, including gravers and end scrapers on flakes, were an important and frequent tool (nearly two-thirds of the total), suggesting carving, butchery, and/or hide processing (Giddings and Anderson 1986:257ff). A handful of organic remains (n=10) from House 22 includes a 5 cm- long, open socketed toggling harpoon head that is very similar in style to a considerably larger piece from Wrangel Island (Dikov 1988:Fig. 2). Several walrus ivory chips or objects reflect an interest in ivory as a medium for artifacts, and may represent trade or travel to obtain walrus (Giddings and Anderson 1986:265). Shallow stone basins were presumably used as lamps. Several wooden items include containers and shafts apparently used for arrows and/or spears. Slate grinding was practiced on a limited scale (less than 2% of total tools). The age of the Krusenstern notched point occupation is contentious and involves 18 radiocarbon ages, two-thirds of which are contaminated with older carbon (Giddings and Anderson 1986:32). Mason and Ludwig (1990) advised that all possibly anomalous older ages be rejected and that only the six C-14 ages on wood that average 2848±57 C-14 yrs B.P. be accepted (Giddings and Anderson 1986:32).

Parallels for the lithic assemblage at Cape Krusenstern are rare and are mostly known from the Interior, especially along the northern foothills of the Brooks Range (Davis et al. 1981). All are undated. Only a handful of notched point sites, termed Northern Archaic tradition sites, lie within 200 km of Cape Krusenstern. Two sites are in the hills just north of Cape Krusenstern: the Palisades site (Giddings 1961b) and NOA-264, on the northern foothills of Mt. Noak (McClenahan and Gibson 1989:264). NOA-264 is a surface site that contains several hundred lithic fragments and includes a microblade technology. Notched points at the Nimiuktuk sites in the middle Noatak valley co-occur with a microblade and burin technology (Anderson 1972:76).

Notched points occur at several widely separated coastal sites from the southern Bering Sea to the eastern Beaufort Sea, a distance of several thousand kilometers. Several artifacts collected from Cape Espenberg are similar to Old Whaling (Giddings and Anderson 1986:227,Pl. 132); the "notches ... formed by unifacial retouch, in a manner characteristic of Old Whaling." The age of the Espenberg notched bifaces can be estimated at 3000-2500 C-14 yrs B.P., based on the age of Choris materials from the same dune ridge (Harritt 1994:133-134), and 3300 to 2000 C-14 yrs B.P., based on geologic age estimates (Mason n.d.). Ackerman (1984) noted similarities between Old Whaling and Wrangel Island but doubted the similarities with early Greenlandic cultures, for example the notched points in Independence II (Fitzhugh 1984). Side-notched points, very similar to Old Whaling, occur at the aceramic seal-hunting Crane site, east of the Mackenzie River, in the Northwest Territories. The Crane site dates to 2600-2400 C-14 yrs B.P. (Le Blanc 1994), is contemporaneous with Choris culture, but it is >2000 km from Kotzebue Sound. Despite the lack of ceramics, the Crane site also resembles Choris and Norton sites from Alaska, especially in terms of its barbed bone points and discoidal scrapers. Affinities with eastern arctic cultures are downplayed by Giddings and Anderson (1986:266), due to the greater width and manufacturing technique of the notches and the greater size of the end scrapers in Alaska. But as Giddings and Anderson (1986:266) note, "because the general forms occur at the same time across the Arctic ... the correspondences are not simply coincidental."

Western arctic coastal notched point sites are also distant from Cape Krusenstern. The Devil's Gorge site on Wrangel Island, 900 km to the northwest, shares several attributes with Cape Krusenstern Archaic sites, principally a reliance on flake tools, a lack of ceramics, and the aforementioned similarities in toggling harpoons. However, bifacial forms at Devil's Gorge are only vaguely notched, and are considered "stemmed" in the brief summary by Dikov (1988). Security Cove, on Kuskokwim Bay, contains a considerable number of side- and corner-notched points from a surface site on a low hill of glacial till that is located about 12 m above the bay (Ackerman 1964:17ff). Security Cove lacks microblade technology, but does contain abundant flake tools and notched cobbles (net sinkers). No radiometric ages are available for Security Cove.

The size of the unnotched bifaces from Cape Krusenstern ridge No. 53 is the most noteworthy characteristic of its technology. In the context of whaling, biface size is assumed to be critical, and leads some to assume that large end blades were necessary for harpoons or lances. Use in butchery cannot be ruled out (Giddings and Anderson 1986:265). Otherwise, the technology at Cape Krusenstern has a predominance of flake tools and notched bifaces that may have some commonality with notched, Northern Archaic bifaces from sites across northern Alaska and the Yukon Territory dating between 6000 and 2000 C-14 yrs B.P. (Clark 1991:48-50). Despite this common ancestry, the Krusenstern notched point technology has been viewed as an intrusive element in the culture history of northwestern Alaska (Giddings 1961b:36). The implication is that sea mammal hunting originated during the Chukchi Archaic period and cannot be derived from elsewhere. Erosion caused by such events as the rapid rise in sea level at 3000 to 2000 C-14 yrs B.P. (Mason and Jordan 1994) may account for the lack of other coastal Archaic sites in northwestern Alaska.

DEFINING CHORIS AS A TECHNOLOGY: LITHICS, LABRETS, AND CERAMICS

In many ways, the Choris culture represents a completely different set of artifact types, including the presence of labrets, ceramics, and barbed harpoon heads. Bifaces found at the type site and other coastal sites tend to be long, comparatively narrow, and stemmed. They generally lack notching. The absence of broad bifaces is also a notable exception, and must account for Giddings' (1967) presentation of the Choris people as non-whalers.

Lithic Technological Attributes

Labrets, oil lamps, inset points, and ceramics are not definitive Choris markers. These artifacts are found only at the Choris Peninsula locality. The more widespread lithics are the hallmark of Choris: diagonally-flaked bifaces, often of substantial length and often with a shouldered hafting at mid-section. Lanceolate spear points, some with concave bases and some stemmed, are common, especially at the famed Choris weapons cache (Feature No. 486) on Cape Krusenstern ridge No. 46 (Giddings and Anderson 1986:215-216, Pl. 124). One of the most interesting finds is the slotted lance head at Trail Creek (Larsen 1968:39ff). Burinization is a critical feature of Choris lithic technology, as is illustrated by Larsen (1968:40ff) at Trail Creek, and presumably was linked with bone insets and, perhaps, the production of microblades. Microblades (or blade-like flakes) occur most prominently within Band 3, Level 2 of the "Old Hearth" of Giddings (1961a:14). Microblades are not mentioned by Anderson (1988), although informed observers report them in the collection (P. Bowers, pers. comm., 1992). The occasional use of slate at the Choris type site appears rather unsophisticated (Giddings and Anderson 1986:Pl. 117), and appears useful as a needle sharpening tool or drill.

Symbolic Culture: Labrets and Lamps

The function of facial insets seems obvious; labrets serve as stylistic markers for kinship groups and as a form of psychic identification with the tusks displayed by the walrus. For, as implied by symbolic anthropologist Lydia Black (1991), if you look like a sea mammal, you must be thinking like one. The use of labrets as a stylistic marker implies, intrinsically, the existence of a population threshold or density that required identification of oneself in relation to foreigners. Further, the uniqueness of the labret as a part of ritual and belief mitigates against casual imitation. However, the early evolution of labrets remains poorly documented along North Pacific coasts.

Labretifery occurred over a wide distribution from the Kuriles to southern British Columbia in the late prehistoric period (Keddie 1981). The basal layers of Chaluka mound on Umnak Island yielded several medial labrets associated with radiocarbon ages of 4000-3800 C-14 yrs B.P. (Aigner 1966, 1978; Laughlin 1981). Labrets of possibly contemporary age, bracketed between 4110±160 (I-1639) and 2810±100 C-14 yrs B.P. (I-3733), are documented at Takli Island, off the eastern Alaska Peninsula coast (Clark 1977:29) and farther afield in southeastern Alaska from Hidden Falls at c. 3200 C-14 yrs B.P. (Davis 1989). The earliest southern Alaskan labrets are broad medial forms common to the areas at European contact. By contrast, the Choris forms are drastically different. These include a bullet-shaped lateral form and a two pronged form (the latter identified tentatively as a bow guard by Giddings and Anderson 1986:198, Pl. 109d), similar to late prehistoric forms used in the lower Yukon-Kuskokwim drainages (Nelson 1899:Pl. xxii). Medial labrets found at Iyatayet within Norton layers dated between 2720±130 C-14 yrs B.P. (M-1260c, Crane and Griffin 1964:21) and 2213±110 C-14 yrs B.P. (average of two runs; Giddings 1964:245).

Significantly, the Aleutian and Alaska Peninsula labret-using cultures differ significantly from Choris, but share some key traits. First, the most profound difference is the absence of ceramics on Aleutian and Alaska Peninsula sites, while slate was worked on the Alaska Peninsula. Certainly, the adoption of pottery could arise far easier than the development of labretifery, and a lack of pottery is explainable by a greater distance from the source area, Siberia. Three key traits indicate a common heritage of unknown, but apparently southern origin: labrets, oil lamps, and dual barbed harpoon heads (cf. Laughlin 1962). The lamp had an extremely important character, both stylistically and psychically, in late prehistoric Eskimo cultures, as the 19th century ethnologist Hough (1898) recognized and as Lucier and VanStone (1991a) have further elaborated. Harpoon head forms may be equally as conservative, but innovation in technological forms is unquantified. In essence, if Choris origins may be placed in southern Alaska, then the entire question of the development of Choris whaling must be recast in terms of the southern Alaskan prehistory.

Ceramics

In view of the importance of ceramics (Arnold 1985), it is advisable to emphasize their occurrence in the Choris horizon. Pottery use by Choris people represents either the appearance of a storage mentality, a cooking technology, or possibly a trade in perishable commodities. The widespread adoption of ceramics implies long-distance connections and, possibly, "colonization" or a population increase (cf. Dumond 1972). Ceramic analysis in Alaska remains a product of the 1950s classificatory impulse, limited to several descriptive efforts by Oswalt (1955), Griffin and Wilmeth (1964), Dumond (1969), and Ackerman (1982). No petrographic, sourcing, or technological studies are available for the earliest pottery. X-ray diffraction studies on Norton pottery from Unalakleet established that local clay sources were used (Lutz 1972). Ethnographic analogies with 19th century manufacture techniques indicate that Alaskan pot-making was rather unsophisticated, and involved patching or coiling clay by hand and setting pots near the fire (Lucier and VanStone 1992). Some pottery was decorated using bone or wood paddles; decorative elements allow the definition of several regional variants. Choris pottery is generally a linear stamp variety, termed "Norton linear stamped" by Oswalt (1955:34). Temper varies from hair on Choris Peninsula to sand and fiber at southern sites. Linear motifs are often crossed to give a check-stamped appearance. A walrus paddle from a Norton site near Point Hope has linear motifs on one side and checks on the other (Oswalt 1955:34). In terms of ceramic classification, Choris is construed by Oswalt (1955) as a variant of the Norton type. The distribution of linear stamp pottery includes the entire Alaska tundra coast from the Yukon Territory to Bristol Bay, coincident with Choris and early Norton (Ackerman 1982:18), including Cape Krusenstern which was erroneously excluded by Ackerman (1982:18; cf. Giddings and Anderson 1986:212). The southern limits of linear design may be the most informative. Linear-impressed potsherds are found on Tununak on Nelson Island, adjacent to the Yukon Delta, and date to 2530±200 C-14 yrs B.P. (GaK-9574), based on examples excavated from a test trench (Okada et al. 1982:6-9). The southern limit of linear stamp ceramics is at Chagvan Bay with the materials from House 107, significantly dated to 2720±80 C-14 yrs B.P. (WSU-3215; Ackerman 1988b:170ff). Linear-stamp sherds on Nunivak Island are as young as 2000 C-14 yrs B.P., and are considered to be Norton products (Nowak 1982). Solely in terms of ceramics, Choris is the first phase of Norton, as Dumond (1977, 1984, and elsewhere) has repeatedly advised. However, there is no Choris stage. Choris is merely the adoption of pottery by preexisting Norton or ASTt peoples.

DISTRIBUTION AND CONTEXT OF COASTAL CHORIS SITES

The most widely accepted typological criteria for Choris sites, those of Anderson (1984), place an abrupt distributional boundary through Seward Peninsula that extends north to near Barrow (Fig. 1). However, if linear pottery is diagnostic, the Choris horizon may have extended from Bristol Bay to the coast of the northern Yukon Territory and throughout the Brooks Range (Anderson 1984). If multi-barbed harpoon heads are diagnostic, then the Engigstciak site in Yukon Territory and the Crane site in the Northwest Territories should be included. A similar point at Nunivak is also a possible site to be included (Dumond 1977:109). Two of the most diagnostic traits, labrets and pottery, are reported primarily along the coast, although both artifacts were indeed carried over 100 km inland into the northern foothills of the Brooks Range, as is evidenced by a medial labret at No-Name Knob (Dale Slaughter, pers. comm., 1993) and the pottery within the Choris tent ring at the Gallagher Flint Station (Peter Bowers, field notes, 1974).

No more than 30 Choris radiocarbon dates are reported from northern Alaska and the Yukon Territory. Most sites are diffuse, surficial scatters of lithic tools atop prominent lookout stations that purportedly were used to observe game. Archaeological sites containing appreciable accumulations of Choris debris include the Onion Portage site on the Kobuk River, the Kayuk site in Anaktuvuk Pass, and the Engigstciak site on the Firth River, northern Yukon Territory (Anderson 1984). However, the greatest concentration of Choris materials (over 180 site loci) is from just three areas: Cape Krusenstern, Cape Espenberg, and Choris Peninsula. The ease of site discovery atop beach ridges and within blowouts quite possibly distorts our view of Choris settlement patterns in these areas. In the following review, we examine the apparent center of Choris culture, Kotzebue Sound.

Kotzebue Sound

The notable decline in Choris traits inland leads most archaeologists to consider the coastal Choris facies as the "true" Choris, or at the very least as the best documented Choris expression. Because of the co-occurrence of Choris remains with whale bone, it is important to understand the specific characteristics of the Choris horizon. Additionally, it is important to define the technology of Choris, because technological attributes provide the basis for inferring whaling at Krusenstern during the Old Whaling/Chukchi Archaic period. Subsequently, we evaluate the relationship between Choris and its near contemporary (or antecedent), the misnamed "Old Whaling" culture. In addition, the sudden appearance of well-fired, decorated ceramics in Choris have traditionally evoked cross-cultural ties to Eurasia, leading to speculations about colonization and migration from the Old to the New World, or speculations about cultural interaction and trait element diffusion (Ackerman 1982). Despite the apparent high number of sites in Kotzebue Sound, the Choris sites can hardly represent a significant number of inhabitants at any one time.

Cape Krusenstern

Most Choris sites at Cape Krusenstern (75 of 96; 77%) are ephemeral scatters (Table 1) on six of 10 ridges (Nos. 44-53), termed the "whalebone beaches" by Giddings and Anderson (1986:209). No sites were found on three ridges (49-51). The Choris ridges lie between two prominent erosional truncations, the older truncation following the Old

Table 1. Cape Krusenstern Choris localities: artifacts and fauna (synthesized from Giddings and Anderson 1986:210ff).

Beach 44 Late Choris
22 localities; four are isolated finds of whale vertebrae or other elements; 13 are hearths, mostly of burned pebbles, some with cut "nearly disintegrated" whale bone (Feature 765); other localities have a considerable amount of chipping debris (>100 within 2 m^2) and some potsherds; some localities are possibly of Norton affinity.

Beach 45 Late Choris
16 localities; four are isolated whale vertebra, six are diffuse lithic scatters with less than 20 artifacts (burin spall cores, weapon points, side or end blade insets); one large lithic scatter, >200 fragments; seven hearths (two unexcavated), one of Norton culture (Feature 780).

Beach 46 Middle Choris
22 localities; five are isolated whale bones; one (Feature 231) is a "pile of whale vertebrae and a jawbone" (Giddings and Anderson 1986:215); one site, not counted in enumeration, is a cache of late Thule or Kotzebue period, i.e., 1500 yrs younger than Choris; the most spectacular find of all Krusenstern (Giddings and Anderson 1986:215): Choris lithics, Feature 486, the total hunting/flaking repertoire of a single hunter, as interpreted by Giddings (1963); the cache includes 73 pieces: 45 oblanceolate spear points, 3 burinated, 10 knives, 8 side blades, 1 flake knife and a notched piece "identical to Old Whaling" pieces.

Beach 47 Early Choris
10 localities; two are isolated whale vertebrae; two are diffuse scatters of tools and debris, including ceramics at one site; a series of closely adjoining hearths (Feature 715-717) contained "with them, under the sod ... bones of small seal and caribou, fragments of whale bones and many chips, artifacts and potsherds" (Giddings and Anderson 1986:219).

Beach 48
2 sites; one containing only a single whale vertebra, the other a lithic scatter of biface, knife, discoids, potsherds (unquantified number).

Beach 52
2 sites; both scatters of lithic debitage (n=>75).

Whaling culture (ridge No. 53) and a younger erosional episode before Ipiutak (ridge No. 35). These are a series of primary, non-reworked ridges, deposited by onshore swells from the south.

Another series of predominantly redeposited ridges (Nos. 54 to 78), with presumed early Choris affinities, lies to the east of the Old Whaling truncation and represents a previous severe erosion cycle under what may be the influence of southerly winds and southwesterly waves (Mason and Ludwig 1990). Twenty-two sites are located on seven of the 25 ridges, although most are diffuse lithic scatters or hearths with few artifacts. Whale bone is reported from one locality on ridge No. 65 (Table 2). Eleven of the early Choris sites are located on ridge No. 78. The location of ridge No. 78 is not provided in Giddings and Anderson (1986), but it may be anomalous in that it antedates the ridge No. 53 disconformity (based on Giddings 1967). Based on the comparatively well-dated Old Whaling ridge No. 53 (Mason and Ludwig 1990), Choris sites should date to less than 3000 C-14 yrs B.P. but to greater than 1600 C-14 yrs B.P., the maximum age of Ipiutak at Cape Krusenstern.

Sites at Krusenstern are thinly distributed. For ridges Nos. 44-51, there are an average of 12 sites per ridge, or one to two sites per 1 km of ridge length. If a single ridge persisted on the beach front for somewhere on the order of 50 to 100 yrs, then the density of Choris occupation was extremely low with perhaps one visit every seven years. One-fifth (n=18) of Choris Krusenstern localities consist only of isolated, perhaps even scavenged, whale bones, since it has never been demonstrated that the Choris people actually hunted whales (Giddings and Anderson 1986:221).

Cultural taxonomists take little notice of the whale bone "caches" reported by Giddings and Anderson (1986:210ff). Isolated finds of whale bone, without artifactual debris (so-called Features 237, 239, 245, 246 on ridge No. 44) should not be considered *prima facie* evidence of cultural activity. Ceramics are also comparatively rare, occurring at only six (6%) of the total Krusenstern sites. Of the 1623 Choris artifacts (excluding potsherds) reported in Giddings and Anderson (1986), nearly half (43%) are from just four lithic scatters. Unfortunately, no tabulation of potsherd frequencies is available from the Krusenstern Choris sites. Nor are any of the Krusenstern localities radiometrically dated. Instead, age is assigned by relative beach ridge position. For example, sites on ridges 44 to 46 are considered contemporaneous with the Choris Peninsula type site that has C-14 dates between 2500 and 3000 C-14 yrs B.P. Thus, sites on ridges 47 to 52 are landward and presumably older than 3000 C-14 yrs B.P. A revised interpretation places ridge No. 53 formation at 3000 C-14 yrs B.P. (Mason and Ludwig 1990). By both interpretations, most Krusenstern Choris sites are younger than 2500 C-14 yrs B.P.

Cape Espenberg

The Choris/Norton "ridge" set at Cape Espenberg is a scarp that was storm-eroded into older (pre-3300 C-14 yrs B.P.) beach/dune ridges, overtopped by dunes, and subsequently dissected by blowouts. The timing of the storms is estimated by geological C-14 ages at 3300-1700 C-14 yrs B.P. (Mason and Jordan 1993). Additional archaeological C-14 ages are associated with buried soils (O horizons), implying that eolian activity ceased and a stabilized surface developed. The important implication here is that Choris/Norton occupations may co-occur with a marked decrease in storminess over the Chukchi Sea (Mason and Gerlach n.d.).

Table 2. Early Choris localities at Cape Krusenstern (n=22) (following Anderson in Giddings and Anderson 1986:268ff, but see Mason and Ludwig 1990 for alternate age estimates).

No sites on beaches 55,56,58,60-64,66-74,76,77.

Beach 53
2 sites; Feature 394 with burinated flake knife, small flake scatter and Feature 460 with three non-diagnostic black chert chips.

Beach 54
1 site; Feature 617, with "Norton-like" adze, chert side blade.

Beach 57
2 sites; Feature 711 with four artifacts, especially asymmetrical bifaces, and Feature 712 with an isolated chert biface of "Norton-like" character.

Beach 59
1 site; isolated biface fragment.

Beach 65
2 sites; one, Feature 632, is a sizable flake scatter (n=85) with no diagnostic pieces; the other is an isolated whale vertebra.

Beach 75
2 sites; two hearths, one with only a burin spall, the other with no artifacts.

Beach 78
10 hearth areas with a "few chert chips and seal bones"; 3 sites recorded, of which only Feature 500 had artifacts, including burinated bifaces, curved shouldered inset blades, and no pottery or microblades.

Choris artifactual debris is scattered but still well-exposed within blowouts, principally on a 2 km stretch of the overbluff dunes of the Espenberg E-14 ridge (and equivalent aged ridges). Only four of the 39 sites identified by Schaaf (1988:Vol.2:186ff) are explicitly termed Choris, four are considered "transitional" to Norton, and five have some Choris traits. One site (KTZ-096) is termed Choris solely on the basis of a 3570±100 C-14 yrs B.P. (Schaaf 1988:Vol.2:281), but it lacks artifacts and should not be defined as Choris. Many of the Choris/Norton sites at Espenberg are associated with culinary by-products such as sea mammal cemented sand, bones, or ceramics. Charcoal was also found at several sites. Radiometric dates fall in the range of 3000-2200 C-14 yrs B.P. but a few must still be corrected for old carbon. A significant number of the Espenberg "sites" are isolated artifacts (14 of 85 loci, or 16%), and a sizable number cover less than 5 m^2 in area (n=22, or 25%). Estimates of site area for the remaining sites are artificial, because of the way that the sites were recorded.

The temporal association between many of the surface finds at Espenberg is unclear. While every flake scatter is potentially a separate behavioral event, the circumstances of discovery within deflation hollows must also be considered. The progressive eolian excavation of surfaces distorts the validity of many of the apparent associations because of blowout formation processes (Mason 1990).

The range of activities represented by artifacts at the sites is limited (Table 3). If contemporary Inupiat gender roles are applicable, many female activities (i.e., lamp tending, cooking, clothing preparation) are infrequently represented. Midden accumulations rarely exceed 5 cm in thickness, implying short-term occupations. However, the high frequency occurrence of charred sea mammal oil, a storable commodity, may indicate short-term camps, probably of only a few days or weeks duration. The consumption of stored goods may also be evidenced by the frequency of potsherds. Unburned cemented sands and associated wooden debris are assumed to represent cache pits. People (perhaps men) did remain at Espenberg long enough to resharpen tools. However, very little evidence of primary reduction processes is observed. Nothing resembling the Choris cache of points has been found at Cape Espenberg.

The range of artifacts is quite restricted at the Espenberg sites, although pottery occurs in a fair percentage (n=28, 33%) of the assemblages. The number of diagnostic pieces within sites is low. Some of the Espenberg sites are scattered over considerable distances, measuring between 85 to 150 m. The total number of recorded lithic and ceramic scatters at Espenberg is estimated at 85, a number derived by ignoring the NPS requirement of a 50 m separation to define an individual site (Schaaf 1988:Vol.1:53).

Table 3. Cultural material in Cape Espenberg sites: Choris/Norton Loci (n=85).

Artifact Type	No. of Sites	Percentage
Flakes-unretouched	54	62.4
Cemented sand	29	34.0
Potsherds	28	32.9
End/side blades	10	11.8
Burins/burinated tools	3	3.5
Scrapers	2	2.4
Lamp	1	1.2

During forays to Espenberg between 1958-1960, Giddings collected a significant number of artifacts from an uncertain number of (unmapped) localities; only 38 diagnostic pieces are illustrated in Giddings and Anderson (1986:Pl.131,132). If the Giddings' collections add another 20+ localities, then only 100 loci are known over the 20 km length of survey.

Considering the 700 yr (likely shorter) range of C-14 assays from Espenberg, a prehistoric visit or occupation may have occurred as rarely as every seven years. The actual dates of occupation may be considerably more restricted, so that annual use may have occurred for several generations or with large gaps between visits. In view of the limited artifactual remains and the nature of site structure, there is no reason to believe that Cape Espenberg was a critical part of the annual subsistence round for Choris or Norton peoples.

Choris Peninsula

Much of the notoriety of the Choris culture derives from the well-preserved organic remains found at the type site on the southwest beach ridge plain of Choris Peninsula. Another Choris/Norton site is located on the eastern beach complex of Choris Peninsula, where a side blade and end blade were found on the oldest ridge (Mason 1990). The Choris housing complex is located on the second oldest beach ridge (A-8; see Mason 1990), only 3 m above sea level and <50 m from a steep slope to the east.

The Choris beach ridges are a mixture of pea gravels and coarse and medium sands with discrete sedimentation units that are inversely graded; larger clasts were deposited at the top as a result of storm intensification. Stratigraphically, each meter thickness of beach ridge shows a succession of >10 cm thick storm beds that are separated by 1 cm thick grass beds, the product of nonstormy intervals. The position of the Choris houses has little to recommend them in terms of site locations, since they are extremely vulnerable to storm attack. Alternatively, the nature of the pole-lattice work suggests the construction of drying racks as much as house structures, a possibility not typically considered by most archaeologists. Other earlier settlements are located on bluffs above the beach ridges.

The completeness of the artifact inventory at the Choris type site has often, it seems, misled archaeologists. A wide range of artifacts occur in two houses. Most prominently, these are bone needles, non-toggling, multiple-barbed harpoon heads, dual barbed points, pecked stone lamps, and bone labrets. The potsherds (n=281) may represent the remains of only two or three pots. The fragmentary projectile points, comprising over 15% of the artifacts, differ considerably from other Choris assemblages, being marked by contracting stems and shoulders. The unappreciated resemblances with Old Whaling and Devil's Gorge forms may not be accidental or coincidental. The faunal assemblage is equally problematic, with a reputed dominance by small-sized caribou, a significant number of ringed seal flippers and beluga whale bones, as well as a wide range of other animal bones.

An intensive mapping program at Choris Peninsula by the Bureau of Indian Affairs (ANCSA office) in 1987 produced little new surficial evidence of core reduction and/or biface manufacture on the Choris ridge, "beach ridge F," according to the BIA enumeration (BIA case file F-22352). By contrast, the Norton ridge at Choris contained many sizable flake scatters, some with >1000 flakes within a 1 m^2 area. No subsurface tests on the Norton ridge were conducted by Giddings.

Southern Seward Peninsula

Several locations on the Agulaak barrier island adjacent to Lopp Lagoon in western Seward Peninsula contain Choris artifacts but lack faunal material. The Agulaak materials derive from recent channel erosion that has exposed a scatter of artifacts below a dune cutbank (Giddings and Anderson 1986:26, 225). Giddings and Anderson (1986:225) observe isolated characteristics of early, "classic," and late Choris. The use of slate implies a late Choris occupation, while the triangular inset blades are even younger and are considered similar to Norton. The presence of microblades, a flake knife, and burinated bifaces indicate an early Choris occupation. Based on apparently mutually exclusive technological categories, three different, temporally separated occupations may have occurred at Agulaak. Until controlled excavations are undertaken, any cultural or temporal associations for the Agulaak materials remain speculative. Sites contemporary with Choris, dated to 2500 C-14 yrs B.P., on nearby Kugzruk Island contain abundant evidence of sealing by Norton inhabitants of Lopp Lagoon (Gerlach et al. n.d.).

THE CHORIS HORIZON: SUDDEN APPEARANCE OR IN SITU EVOLUTION

Radiometric Ages for Choris Culture

Any discussion of whaling origins at Cape Krusenstern involves determining the age of the its occupation, assessing faunal data from contemporary, related communities, and inferring the antecedents of the residents on the whale bone ridges. In contrast to many arctic cultural taxonomists, we consider Choris to be a horizon, a geographically-widespread but thin cultural "veneer," rather than a longer-term cultural phenomena. Only two Choris sites are usually considered in the standard chronologies: Choris Peninsula and Onion Portage. The dating of the Choris Peninsula site is reasonably consistent, despite the large standard deviations (150-180 yrs) of the three C-14 determinations which average 2492±81 C-14 yrs B.P. General agreement exists between most of the seven Choris dates from Onion Portage, Band 3 (Mason and Gerlach n.d.), as reported in Anderson (1988:48). The Choris occupations in Band 3, Level 2, are in the range of 2700-2350 C-14 yrs B.P. A single, non-cultural driftwood date of 3170±120 C-14 yrs B.P. (K-835) from Level 5 below the Choris levels (Lawn 1975) has been ill-advisedly used to define the start of Choris by Anderson (1984, 1988), Giddings and Anderson 1986), Schaaf (1988), and Harritt (1994). Most of the coastal sites are undated, including all of the Cape Krusenstern and most of the Cape Espenberg localities. One Cape Espenberg site, KTZ-127, with definitive Choris lithic characteristics, has two dates within the range 2900-2700 C-14 yrs B.P. (Harritt 1994:134), while an Espenberg site containing cord-marked potsherds is comparatively young, with a date of 2285±90 C-14 yrs B.P. (Beta-17968; Schaaf 1988:Vol.2:312). The Lopp Lagoon sites are as old as 2570±50 (P-598) on "charcoal and sand" and as young as 2300±40 C-14 yrs B.P. (P-629) on wood (ages are rounded off from those presented in Giddings and Anderson 1986:30). The charcoal/sand amalgam may represent a sea mammal oil clinker containing old carbon; an age of 2400-2200 C-14 yrs B.P. is probably a close estimate for Kugzruk and Agulaak.

Siberian Pottery Complexes: Choris Origins?

Unfortunately, ceramic history and distribution is poorly documented for northeastern Chukotka, at least for a non-Russian audience. Ackerman (1982:20-21) briefly reviewed this literature (Dikov 1977, 1979), and concluded that "it would be best to consider Chukotka and western Alaska a single culture area." The linear-stamp-using ceramic cultures of the middle Anadyr River basin appear to be the most likely source area for Choris pottery. Significantly, the Ust-Bel'skaia culture dates from 2920±95 to 2860±95 C-14 yrs B.P., just older than the earliest Choris sites in Kotzebue Sound (Ackerman 1982). Because many Alaskan archaeologists have accepted the Onion Portage limiting age (3200 C-14 yrs B.P.) for the start of Choris, the antecedent ages of Ust-Bel'skaia have not been widely appreciated. Another inland, eastern Chukotsk Peninsula pottery site is even closer to Alaska, the site of Paliakvyn which yielded linear ceramics and bifacial lithics within an oval outlined house containing charcoal dated to 3100±100 C-14 yrs B.P. (MAG-918; Lozhkin and Trumpe 1990). The Paliakvyn site is 50 km upriver from the Bering Sea; unfortunately, no coastal sites dating to c. 3000 C-14 yrs B.P. are presently reported for the northern and eastern Chukotkan coasts (T. Goebel, pers. comm., 1992). This negative evidence might be easily rectified if beach ridge archaeological surveys were undertaken on the extensive barrier island/spit complexes at the entrances to Kolchinskaya Guba and other locations. Nonetheless, we cannot link either the Chukchi Archaic or Choris to an antecedent or contemporary whaling culture in Siberia.

CHUKCHI ARCHAIC AND CHORIS SUBSISTENCE PATTERNS AND RESOURCE USE BASED ON RECOVERED FAUNAL REMAINS

Whaling in the Old Whaling Phase of Chukchi Archaic: An Exegesis of Giddings' Work

The first flush of discovery by Giddings continues to dominate any discussion of Old Whaling. After just a few days of digging, Giddings (Giddings and Anderson 1986:232) felt certain of tool function:

> Whaling implements in particular excited us, and as more were uncovered and we observed the uses to which whale bone had been put, as well as the quantities of it on the beach, we could only conclude that at an unexpectedly early time there had been a settlement of whaling people here, whose culture we have designated Old Whaling.

Unfortunately, no information on the "uses" of whale bone or on its "quantities" are provided in the final report on Cape Krusenstern by Giddings and Anderson (1986). A focus on sea mammal hunting in the Chukchi Archaic tradition is clear, based on the sheer amount of oil-soaked clinker encountered in the Krusenstern houses (Giddings and Anderson 1986:233): "... fireplaces were invariably thick and consolidated. A great deal of grease or oil appears to have been used in connection with them. If the houses were occupied for a relatively short time, the remarkable thickness and density of the hearths indicate great fuel consumption." Although lamp use is reported, box-enclosed fireplaces appear to be the norm. It is unclear what type of culinary, disposal, or heating process is represented by the sea mammal-soaked clinkers. Krusenstern house supports were of small, uniformly-sized spruce poles, lacking a substantial roof. Giddings (1967:237-8) hypothesized purposeful, laborious journeys to select timber from living stands along the Noatak, instead of the casual use of driftwood, possibly not available.

Giddings' writings provide an instructive case study of argument by assertion rather than from data. As early as six months after the first Old Whaling discoveries, in the spring of 1960, Giddings announced: "The makers of the new flint forms were primarily seal hunters and whalers," while admitting his lack of "careful analysis" (Giddings 1960:127). One year later, Giddings (1961a:164) briefly mentioned a possible butchery site containing a "dismembered whale skull in the crest of the beach [ridge No. 53] together with a number of large, broad flints believed to be the blades of whaling harpoon heads." This statement implies human dismemberment rather than disarticulation which could have resulted from several taphonomic processes. Further, the use of "together" implies cultural association. A few details are available in the Giddings (1967:237) popular account, *Ancient Men of the Arctic*, published after Giddings death in 1964. The whale skull at the "winter village" was found "30 ft along the crest of the ridge beyond House 23 and only a few inches below the surface ... It appeared to have been buried intentionally ... and few flint flakes lay about its base. Part of one ear bone and another segment of bone had been removed as though needed for manufactures of some kind." Unfortunately, no further data are mentioned in the final report by Giddings and Anderson (1986:231-267). The location of the skulls is described by Giddings and Anderson (1986:270) in the context of the many whale bones on beach ridge No. 53, "though not all were marked on our map. If our projection of this beach eastward is correct, there were two whale skulls there as well as the one found in the Old Whaling village. It is likely that at least some of the whale bones were associated with the Old Whalers rather than the early Choris occupants of beach 53." The near contemporaneity of Choris and Old Whaling remains controversial.

Two arguments are used by Giddings (1967) to infer whale hunting at Krusenstern: tool morphology/size and location of the whale bone atop the crests of the beach ridge. The large bifaces must have been used for whaling, because of their similarity to late prehistoric whaling forms, so says Giddings. A large and distinctive lanceolate form, 21 cm in length, is offered as further proof. This form is nearly identical to an early Holocene biface at the Panguingine site (Powers and Maxwell 1986: Fig. 3D) in central Alaska. Hence, shall we propose whaling in the Nenana Valley? Consider Bowers's (1982:102) cautionary statement when he used a single biface to extend Old Whaling to the Brooks Range foothills: "one need not conjure up images of whales swimming up Iteriak Creek" to postulate an interior caribou hunting facies in "Old Whaling." On the second point, Giddings (1967:237) inferred human transport of the whale bones; they "could scarcely have been washed all the way up to its crest but must have been hauled there by men." This point will be discussed below as a taphonomic problem.

The recovery of whale vertebra "seats" adjacent to fire pits in Houses 21 and 24 (Giddings and Anderson 1986:238, 246) remains the most substantial evidence of an early human use of whale in the Western Arctic, a not overly convincing testament for the hunting of bowheads. In addition, a whale vertebra "lay atop the fallen wall or roof timbers" of House 23, "indicating that it had been laid on the roof or had been part of roof construction" (Giddings and Anderson 1986:243). Similar evidence is that of a whale vertebra in a shallow (hence, "summer") House 201 (Giddings and Anderson 1986:250).

Other references to faunal remains on the Old Whaling ridge are vague but entirely exclude whale bone. Unidentified, well preserved animal bones were found adjacent House 201 with the presence of "well preserved seal bones and a few other animals" scattered throughout House 202. In locating Old Whaling "summer" houses, Giddings and Anderson (1986:248) observed that "it was obvious that the hearths were of the Old Whaling culture. Their consistency as well as their *seal bones* and characteristic flints left no question as to cultural affinity" (our italics). Why not Old Sealers, then?

Giddings (1967:224) readily admits his preconceptions about whaling in the first pages of his Old Whaling chapter in *Ancient Men:* "On Beach [No.] 53, *even before* we began looking for houses we noticed the uncommon number of whale bones there and decided that the people of its day had hunted the greatest game of all — whales" (our italics).

Taphonomy and Tangible Evidence of Whaling

Research in the Eastern Arctic (Savelle and McCartney 1991; McCartney and Savelle 1993; Whitridge 1993) shows the utility of comparing whale demographic parameters to define human predation from natural mortality. Unfortunately, demographic data on Krusenstern whale bones is lacking. One agency selected as proof by Giddings (1967:237) is the inability of waves to transport whale bone onshore. However, the buoyancy of decomposing, gas-filled, oil-rich sea mammals is well attested as a marine transport mechanism (Schäfer 1972:20ff). The altitude reached by a particular carcass is a function of relative sea level which, in turn, is dependent on wave energy and the storm surge elevation of sea level. In the Chukchi Sea, storm-elevated sea levels of up to 5 m are reported (Wise et al. 1981). The possibility of mass strandings of whales in shallow areas should also be considered. Possibly, unusual ice conditions or strong gale force winds forced bowheads closer to shore during Choris and/or Old Whaling times, as indicated by the transgressive nature of the ridge No. 53. The fact that the whale skull was covered by gravel is probably equally good evidence of storm-induced deposition.

Walrus and Seal Hunting in the Northern Chukchi Sea: Wrangel Island

Archaeological data on the western and northern shores of the Chukchi Sea, although limited, provides some hints to Alaskan culture history (Gerlach and Mason 1992). About 3000 C-14 yrs B.P., a sizable "paleo-Eskimo" community thrived at Devil's Gorge on southern Wrangel Island. With shouldered bifaces inset into toggling harpoons, these people hunted walrus, bearded seal, and small seals (Dikov 1988). Unifacial tools form a significant portion (15%) of the small 1975 artifact collection (n=54), although a "great number" of artifacts and flakes were reported by Dikov (1988:81) within the bone-rich, 60 cm deep midden pits. Similarities in the toggling harpoon and the hafting styles led Ackerman (1984, n.d.) to hypothesize direct links (i.e., descent) between the Alaskan Old Whaling culture and the community at Devils Gorge on Wrangel Island. Significantly, no pottery was found on Wrangel Island.

The five radiocarbon dates from Devil's Gorge imply, at face value, a lengthy occupation: the oldest is 3360±155 C-14 yrs B.P. (MAG-198) and the youngest is 2851±50 C-14 yrs B.P. (MAG-415). On the basis of a few brief contextual comments in Dikov (1988:85), it is difficult to ascertain whether the same cultural layer is dated and whether there is any possibility of contamination with sea mammal old carbon. Using the two sigma range, the ages from the C-14 dates at Devil's Gorge are mutually exclusive and the occupation may be as old as 3700-3000 C-14 yrs B.P. or as recent as 2950-2750 C-14 yrs B.P. The latter range would be contemporaneous with the Krusenstern notched point occupation. Several alternative explanations for the differing date ranges are possible and involve either taphonomic processes or human behavior. First, the materials used for dating may not reflect the time of occupation. Dates on sea mammal materials can be in error because sea mammals may have consumed deep water carbon with the surface water. This could add 400 to 800 yr to an age estimate (Taylor 1987). The use of long-lived trees can introduce an error, if large samples are used or if old wood were scavenged from the beach. Both possibilities could add 250-400 yrs. The date range could reflect repeated but isolated visitations over a 1000 year period. The least likely scenario is an intensive, thousand year occupation, especially in light of the fact that there is only a handful of artifacts from the site.

Faunal remains at Devil's Gorge are not yet analyzed (Dikov 1988:81), but include bones of walrus, bearded seal, "small" seal, and bird ("ducks" as well as other unidentified birds). All observers have noticed that Devil's Gorge is located at the northern extreme of the walrus migration range toward the polar ice. Significantly, no trace of whale bone is evident on Wrangel Island. Large bifaces are reported from the 1981 excavations by T. S. Tein, but otherwise the tool inventory is limited. The harpoon head found is certainly large enough for a whaling lance head, but lacking faunal data, no claims are made for whaling by Devil's Gorge people.

Seal Hunters of Choris Peninsula and Lopp Lagoon

Gerlach and colleagues (n.d.) reanalyzed the 1950s non-avian faunal collections (n=1500) of Giddings from Choris and Seward peninsulas. A wide variety of species were harvested, including fox, bear, moose, rabbit, beaver, musk-ox, and birds. However, caribou and ringed seal provided the focal point of subsistence harvest for Choris people. Seal represented only one-third of the faunal remains at Choris Peninsula, but 43% and 51%, respectively, at the western Seward Peninsula sites of Singauruk and Kugzruk. The disproportionate amount of seal appendicular and axial elements indicates the operation of cultural selection or possibly dog consumption. Skulls predominate in terms of minimum animal units (MAUs), and may reflect ritual behavior rather than disposal. Measurements on Choris seal mandibles are identical to Baffin Island adult seals from pack ice. The high number of adult seals indicates hunting at breathing holes, as seals segregate by age on the ice. Hunting occurred on the ice, and delicacies such as flippers were removed for consumption at the camp. All told, the hunting strategy resembled a late winter ice hole-hunting episode as described by Lucier and VanStone (1991b) for the 19th century. Caribou bones at Choris sites reflect an opposite selection process than that for seal. Bones of low use-value (mandible, ribs, sternum, vertebrae, extremities) occur in low frequencies, suggesting that caribou were transported over considerable distances. No evidence of caribou herding can be supported by the data. Because the minimum number of individuals (MNIs) of caribou are similar to that of seal, Gerlach et al. (n.d.) argue that a significant part of the subsistence cycle (fall to spring) is represented at the three Choris sites; missing is the possible summer fish camp, if 19th century analogies are appropriate. At Cape Espenberg, the only bone dated to Choris times is burned seal bone (Schaaf 1988:2:236), and the occurrence of the residue of sea mammal oil, charcoal clinkers, is good proxy evidence for sea mammal consumption at Capes Espenberg and Krusenstern.

Whaling at Cape Krusenstern During Choris?

The Cape Krusenstern Choris ridges contain such abundant whale bone (presumably from bowheads) that Giddings singled them out as "the" whale bone ridges, implying that the amount of bone debris exceeded all other ridges. What are the implications of this for prehistoric subsistence and for paleoclimatology? First, as yet there is no definitive faunal evidence of human hunting of whales, or, in the case of the Choris culture, even of artifacts whose size indicates whaling to some workers (Giddings and Anderson 1986). The circumstantial evidence of isolated vertebrae and other elements suggests a selective taphonomic agent, possibly, although not demonstrably, human. The occurrence of isolated whale elements on the ridge crest is just as suggestive of sorting and redistribution by wave action. However, the relative abundance of whale bone during Choris and Chukchi Archaic times remains a subject of concern. Until controlled faunal collections and analyses are available, several working hypotheses with respect to the whaling problem are suggested: (a) higher bone concentrations reflect higher non-cultural mortality or higher absolute numbers of whales; (b) higher whale mortality was due to unsuccessful hunting; (c) less human interest was expressed in dead whales as compared to later times; and (d) more whale bones on the beach because human populations were lower and there was less scavenging activity.

Little is known about the climatic interrelationships of whale numbers and zooplankton, but this seems a promising area for future investigations. Bowheads in the Beaufort Sea require high densities of zooplankton, which are found at or near oceanic fronts and in nutrient upwelling conditions such as those caused by water column turbulence (Fraker 1989). Perhaps, during the onset of the Neoglacial period, zooplankton numbers increased due to increased water column turbulence, less fresh water inflow, and greater ice cover. The occasional association of whale bones and human artifacts or habitations would argue against hypothesis (c) above. Given what is known about Choris site numbers, it is most likely that the high density of whale bones reflects a combination of low human populations and natural taphonomic processes.

In assessing the role of natural factors, recall that longshore drift during Choris times, as today, followed a north to south pathway (Moore 1966; Mason 1990). Thus, the source of beached whales lies to the north of Krusenstern, in the lee of Cape Thompson. Cape Krusenstern served as a "sink" for whale bones that were ultimately derived from offshore. The whale population in question is that of the central Chukchi Sea, one that was severely depleted by European whalers (Fraker 1989). Beach ridge geomorphology may partially explain whale bone abundance estimates. For example, if the nine whale bone ridges date to 3000-2300 C-14 yrs B.P. at Cape Krusenstern, then one ridge reflects over 100 yrs of accumulation. Conversely, the 20 Ipiutak ridges reflect only 400 yrs or 25 yrs of accumulation per ridge during the period dated to 1600-1200 C-14 yrs C-14 yrs B.P. (Gerlach and Mason 1992). The implication here is that four times as many bones would be expected on any one Choris ridge as on an Ipiutak ridge.

CHUKCHI ARCHAIC AND CHORIS SITE DISTRIBUTION: A PROXY FOR LOW POPULATION DENSITY?

Both Old Whaling and Choris people arrived on an Alaskan shoreline that we suspect was virtually uninhabited. A considerable temporal gap exists between the youngest ASTt sites (3500-3700 C-14 yrs B.P.) and the oldest Choris sites (c. 3000 C-14 yrs B.P.). This is particularly apparent on the coast, especially if one accepts the chronometric revisions of Mason and Ludwig (1990) for ridge Nos. 54-78 at Cape Krusenstern. If the ASTt migration thesis of the Canadian Arctic is accepted (discussed in Maxwell 1985), then the apparent abandonment of Alaska must be tied in some way to subsistence resource fluctuations, as originally proposed by McGhee (1981). The Old Whaling/Chukchi Archaic tradition was possibly a more widespread phenomenon, especially if the definition is expanded to include the Palisades and Security Cove sites (Ackerman 1964). Most archaeologists are not comfortable with a classification of these sites with the "interior-based" Northern Archaic tradition. Without more solid evidence to argue from, it is possible that intersocietal relationships were considerably more fluid than we presume, that riverine groups had sufficient maritime capabilities if climatic and ecological conditions changed, and/or, alternatively, that Chukchi Archaic represents successful sea mammal hunting. In this scenario, a reversal in hunting success may have led to the demise and migration of the population northward to Canada, as witnessed by the notched point assemblage of the Crane site (Le Blanc 1994). It is a distinct possibility that Choris expanded into and consumed the niche previously occupied by Chukchi Archaic, again noting the co-occurrence of Choris-like barbed harpoons with notched points at the Crane site. As Willey (1966: 431) observed, "Old Whaling ... in no way provides a transition from Denbigh to Choris-Norton," it apparently was not "involved directly in the evolution of either the Choris-Norton-Near Ipiutak or Northern Maritime subtraditions but suggests another line of development," and "it points up the historical complexity of events in the western Arctic during the second millennium B.C." Although not relevant to the question of maritime adaptations, side-notched points are dated as recently as 2000 yrs B.P. in the upper Tanana valley (Clark 1991:50).

Several critical data points are lacking to link northern Chukotka and Kotzebue Sound into a single *oikoumene* or even a trade network (*sensu* Caldwell 1964). Any trade appears to have been in perishable goods or in clay. Few of these goods, it seems, reached any distance inland and no site can be ascribed as a market place or *entrepôt* for goods. The distribution of pottery is extremely limited, and does not suggest a substantial movement of goods or services. No evidence of warfare or of trade in exotic lithics is reported. Overall, Choris site density, hence population density, appears too low for a consumer population requiring elaborate redistributive systems. Labretifery may suggest boundary maintenance, status differences, differential hunting prowess, if not access to goods, or, alternatively, exclusively mystico-religious beliefs. In view of the low population numbers for Choris in Kotzebue Sound, little need is apparent for boundary markers. Choris societies, then, may represent the early colonizers, the leading edge of a population expanding into a new area *from* one with a high population that used boundary markers. A number of key traits, labretifery for example, support the original proposal by Laughlin (1962) for a southern origin for Choris, termed the "boat load of Aleuts" model as followed up by Dumond (1990). Again, too little data exist to assess this possibility.

CONCLUSIONS

The high frequency of whale bone on the Choris ridges is noteworthy, although no claims are made for clear cultural association or use of the bone assemblages. The occurrence of whale bone probably reflects a low human population density rather than the efficiency of an incipient whale hunting technology. The same case may be offered for the Old Whaling phase of the Chukchi Archaic, despite the presence of putative whaling harpoon blades of considerable size. Although the lithic tools of Chukchi Archaic and Choris people are sufficiently large for dispatching whales, no definitive harpoon heads have been found to substantiate this inference. The large bifaces are commonly termed lance heads, and the agility in using such an implement presumes a facility with bowhead whale hunting not commonly reported in the ethnographic literature. Further, the limited associations of whale bones and Choris or Old Whaling hearths are insufficiently documented for clear use to be established.

The preceding description of Chukchi Archaic and Choris site density is deceptive. Although the greatest concentration of sites lies on the Chukchi Sea coast, does this reflect prehistoric subsistence and settlement patterns? The handful of side-notched point sites that occur on the coast (Security Cove on Kuskokwim Bay, Cape Krusenstern, the Crane site near Cape Bathurst in the Beaufort Sea) are separated by thousands of kilometers. The nearly 200 Choris sites at Cape Krusenstern and Cape Espenberg are extremely transient phenomena, based on limited artifactual remains and site size. At Cape Espenberg, a Choris visit may have occurred as infrequently as every seven years, because all midden accumulations are less than 5 cm thick. Pottery implies that food storage or cooking played an important role, but this may be inconsistent with the suspected nature of occupational function. Similarly, the Choris sites at Krusenstern occur on one-third (13) of the 34 beach ridges deposited during an estimated 750 years of potential Choris occupations, and, hence, also may record visits only every seven years. Five or six discrete tool-making events represent the bulk of the artifactual remains at Cape Krusenstern.

The recent ethnohistorical article by Lucier and VanStone (1991b) may prove to be one of the most critical building blocks for understanding western arctic prehistory. Discussing the settlement pattern of Kotzebue Sound Inupiat, Lucier and VanStone (1991b) recount an annual round that could potentially leave little or no trace on the terrestrial landscape. Winter and early spring sealing was conducted from snow houses placed on the ice, in contradistinction to most of historic Alaska. Ice in Kotzebue Sound is peculiarly stable in most years, a fact that probably was even more pertinent in the frigid 19th century. An inland focus during the summer and fall required that people remain fairly mobile in the winter.

Both the Choris and Chukchi Archaic occupations are a particularly thin overlay on the Alaskan landscape, so thin that it is the favorable geomorphology for site discovery rather than occupational intensity that must account for their archaeological detection at all. Notched points are comparatively rare in northwestern Alaska, although no detailed inventory is yet available. Apparently, Choris people were highly mobile, ranging far inland in search of caribou and other terrestrial resources. The low densities of Choris sites in the interior is probably a better reflection of Choris occupational density at any one point in time.

We recommend that analysis of existing collections be undertaken in order to distinguish blood residue traces on artifacts (Eisele 1994; Gerlach et al. n.d.; Newman et al. n.d.) and finding microwear patterns indicative of butchery. Procedures must be refined to better distinguish whale hunting from the scavenging of whale carcasses by using demographic patterns reconstructible from faunal data (cf. McCartney and Savelle 1993). In view of the dearth of faunal collections, it is urgent that any remaining, unexcavated Cape Krusenstern whale bone sites be excavated in order to distinguish natural whale mortality from human predation. Archaeologists must enlist native cooperation and, hopefully, inculcate a sense of the value of archaeology as a historic pursuit. Surely, the undertaking of archaeological research by Native Alaskans will produce an alternative point of view and, perhaps, unexpected syntheses. Archaeology in northwestern Alaska is in a crisis: sites are in the process of destruction by erosion and by pot-hunting. Lack of problem orientation can stifle research in the coming decades, along with the increased politicization of archaeology. Moreover, we need to continually reexamine the nature of archaeological data, with an eye towards questioning interpretations that have outlived the data upon which they are made and, thus, their usefulness in reconstructing the past.

REFERENCES

Ackerman, Robert E.
- 1964 Prehistory in the Kuskokwim-Bristol Bay Region, Southwestern Alaska. *Washington State University Laboratory of Archaeology, Reports of Investigations* No. 26.
- 1982 The Neolithic-Bronze Age Cultures of Asia and the Norton phase of Alaskan Prehistory. *Arctic Anthropology* 19(2):11-38.
- 1984 Prehistory of the Asian Eskimo Zone. In: *Handbook of North American Indians, Vol. 5, Arctic,* edited by D. Damas, pp. 106-118. Smithsonian Institution, Washington, D.C.
- 1988a Settlements and Sea Mammal Hunting in the Bering-Chukchi Sea Region. *Arctic Anthropology* 25(1):52-79.
- 1988b Late Prehistoric Settlement at Chagvan Bay. In: The Late Prehistoric Development of Alaska's Native People, edited by R. D. Shaw, R. K. Harritt, and D. E. Dumond, pp. 169-188. *Aurora: Alaska Anthropological Association Monograph Series* No. 4.
- n.d. Development of the Northern Maritime Tradition: The Bering Strait Region. Paper presented at the U.S-Japan Joint Seminar on Origins, Development, and Spread of Prehistoric North Pacific-Bering Sea Maritime Cultures, Honolulu, 1993.

Aigner, Jean S.
- 1966 Bone Tools and Decorative Motifs from Chaluka, Umnak Island. *Arctic Anthropology* 3(2):57-85.
- 1978 Activity Zonation in a 4000 Year Old Aleut House, Chaluka Village, Umnak Island, Alaska. *Anthropological Papers of the University of Alaska* 19(1):11-25.

Anderson, Douglas D.
- 1972 An Archaeological Survey of Noatak Drainage, Alaska. *Anthropological Papers of the University of Alaska* 9(1):66-117.
- 1983 Changing Prehistoric Eskimo Subsistence Patterns: A Working Paper. In: *Cultures of the Bering Sea Region, Papers from an International Symposium* (Moscow), edited by

H. N. Michael and J. W. VanStone, pp. 62-83. International Research and Exchanges Board, New York.

1984   Prehistory of North Alaska. In: *Handbook of North American Indians*, Vol.5, *Arctic*, edited by D. Damas, pp. 80-93. Smithsonian Institution, Washington, D.C.

1988   Onion Portage: The Archaeology of a Stratified Site from the Kobuk River, Northwest Alaska. *Anthropological Papers of the University of Alaska* 22(1-2):1-163.

Arnold, Dean

1985   *Ceramic Theory and Cultural Process*. Cambridge University Press, Cambridge.

Black, Lydia

1991   *Glory Remembered: Wooden Headgear of Alaska Sea Hunters*. Alaska State Museum, Juneau.

Bockstoce, John R.

1973   A Prehistoric Population Change in the Bering Strait Region. *Polar Record* 16:793-803.

1979   The Archaeology of Cape Nome. *University of Pennsylvania, University Museum Monograph* 38.

Bowers, Peter M.

1982   The Lisburne Site: Analysis and Cultural History of a Multi-Component Lithic Workshop in the Iteriak Valley, Arctic Foothills, Northern Alaska. *Anthropological Papers of the University of Alaska* 20(1-2): 79-112.

Caldwell, J. R.

1964   Interaction Spheres in Prehistory. *Illinois State Museum Scientific Paper* 12(6):133-143.

Clark, Gerald

1977   Archaeology on the Alaska Peninsula: The Coast of Shelikof Strait, 1963-1965. *University of Oregon Anthropological Papers* No.13.

Clark, Donald W.

1991   *Western Subarctic Prehistory*. Canadian Museum of Civilization, Hull.

Coachman, L. K.

1993   On the Flow Field in the Chirikov Basin. *Continental Shelf Research* 13(5/6):481-508.

Coachman, L. K. and D. A. Hansell (editors)

1993   ISHTAR: Inner Shelf Transfer And Recycling in the Bering and Chukchi Seas. *Continental Shelf Research* 13(5/6):473-704

Collins, Henry B., Jr.

1937   Archaeology of St. Lawrence Island, Alaska. *Smithsonian Miscellaneous Collection* 96(1).

Crane, H. R. and James B. Griffin

1964   University of Michigan Radiocarbon Dates IX. *Radiocarbon* 6:21-22.

Davis, Craig W., Dana C. Linck, Kenneth M. Schoenberg, Harvey M. Shields

1981   Slogging, Humping and Mucking Through the NPR-A: An Archaeological Interlude, Vol. 2. *Cooperative Park Studies Unit Occasional Paper* No. 25. University of Alaska Fairbanks.

Davis, Stanley

1989   Cultural Component III. In: The Hidden Falls Site, Baranof Island, Alaska, edited by S. Davis, pp. 275-344. *Aurora: Alaska Anthropological Association Monograph Series* No. 5.

Dekin, Albert A.
   1984   The Archaeology of a Contemporary Whaling Society: The Later Prehistory, Ethno-archaeology and Recent Ethnohistory of Whaling in the Barrow, Alaska area. In: *Arctic Whaling: Proceedings of the International Symposium on Arctic Whaling*, February, 1983, edited by H. K. s'Jacob, K. Snoeijing, and R. Vaughan, pp. 113-120. Arctic Centre, University of Groningen.

Dikov, Nikolai N.
   1977   *Arkheologicheskie Pamiatniki Kamchatki, Chukotki i Verkhnei Kolymy (Aziia na styke Amerikoi v Drevnosti)*. (Archaeological Sites of Kamchatka, Chukotka and the Upper Kolyma: Asia Joining America in Ancient Times). Izdatel'stvo, Nauka, Moscow.
   1979   *Drevie Kul'tury Severo-vostochoi Azii: Aziia na Styke s Amerikoi v Drevosti* (Early Cultures of Northeast Asia: Asia Joining America in Ancient Times). Izdatel'stvo, Nauka, Moscow.
   1988   The Earliest Sea Mammal Hunters of Wrangell Island. *Arctic Anthropology* 25(1):80-93.

Dumond, Don E.
   1969   The Prehistoric Pottery of Southwestern Alaska. *Anthropological Papers of the University of Alaska* 14(2):19-42.
   1972   Prehistoric Population Growth and Subsistence Change in Eskimo Alaska. In: *Population Growth: Anthropological Implications*, edited by B. Spooner, pp. 311-328. MIT Press, Cambridge.
   1977   *Eskimos and Aleuts*. Thames and Hudson, London.
   1984   Prehistory of the Bering Sea Region. In: *Handbook of North American Indians, Vol. 5, Arctic,* edited by D. Damas, pp. 94-105. Smithsonian Institution, Washington, D.C.
   1990   A Southern Origin for Norton? Paper presented at the 17th Annual Meeting of the Alaska Anthropological Association, Fairbanks.

Eisele, Judith
   1994   Survival and Detection of Blood Residues on Stone Tools. *Department of Anthropology, University of Nevada, Reno, Technical Report* 94-1.

Fitzhugh, William W.
   1984   Paleo-Eskimo Cultures of Greenland. In: *Handbook of North American Indians*, Vol. 5, *Arctic,* edited by D. Damas, pp. 528-539. Smithsonian Institution, Washington, D.C.

Fraker, Mark A.
   1989   Aspects of the Ecology of the Bowhead Whale (*Balaena mysticetus*) in the Western Arctic. In: *Proceedings of the 6th Conference of the Comité Arctique International*, edited by L. Rey and V. Alexander, pp. 252-279. E. J. Brill, Leiden.

Gerlach, S.Craig and Owen K. Mason
   1992   Calibrated Radiocarbon Dates and Cultural Interaction in the Western Arctic. *Arctic Anthropology* 29(1):54-81.

Gerlach, S.Craig, Mark Diab, and Owen K. Mason
   n.d.   The Bones that Didn't Sink: A Re-Analysis of the Faunal Assemblage from Choris Peninsula. MS. on file, Department of Anthropology, University of Alaska Fairbanks, pp. 32.

Giddings, J. Louis
   1960   The Archaeology of Bering Strait. *Current Anthropology* 1(2):121-138.
   1961a   Cultural Continuities of Eskimos. *American Antiquity* 27(2):155-173.
   1961b   Onion Portage and Other Flint Sites of the Kobuk River. *Arctic Anthropology* 1(1):6-27.

1962 Side-Notched Points Near Bering Strait. In: Prehistoric Cultural Relations Between the Arctic and Temperate Zones of North America, edited by J. M. Campbell, pp. 35-38. *Arctic Institute of North America Technical Paper* No. 11. University of Toronto Press, Toronto.

1964 *The Archeology of Cape Denbigh.* Brown University Press, Providence.

1967 *Ancient Men of the Arctic.* Alfred A. Knopf, New York.

Giddings, J. Louis and Douglas D. Anderson

1986 Beach Ridge Archaeology of Cape Krusenstern: Eskimo and Pre-Eskimo Settlements Around Kotzebue Sound, Alaska. *National Park Service, Publications in Archeology* No. 20.

Gillispie, Thomas E.

1985 Radiocarbon Dated Notched Biface Sites in Alaska. Paper presented at the 12th Annual Meeting of the Alaska Anthropological Association, Anchorage.

Griffin, James B. and Roscoe H. Wilmeth, Jr.

1964 The Ceramic Complexes at Iyatayet. In: *The Archeology of Cape Denbigh*, J. Louis Giddings, pp. 271- 303. Brown University Press, Providence.

Harritt, Roger K.

1991 Relationships Between Whale Hunting, Human Social Organization, and Subsistence Economies in Coastal Areas of Northwest Alaska During Late Prehistoric Times. In: *Proceedings of the International Conference on the Role of Polar Regions in Global Change*, Vol. 1, edited by G. Weller et al., pp. 401-405. Geophysical Institute, Fairbanks.

1992 Continuity and Change in Prehistoric Eskimo Socio-Territorial Patterns in Bering Strait. Paper presented at the 22nd Arctic Workshop, Institute of Arctic and Alpine Research, Boulder.

1994 Eskimo Prehistory on the Seward Peninsula. *National Park Service Research Report AR* 21, Alaska Regional Office, Anchorage.

Hough, Walter

1898 The Lamp of the Eskimo. *U.S. National Museum Annual Report for 1896*, pp. 1027-1057. Washington, D.C.

Keddie, Grant

1981 The Use and Distribution of Labrets on the North Pacific Rim. *Syesis* 14:59-80.

Krupnik, Igor

1987 The Bowhead vs. the Gray Whale in Chukotkan Aboriginal Whaling. *Arctic* 40(1):16-32.

1993 *Arctic Adaptations: Native Whalers and Reindeer Herders of Northern Eurasia.* Dartmouth College, University Press of New England, Hanover, NH.

Larsen, Helge

1968 Trail Creek: Final Report on the Excavation of Two Caves on Seward Peninsula, Alaska. *Acta Arctica* 15. Munksgaard, Copenhagen.

Larsen, Helge and Froelich Rainey

1948 Ipiutak and the Old Whaling Culture. *Anthropological Papers of the American Museum of Natural History* 42.

Laughlin, William S.

1962 Bering Strait to Puget Sound: Dichotomy and Affinity Between Eskimo-Aleuts and American Indians. In: Prehistoric Cultural Relations Between the Arctic and Temperate Zones of North America, edited by J. M. Campbell, pp. 113-125. *Arctic Institute of North America Technical Paper* No. 11. University of Toronto Press, Toronto.

    1981   Anangula and Chaluka Investigations of 1972. *National Geographic Society Research Reports* 13:365-379.

Lawn, Barbara
    1975   University of Pennsylvania Radiocarbon Date List XVIII. *Radiocarbon 17(2):210.*

LeBlanc, Raymond
    1994   The Crane Site: A Paleo-Eskimo Site in the Western Canadian Arctic. *Canadian Museum of Civilization, Mercury Series, Archaeological Survey of Canada Paper* No. 148.

Lozhkin, A. V. and M. A. Trumpe
    1990   Sistematizatsiia Radiouglerodnykh Datirovok Arkheologischeskikh Pamiatnikov Magadanskoi Oblasti. In: *Drevie Pamiatniki Severa Dal' Nego Vostoka*, pp. 176-179. Nauka, Magadan.

Lucier, Charles V. and James W. VanStone
    1991a   The Traditional Oil Lamp Among Kangigmiut and Neighboring Inupiat of Kotzebue Sound, Alaska. *Arctic Anthropology* 28(2):1-14.
    1991b   Winter and Spring Fast Ice Seal Hunting by Kangigmiut and Other Inupiat of Kotzebue Sound. *Etudes/Inuit/Studies* 15(1):29-49.
    1992   Historic Pottery of the Kotzebue Sound Inupiat. *Fieldiana: Anthropology* (New Series) No. 18.

Lutz, Bruce J.
    1972   *A Methodology for Determining Regional Intracultural Variation Within Norton, An Alaskan Archaeological Culture.* Unpublished Ph.D. dissertation, Department of Anthropology, University of Pennsylvania.

McCartney, Allen P.
    1984   History of Native Whaling in the Arctic and Subarctic. In: *Arctic Whaling: Proceedings of the International Symposium on Arctic Whaling*, February, 1983, edited by H. K. s'Jacob, K. Snoeijing, and R. Vaughan, pp. 79-112. Arctic Centre, University of Groningen.

McCartney, Allen P. and James M. Savelle
    1993   Bowhead Whale Bone and Thule Eskimo Subsistence Patterns in the Central Canadian Arctic. *Polar Record* 29:1-12.

McClenahan, Patricia and Douglas E. Gibson
    1990   Cape Krusenstern National Monument: An Archaeological Survey, Vol. 2, *National Park Service Research/Resource Management Report* AR-17, Alaska Regional Office, Anchorage.

McGhee, Robert
    1969   Speculations on Climatic Change and Thule Culture Development. *Folk 1970* 11-12:173-184.
    1972   Climatic Change and the Development of Canadian Arctic Cultural Traditions. In: Climatic Change in Arctic Areas During the Last Ten Thousand Years, edited by Y. Vasari et al., pp. 39-60. *Acta Universitatis Ouluensis*, Series A, *Geologica*, Oulu, Finland.
    1981   Archaeological Evidence for Climatic Change During the Last 5000 Years. In: *Climate and History*, edited by T. M. L. Wigley et al., pp. 162-179. Cambridge University Press, Cambridge.

Mason, Owen K.
    1990   *Beach Ridge Geomorphology of Kotzebue Sound: Implications for Paleoclimatology and Archaeology.* Unpublished Ph.D. dissertation, Quaternary Science, University of Alaska Fairbanks.

Mason, Owen K. and S. Craig Gerlach
n.d. Chukchi Sea Hot Spots, Paleo-Polynyas and Caribou Crashes: Climatic and Ecological Constraints on Northern Alaska Prehistory. *Arctic Anthropology*. In press.

Mason, Owen K., David M. Hopkins, and Lawrence Plug
n.d. Chronology and Paleoclimate of Storm-Induced Erosion and Episodic Dune Growth Across Cape Espenberg Spit, Alaska, U.S.A. MS. on file, Alaska Quaternary Center, University of Alaska Fairbanks, pp. 55.

Mason, Owen K. and James W. Jordan
1993 Heightened North Pacific Storminess and Synchronous Late Holocene Erosion of Northwest Alaska Beach Ridge Complexes. *Quaternary Research* 40(1):55-69.
1994 Eustatic Sea-Level Changes in Northwest Alaska During the Last 3000 Years, Abstracts with Programs, Annual Meeting, Geological Society of America, p. A309.

Mason, Owen K. and Stefanie L. Ludwig
1990 Resurrecting Beach Ridge Archaeology: Parallel Depositional Records from St. Lawrence Island and Cape Krusenstern. *Geoarchaeology* 5(4): 349-373.

Maxwell, Moreau
1985 *Prehistory of the Eastern Arctic*. Academic Press, New York.

Minc, Leah and Kevin P. Smith
1989 The Spirit of Survival: Cultural Responses to Resource Variability in North Alaska. In: *Bad Year Economics: Cultural Responses to Risk and Uncertainty*, edited by P. Halstead and J. O'Shea, pp. 8-39. Cambridge University Press, Cambridge.

Moore, George W.
1966 Arctic Beach Sedimentation. In: *Environment of the Cape Thompson Region, Alaska*, edited by N. Wilimovsky and J. N. Wolfe, pp. 587-608, Atomic Energy Commission, Oak Ridge, TN.

Nelson, Edward W.
1899 Eskimo About Bering Strait. *18th Annual Report of the Bureau of American Ethnology for the Years 1896-1897*. Washington, D.C.

Newman, M. E., H. Ceri, and B. Kooyman
n.d. The Use of Immunological Methods to Detect Archaeological Residues: A Reply to Eisele. MS. on file, pp. 11.

Nowak, Michael
1982 The Norton Period of Nunivak Island: Internal Change and External Influence. *Arctic Anthropology* 19(2):75-92.

Okada, H., A. Okada, K. Yajima, O. Miyaoka, and C. Oka
1982 *The Qaluyaarmiut: An Anthropological Survey of the Southwestern Alaska Eskimo*. Department of Behavioral Science, Hokkaido University, Sapporo.

Oswalt, Wendell
1955 Alaskan Pottery: A Classification and Historical Reconstruction. *American Antiquity* 21(1):32-43.

Rudenko, S. I.
1961 The Ancient Culture of the Bering Sea and the Eskimo Problem. *Arctic Institute of North America, Anthropology of the North, Translations from Russian Sources* No. 1. University of Toronto Press, Toronto.

Powers, W. Roger and Howard E. Maxwell
1986 Lithic Remains from Panguingue Creek: An Early Holocene Site in the Northern Foothills of the Alaska Range. *Alaska Historical Commission Studies in History* No. 189.

Savelle, James M. and Allen P. McCartney
   1991   Thule Eskimo Subsistence and Bowhead Whale Procurement. In: *Human Predators and Prey Mortality*, edited by M. C. Stiner, pp. 201-216. Westview Press, Boulder.

Schaaf, Jeanne
   1988   The Bering Land Bridge: An Archaeological Survey. 2 vols. *National Park Service Resources Management Report* 14. Alaska Regional Office, Anchorage.

Schäfer, Wilhelm
   1972   *Ecology and Palaeoecology of Marine Environments*. University of Chicago Press, Chicago.

Sheehan, Glenn W.
   1985   Whaling as an Organizing Focus in Northwestern Alaskan Eskimo Society. In: *Prehistoric Hunter-Gatherers: The Emergence of Cultural Complexity*, edited by T. D. Price and J. A. Brown, pp. 123-154. Academic Press, New York.

Stanford, Dennis J.
   1973   *The Origins of Thule Culture*. Unpublished Ph.D. dissertation, Department of Anthropology, University of New Mexico.

Springer, A. M. and C. P. McRoy
   1993   The Paradox of Pelagic Food Webs in the Northern Bering Sea-III. Patterns of Primary Production. *Continental Shelf Research* 13(5/6):57 5-599.

Taylor, R.
   1987   *Radiocarbon Dating: An Archaeological Perspective*. Academic Press, New York.

Taylor, William E., Jr.
   1963   Hypotheses on the Origin of the Canadian Thule Culture. *American Antiquity* 28(4):456-464.

Walsh, J. J., C. P. McRoy, L. K. Coachman, J. J. Goering, J. J. Nihoul, T. E. Whitledge, T. H. Blackburn, P. L. Parker, C. D. Wirick, P. G. Shuert, J. M. Grebmeier, A. M. Springer, R. D. Tripp, D. A. Hansell, S. Djenidi, E. Deleersnijder, K. Henriksen, B. A. Lund, P. Andersen, F. E. Müller-Karger, and K. Dean
   1989   Carbon and Nitrogen Cycling Within the Bering/Chukchi Seas: Source Regions for Organic Matter Effecting AOU Demands of the Arctic Ocean. *Progress in Oceanography* 22:277-359.

Whitridge, Peter
   1993   The Spatial Analysis of Thule Whale Bone Distributions: A Case Study from the Central Canadian Arctic. Paper Presented at the 58th Annual Meeting of the Society for American Archaeology, St. Louis.

Willey, Gordon R.
   1966   *An Introduction to American Archaeology*, Vol. I, *North and Middle America*. Prentice Hall, New York.

Willey, Gordon R. and Philip Phillips
   1958   *Method and Theory in American Archaeology*. University of Chicago Press, Chicago.

Wise, J. L., A. L. Comiskey, and R. Becker
   1981   *Storm Surge Climatology and Forecasting in Alaska*. Arctic Environmental Information and Data Center, Anchorage.

The Development and Spread of the Whale Hunting Complex in Bering Strait: Retrospective and Prospects

Roger K. Harritt
National Park Service, Alaska Regional Office
2525 Gambell Street
Anchorage, AK 99503-2892

Abstract. *The origins and development of Eskimo whaling in western Alaska, the Bering Sea islands, and Chukotka are examined in this paper. Evidence for whaling subsistence/technology is reviewed within the major western and northwestern Alaskan archaeological cultures, beginning with the 5500-year-old Denbigh Flint complex and especially within Birnirk, Punuk, and Thule contexts of the past 1500 years. Historic Alaskan Eskimo whaling appears to have arisen from the intrusion of Punuk influences, centered on the coasts of Chukchi Peninsula, into Birnirk territory in western and northwestern Alaska at c. A.D. 600-800. Relatively large coastal villages with an effective sociopolitical organization, the regular presence of whales, and an effective boat-harpoon technology were pre- or co-requisites of such a whaling complex. It is suggested that the Alaskan bilaterally descended local family formed the basis for development of the patrilineal clan organization on the Siberian side of Bering Strait, where large, ranked, clan-like social structures initially developed that facilitated whale hunting and butchering. Such whaling supported large settlements with an abundance of meat and blubber. Given limitations of identifying precontact whaling solely on technological evidence or settlement patterns, it is suggested that Bering Strait coastal societies be studied through human skeletal remains and through language studies, in order to reconstruct population movements and cultural diffusion over the past two millennia, during which time the whaling complex arose.*

INTRODUCTION

This paper addresses the development of traditional, historic Eskimo whale hunting methods found in coastal areas of western Alaska, the Bering Sea islands, and coastal areas of Siberia, as viewed from Alaskan shores (Fig. 1). Elements of the whaling complex include hunting techniques utilizing *umiaks* manned with crews of hunters led by whaling captains, or *umialiq*, related technology such as the drag float and large toggling whaling harpoon head, and large whaling villages (cf. Ackerman 1984; Worl 1980). Although this

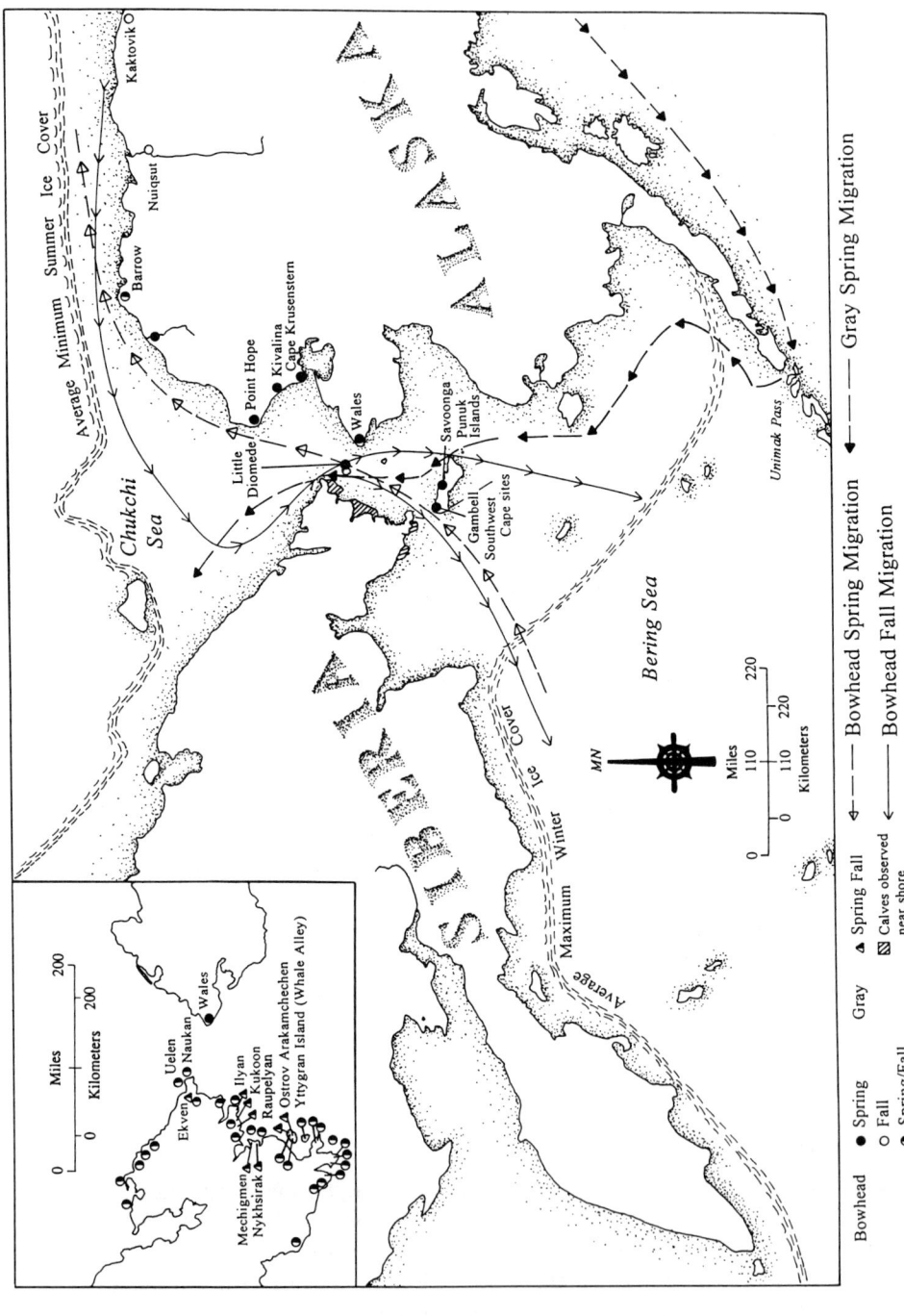

Figure 1.
Seasonal whale migration routes and primary prehistoric and historic whaling villages in Bering Strait and the western Beaufort Sea.

treatment subsumes two language groups, Yupik and Inupiaq, it nevertheless focuses on the important shared theme of traditional arctic culture generally conceived as Eskimo (see Oswalt 1967:258-259). The primary focus is on historic variation in Eskimo socioterritorial and subsistence patterns, and related elements of whale-hunting technology. Prehistoric human biology and language in Bering Strait are also identified as critical areas of investigation of the relationships between whale-hunting Eskimos and those with alternative subsistence foci.

Historic examples of ecological adaptations, expressed as use of coastal and interior areas, provide analogues for inferring types of socioterritorial organization in operation during prehistoric Denbigh Flint complex times, dating to as early as 5500 years ago (Harritt 1994a; Larsen 1972). It is inferred that prehistoric occupants of areas such as the Seward Peninsula pursued a mobile life style of hunting and gathering, represented by small settlements whose inhabitants concentrated on hunting small sea mammals and caribou. This trend was pursued in varying degrees, beginning with the Denbigh Flint complex and continuing through Choris, Norton-Near Ipiutak, Ipiutak, and Thule times. The primary evidence for this prehistoric lifeway pattern are settlement settings, technological assemblages, and, in later prehistoric times, the terrestrial and marine species represented in archaeological faunal assemblages. It is seen historically in groups such as those of Kawerak and Shishmaref (Ray 1983).

RETROSPECTIVE: PREHISTORIC ESKIMO ECOLOGY

While it may never be possible to establish the specific elements of Denbigh socioterritorial patterns, broad-based, cross-cultural studies such as those by Birdsell (1968:235) and Eskimo sociohistorical organization by Ray (1983) provide analogues relevant to Denbigh culture and ethnographic examples of traditional native use of the environment. These analogues indicate the type of human distribution necessary to maintain viable, self-perpetuating populations in northwestern Alaska. Denbigh people lived in Alaska for at least two millennia, thereby reflecting a fully-developed arctic adaptation. The similarity of the Denbigh Flint complex lithic technology to those of Europe was noted in the original definition of the culture (Giddings 1964:201-202). Particular affinities were observed between Denbigh and the European Mousterian-Levalloisan approach in removing blade-like flakes by striking one end of a pebble, the production of Mesolithic microblades, and burin production of Aurignacian times. This type of technology had spread to eastern Siberia, as reflected in the Belkachi culture, by around 4000 B.C. The earliest Denbigh remains presently known in Alaska were discovered at Kuzitrin Lake in the interior of Seward Peninsula, dating to c. 5500 B.P. (calibrated age; Harritt 1994a). Subsistence pursuits of northwestern Alaskan Denbigh people were focused on seals and caribou, reflecting some balance between marine and terrestrial resources. Sites have also been found on salmon streams, and the presence of *Salmonidae* remains in Denbigh contexts in other regions indicates the use of this resource as well (cf. Harritt 1994a:35-36). Denbigh sites in northwestern Alaska include temporary camps located on the coast, such as at Cape Espenberg, and as mentioned, at inland locations such as Kuzitrin Lake. Denbigh winter houses are also found at Onion Portage on the Kobuk River (Anderson 1988; Harritt 1994a:35-36).

On the basis of cross-cultural studies, it can be asserted that small semisedentary local bands of Denbigh individuals occupied territories with known resource locations and seasonality. Although the actual number of individuals comprising a band may have varied, populations probably ranged between 1-25 individuals per 100 mi^2 (Lee and DeVore 1968:10-11). Bands were likely allied through some form of exogamous marriage and other types of social arrangements, providing a flexible organization perhaps approximating that of some historic Eskimo groups (see Binford 1983; Birdsell 1968:234; Giddings 1954:86-88; Spencer 1959).[1] Cross-cultural examples of this type of pattern indicate that prehistoric Eskimo bands were united through geographic proximity, social and biological ties, and, presumably, language (see Birdsell 1968; Burch 1976; Lee and DeVore 1968). Tribes, or aggregations of bands, consisted of approximately 500 individuals who occupied a territory defined by its relationship with other, equivalent tribal territories (Lee and DeVore 1968:11; Wobst 1974:153). The size of the tribe may be related to proportionate numbers of males within a group, a relationship that has direct bearing on availability of marriage partners and maintenance of social contact between tribal members (compare Krupnik 1985:120 and Wobst 1974:152). It is likely that more than one such territory and group existed in an area such as Seward Peninsula during Denbigh times, based on analogy with ethnographic Eskimo groupings.

VARIATION IN SUBSISTENCE FOCI

Denbigh people practiced a balanced use of terrestrial and marine resources, that included caribou and seals. The advent of Denbigh culture in northwestern Alaska marks the first known, routine use of the coast in this area. This pattern continued through Choris times, as reflected in both types of faunal remains found in Choris assemblages (Giddings and Anderson 1986:228, 230, Fig. 126). This pattern shifted during Norton-Near Ipiutak times to an emphasis on marine subsistence, such as sealing, but use of terrestrial resources, especially caribou, continued (Dumond 1980:34-35, 1987:110; Giddings 1964:Table 13, 242; Giddings and Anderson 1986:Fig. 126, 320). A casual interest in whaling may have been present among the Norton occupants of the coasts, but there are no indications that the methods for taking the animals resembled those of the late prehistoric whale hunters (Giddings 1964). Norton people occupied coastal areas more continuously than had people of the preceding cultures in western Alaska (Dumond 1982).

In the succeeding Birnirk and Ipiutak cultures, use of caribou and seals, represented by their skeletal remains, indicates that both coastal and inland resources continued to be used. Although Birnirk culture appeared in the Bering Sea at around A.D. 300 and preceded the appearance of Punuk culture by approximately 300 years, there is ample evidence that the two entities were contemporaneous during parts of their respective tenures in the area (Ackerman 1962; Gerlach and Mason 1992; Rainey and Ralph 1959). There is evidence of interaction between Birnirk and Ipiutak peoples, but the distinctiveness of each assemblage — especially the absence of ceramics in Ipiutak — indicates that each represents a distinctive culture history. Inasmuch as Ipiutak culture did not include

---

1. Riches (1982:15, 127) suggests that the band is "... the largest group of which people themselves consider they are members ..." and uses the term in references to groups of 150-500 individuals, a collective of the size Ray (1983) and Burch (1980) would ascribe to a tribe or society.

a whaling focus in its economy, and was not directly related to the development of the northwest Alaskan whaling complex, it is not considered further in this review.

The development of whaling in northwestern Alaska is related to the evolution of Birnirk culture augmented by Punuk influences, a well-known scenario that was initially advanced by Collins (1937) and later reaffirmed by Ford (1959) and Stanford (1976:113). Alaskan Birnirk sites are distributed in coastal areas from Norton Sound to Point Barrow (Ford 1959; Gerlach and Mason 1992:64). Birnirk settlements range in size from two or three houses to as many as 16 houses at Point Barrow (Anderson 1984:90-91). Birnirk houses at the Birnirk site average approximately 3.05 x 3.12 m for the dimensions of the room; these fall within the dimensional range of Krupnik's (1983:89) type 3 houses, a size he suggests served as short-term, single-family winter dwellings.[2] In contrast, Siberian Punuk settlements were larger and more densely distributed than Birnirk settlements, and usually had very large middens and whale bones in abundance (Ackerman 1984:112-113). The large Siberian Punuk houses range from 6-20 x 8-25 m for dimensions of the room, and for this reason and because of other house features, they are interpreted as communal dwellings that may have accommodated as many as 30-80 inhabitants. These houses are thought to reflect an overall increase in population in the central and western regions of Bering Strait (Krupnik 1983:90-92, Table 1). In Siberia, elements that appeared in coastal cultures by 300 B.C. persisted into historic times, with an important shift to single family dwellings sometime around A.D. 1500 (cf. Krupnik 1983).

Feature remains encountered at Nunagiak that are clearly attributable to Punuk culture were interpreted by Ford (1959:61-66) as those of a house, although outlines of a dwelling were not discovered nor were probable dimensions recorded. Other Punuk remains found on Alaskan shores are restricted to portable implements such as particular types of harpoon heads that are usually found in association with Birnirk remains. Sites with some combination of Ipiutak, Birnirk, and Punuk cultural elements include Kurigitavik, with Birnirk and Punuk elements, Cape Krusenstern, with Birnirk and Ipiutak elements, and Point Hope, with Birnirk and Ipiutak elements (Gerlach and Mason 1992; Larson and Rainey 1948).

Changes that occurred after A.D. 600 in Alaska did not substantially alter the basic technologies that had been developed by Birnirk and Ipiutak people. The appearance of Punuk art and tools at Kurigitavik dates to around A.D. 600, and Punuk whaling harpoon heads appear in the vicinity of Point Barrow at around A.D. 800, marking the advent of Siberian whaling techniques on Alaskan shores (Stanford 1976).

---

2. Birnirk site house dimensions are based on Ford's (1959) measurements.

HISTORIC SOCIOTERRITORIAL ORGANIZATION

Although Sheehan (1985) asserts that the Alaskan whaling villages had developed into ranked societies at least in part because of the abundance and stability of food obtained from whale hunting, I have noted elsewhere (Harritt 1992) that a number of issues remain regarding socioterritorial structure of the Alaskan whaling village. Socioterritorial organization is a critical element of the whaling complex (see Worl 1980), and therefore must be addressed as part of any review of traditional whaling villages (Fig. 1). Descriptions of historic Alaskan Eskimo socioterritorial organization vary widely with respect to the nature of affinal relationships that may have influenced their structure (Harritt 1992). In these cases, additional analysis will be necessary to establish the range and variation of organization that were present in Alaska (cf. Burch 1980; Ray 1983; Sheehan 1985). Although the present review should be regarded as provisional and is intended only as a general outline of current knowledge, the key elements in any interpretation are the movements of individuals and nuclear families between villages and territories and territorial ties perceived by migrants and host occupants (Binford 1983; Spencer 1959:62-64).

The smallest social unit — the nuclear family — is the most basic social division that can be characterized with respect to geographically specific residence and economic importance. Spencer (1959:64-65) notes that the nuclear family unit operated autonomously on the level of day-to-day living in procuring and preparing food and in other aspects of routine habitation.

The next larger social segment has been referred to as the extended or local family or band (Burch 1980; Riches 1982; Spencer 1959).[3] In the case of a nuclear Eskimo family which reckons relationships bilaterally, the extended family includes the wife's and husband's parents, siblings, aunts, uncles, and cousins. Spencer (1959:65) suggests that the extended family provided support in times of stress, but it was not a formal institution beyond relationships that were established between individuals. By way of clarification, Spencer (1959:65) indicates that charms, amulets, and even property marks might be inherited, but only between individuals. They were not symbols of particular families or whole families.

Evaluations of the primary leadership in traditional Alaskan Inupiaq Eskimo bands indicate that the role is expressed in at least two forms, representing different levels of socioterritorial organization. Of these, Ray (1983:150, 154) suggests that each *kazgi* in a village had a chief, and in the case of a village with more than one *kazgi*, a "principal chief."[4] This man emerged as village leader by the strength of his personality and talents. A chief sometimes had a protege who assumed some leadership responsibilities based on his association with the chief and, because of this training, sometimes succeeded the chief upon his death. Ray (1983:154) suggests that "ideally" the role of chief was hereditary,

---

3. Burch (1980:263, 265) suggests that the size of the local family for traditional Eskimos may be as small as 12 individuals in resource-poor interior areas, and as large as 100 individuals in productive whaling villages such as Wales; the average local family size was 30-60 individuals. Binford (1983:36) suggests that the average size of a "band" — an equivalent of Burch's local family — was approximately 35 individuals for the interior-based Nunamiut.
4. The men's ceremonial house, the *kashim* or *kazgi*, is notably absent in Siberian settlements (see Hughes 1984a; 1984b). In Alaska, this structure is associated with mens" activities in general, and at least indirectly with leadership of the extended family.

or that the position belonged to a specific family. Burch's (1980:264) alternative description of the chief's role is that of the head of an extended family, a so-called *umealik*, who served in this capacity over some period of his adult life. Burch (1980:264) points out that in circumstances where a village was occupied by a single extended family, the *umealik* may appear to function as a so-called chief, as Ray suggests, but that assigning the *umealik* leadership authority beyond his extended family was inappropriate (see also Riches 1982).

Views on leadership and organization on the tribal level diverge more widely than those of the single extended family. Ray (1983:151, 174) describes tribal settlement and social patterns as a large principle village surrounded by smaller, outlying satellite villages. Burch (1980:264) suggests that traditional Eskimo tribes were "segmental" in terms of being made up of units — extended families — that were more or less equivalent in rank. However, he further suggests that tribes may have been composed of stratified local families, owing primarily to their relative sizes and wealth (Burch 1980:264-265). But times of diminished resource availability or unsuccessful hunts resulted in fissioning of local families, and the absence of a strong interlocal family political hierarchy enabled the dispersal of smaller family segments, even at the level of the nuclear family or individual during stressful times (Burch 1980:265-266; Harritt 1992). It is apparent from Burch's and Ray's descriptions that the highest rank one could achieve in traditional Eskimo society was head of an extended family; in the local group winter settlement, this role would also be expressed as chief of the village *kazgi*.

Traditional Yupik Eskimo socioterritorial organization on Siberian shores and the Bering Sea islands was based on the patrilineal clan. Lineage was reckoned from a common clan ancestor, even though more immediate ancestors may be forgotten or unknown (Hughes 1984a:244). Siberian society revolved around the patrilineage and the clan, insofar as marriage, composition of whaling boat crews, settlement patterns, burial patterns, and other conventions were related to these two elements (Hughes 1984a:244). Hughes suggests that Siberian social organization may have originated in the naming of territorial groups that were presumably some version of the type found on the Seward Peninsula.

Siberian natives practiced patrilineal clan exogamous marriage within a village, in contrast to Seward Peninsula Eskimos who practiced a form of local, bilaterally-descended group exogamous marriage. That is, Alaskan groups were endogamous on the level of the *tribe* (Burch 1980; Hughes 1984a). Siberian residential areas within villages bore a strong resemblance to the "family compounds" found in large Alaskan coastal villages (Burch 1980:266; Hughes 1984b:254), the primary difference between the two being spatial organization based on the patrilineal Siberian clan versus the bilateral, extended Alaskan family.

Leadership in the Siberian clan was usually provided by an elderly man, designated the *nuna'-laxtaq*. Duties of this position included leadership in trade with other clans and hunting, but also included making judgments in settling disputes and in carrying out clan religious ceremonies (Hughes 1984b:254). Although the general clan leadership role strongly resembles the role of the *umealik* in Alaskan villages, Siberian leadership differed in being hereditary, and it was always passed down in the patrilineage from father to son (Hughes 1984b:254).

Leadership of the collective Siberian village fell to the leader of the "most powerful and respected" village clan (Hughes 1984b:254-255), also a pattern that echoes tendencies of Alaskan groups, but was formalized in Siberian societies. In Alaskan villages, the proximity of family groups, each with its own *umealik,* produced stressful situations. This tendency suggests that there was no integrating social mechanism in operation to support a collective sociopolitical structure that cross-cut more than one family group. Family compounds here maintained their political autonomy (Burch 1980:266) and reflect a clear parallel with tribal territory patterns.

Although Burch suggests that local Alaskan family groups were stratified, this interpretation is predicated on consistent availability of sufficient resources to support the larger groupings within a territory, and it downplays effects of fluctuations in resource abundances that were chronic in many Alaskan villages. In addition, the lack of a pan-village political structure integrating more than one extended family is inconsistent with the conventional application of the "stratified" society designation (Fried 1967).

GENERAL TRENDS

Based on the preceding, it can be seen that Alaskan Eskimo groups did not develop a form of social organization that could reasonably be termed "ranked" (cf. Harritt 1992 and Sheehan 1985). The possibility that local families were "stratified," as Burch suggests, appears to lack substantial support, insofar as the appearance of a successful *umealik* leading a large, wealthy family could simply reflect temporary circumstances rather than a long-term, formalized type of organization. As mentioned, diminishing resources would result in the splitting up of a local family, a process that even an influential *umealik* could not forestall. The tendency for an aggregate group of families to fission militates against maintaining a strong social hierarchy within a local family, and precludes development of a formal sociopolitical hierarchy on the level of the tribe (Burch 1980:266; Hassan 1981:182).

In contrast, traditional Siberian organization is structured so that patrilineal clans provide well-defined social segments on the level of the extended family. These segments provide a basis for social ranking with respect to hereditary leadership positions, mentioned above, as well as forming a basis for stratification of clans based on relative size and wealth (cf. Burch 1980). The formalization of this organization is reflected by the hereditary aspect of the clan leadership, as it is passed down within a particular patrilineage (Hughes 1984b).

SOME CONSIDERATIONS

All traditional Eskimo social organizations are undoubtedly effective adaptations to their environments, *but the development of the different socio-territorial forms cannot presently be attributed to a direct ecological cause and cultural effect relationship* (see Beardsley et al. 1956; Chang 1962; Damas 1967:117; 1969b:57-58; Hassan 1981:179-180, 182, 184; Riches 1982; Steward 1968:323-329). Specific considerations in this respect include:

(a) Alaskan and Siberian Eskimo groups living in virtually the same Bering Strait environments and hunting the same marine game animals possessed different forms of social organization. The persistence of the caribou and small sea mammal patterns from Denbigh times as early as 5500 years ago up to early historic times and a late (c. 1200 year old) development of the whaling pattern in western Alaska also indicate that the development of Siberian whaling techniques and socioterritorial organization did not proceed as a collective, linear development or progression across the Bering Strait area.

(b) If development of the Siberian patrilineal clan organization is attributed to a long-term dependence on sea mammals, then an explanation for the development of the pattern over the course of two millennia is needed. Although it is tempting to attribute the Siberian development to a long-term, reliable sea mammal base, explanations of how the changes progressed through time are necessary to develop a complete culture history.

(c) The relatively late development of Alaskan whaling villages suggests that threshold conditions for change in social organization on Alaskan shores to the more complex Siberian form were not attained as early as they were on the Siberian side of Bering Strait. Furthermore, achieving the postulated threshold conditions may have been requirements for Alaskan groups to adopt the Siberian form of open-water whaling. Although it can be suggested that pertinent threshold factors may include a requisite population size, technology, and ecological factors such as proximity of whale migration routes, identification of specific factors will be necessary to explain the development (see Bockstoce 1976, 1979).

New insights into the factors that contributed to the development of Siberian and Alaskan socioterritorial patterns may be provided by careful examination of the similarities and differences of ecologies from one tribal territory to the next. Although it may not be possible to establish that ecological factors contributed to some of the cultural developments in the region, it is nevertheless important to consider other probable causes, as Damas (1969b:40) has suggested for the Central Canadian Arctic.

PARADIGMS FOR FURTHER INVESTIGATION

Environmental Studies

Environmental aspects of particular interest are those related to the proximity of villages to whale migration routes and the locations of deep, near-shore ocean trenches. A cursory review of Siberian and Alaskan whaling village locations, locations of polynyas, deep ocean trench locations, and whale migration routes suggests that there are correspondences between these phenomena in Bering Strait (Harritt 1994b; Mason and Gerlach n.d.). Village locations near polynyas that apparently persisted over many decades have been identified in the Canadian Arctic by Schledermann (1980).

Biological Studies

Differences between Bering Strait cultures of the last 2000 years are represented in the physical human remains. Debets (1975; see also Arutyunov 1979:30) suggests that two distinct populations are represented by the early Old Bering Sea dolichocephalic human remains at East Cape and, by comparison, contemporaneous mesocephalic Ipiutak remains at Point Hope. According to Debets, the later Birnirk occupants of Point Barrow have the same characteristics as their contemporaries at East Cape, a circumstance he preferred to regard as localized population change, even though he had considered the possibility of migrations from East Cape to northwestern Alaska. Arutyunov (1979:30) suggests that Punuk was a biological and cultural derivation from Okvik/Old Bering Sea predecessors and that Birnirk was intrusive in Punuk areas of Bering Strait. In contrast, Utermohle (1988) suggests that Old Bering Sea and Birnirk people were genetically close groups, and that Ipiutak and Birnirk remains represent two distinct populations (Utermohle 1988:Fig 3, 45; see also, Turner 1988:35-36). Utermohle's sample did not include Punuk examples (see Utermohle 1984:Tables 2 & 3, 1988).

Linguistic Studies

Alternative interpretations of language development in Bering Strait over the past 2000 years primarily focus on differences in the timing for the split between Yupik and Inupiaq, the two main branches of the Eskimoan language. Woodbury (1984:61) suggests that the split between the two occurred sometime around the middle of the first millennium B.C., a time when Okvik/Old Bering Sea and Norton cultures appeared in Bering Strait. In an alternative interpretation, Dumond (1988) reasserts a prior suggestion by Collins (1954) in placing the split approximately 1000 years later, at around A.D. 500, a time that roughly coincides with the appearance of Punuk and Ipiutak cultures in Bering Strait (cf. Anderson 1984:88-90). Along similar lines, Utermohle (1988:43-46) concludes that the spread of Birnirk people into northwestern Alaska also signaled the arrival of the Inupiaq language, based on his assessment that Inupiaq language distribution corresponded with distributions of prehistoric genetic and cultural groups. Attempts to reconstruct distributions of late prehistoric groups based on language also include an approach suggested by Burch (1980:279), in which dialect areas are analyzed. Burch defines dialectic zones that include more than one society in intersocietal groupings. In this interpretation, Barrow and Wales are relegated to different dialectic zones, even though no significant differences were found by Hirsch (1954) and Dumond (1965:1236-1237) in the languages spoken by members of these settlements.

A SYNTHESIS

For the present discussion, it is presumed that some form of seal and caribou subsistence focus provided an economic baseline that supported further cultural developments and changes that led, in turn, to the whaling focus for some groups. Other groups continued to pursue the prior seal and/or caribou foci up to historic contact. It is further presumed that the Birnirk inhabitants of the Alaskan coastal promontories, such as those at Point Barrow, were the immediate cultural antecedents to the later Thule inhabitants, and were the receptors of Punuk influences that led to the development of the Alaskan whaling village. These suppositions enable us to postulate how the Alaskan whaling village organization developed. These postulates are as follows:

(a) Historic Alaskan whaling villages were located within territories that were occupied by Birnirk groups. Punuk influences on Alaskan Birnirk groups were established over the course of their shared history in the Bering Sea region. Punuk influences at Barrow were in the form of the "trait-unit intrusions" that are described by Ackerman (1962:33) for St. Lawrence Island Punuk influence on contemporaneous Birnirk occupants.

(b) Punuk and Birnirk groups around Bering Strait spoke dialects that were sufficiently mutually intelligible to enable transfers of technological units (compare Nelson 1983:228-232 and Ray 1975:61). The history of the close relationship between the Birnirk and Punuk cultures on St. Lawrence Island, for example, suggests that a mutually intelligible dialect for the two cultures existed. The possibility that the languages spoken at Wales and Barrow were closely related (Hirsch 1954) suggests that the Birnirk occupants of these locations may have communicated with Punuk neighbors in the same way, and had the same type of relationship as the Birnirk and Punuk occupants of St. Lawrence Island.

(c) Although Ford (1959:64) suggests that a small family group of Punuk people inhabited the lower level house at Nunagiak, it seems unlikely that substantial, intact Punuk socioterritorial groups migrated to Alaskan shores. Such an incursion would be one with a high profile in the archaeological record, insofar as a Siberian type of Punuk settlement in northwestern Alaska would undoubtedly consist of substantial remains with clearly recognizable Punuk traits. Based solely on the distribution of Birnirk and Punuk traits in the Bering Strait region, it is suggested that Punuk culture was centered on the coastal Chukchi Peninsula, and that Birnirk culture centered in the Bering Sea islands or northwestern Alaska. In following this reasoning, it is possible that an admixture of Birnirk and Punuk traits, such as those found at the S'keliyuk site on St. Lawrence Island, occurred at other locations as well. Rather than belonging to one, they may represent eastern and western Bering Sea influences on the 5th and 6th century inhabitants of the islands that lay between the continents. Sites such as Kurigitavik (Wales) and Nunagiak (Barrow) may contain information about how the whaling complex was acquired and used by Alaskans from A.D. 800 to the 19th century. Remains at these locations should reveal the extent and type of Siberian influences on Alaskan shores. Specifically, we would hope to determine when whaling became the primary focus of the occupants, or why seals and caribou remained the primary foci in the economies over the course of the past 1200 years.

(d) Given the present lack of Punuk human biological data, it is not known if Punuk people were biologically different from the Birnirk inhabitants of Alaska to any significant degree. Although distinctive groups were responsible for the development and spread of each culture, biological traits may represent an admixture of Punuk and Birnirk people. A possibility that should be considered is that some type of systematic marriage exchange was in operation between Punuk and Birnirk groups. This type of social mechanism could result in an initial spread of Punuk traits in Alaska, with subsequent increasing affinities between the two cultures at certain locations on Alaskan shores. Of particular interest are human remains from Siberian sites and Bering Sea island sites that may eventually yield evidence of the type of exchange described above.

(e) Although the proclivity of Birnirk to receive Punuk influences indicates a history of interaction between the cultures, it nevertheless does not explain the differences between them over most of their histories. The technology and organizational knowledge for taking large whales were present in the area by A.D. 600, but unequivocal indications of the arrival of whale hunting are absent until the appearance of the Sicco harpoon head, a Punuk type, in the Barrow area at around A.D. 800 (Bockstoce 1979; Stanford 1976). By that time, a few groups in northwest Alaska must have attained the conditions necessary for the whaling pattern to operate (Harritt 1992; see also Bockstoce 1979). The requirements for implementing the Punuk whaling techniques must have included the presence of a sufficiently large human population, proximity of whales, and effective technology, including *umiaks*, harpoons, and storage facilities for the whale products. The organizational capabilities necessary for hunting and butchering the huge animals may not require more than basic organizational elements present in hunter-gatherer societies (Maschner 1992).

(f) The presumed transition from the ethnographic Alaskan whaling village form to the Siberian form requires further examination. In following Hughes" (1984a) suggestion about the transition, it is important to base comparisons on equivalent socioterritorial units for each of the areas. In this respect, although Hughes (1984a:244) suggests that the naming of territorial groups was a significant step in the development of the Siberian clan-based territorial organization, he does not specify whether territorial groups are defined on the level of the local group, the extended family (or clan), or the tribe. Insofar as Alaskan territories are generally defined on the level of the tribe (i.e., Burch 1988; Ray 1983), it is presumed that Hughes is referring to development of territory ownership on the level of the clan, an organizational characteristic that would parallel territorial organization in the American Northwest Coast culture area (i.e., Donald and Mitchell 1975; Maschner 1992:25-39). A productive avenue for investigating the development of named territorial groups is the postulation that population sizes and available land and resources are key elements in their development. In this respect, it is suggested that larger populations and the limited amount of land available on Bering Sea islands such as St. Lawrence may have engendered a need for formalizing relationships between social groups and the areas occupied by them (Price and Brown 1985:16). The developments that led to the Siberian Eskimo social complexity, therefore, may have occurred in insular environments such as St. Lawrence Island, because of large human populations and limitations of available land and resources.

(g) The correlation between these attributes of the society and the distribution of the Yupik language may reflect a heretofore unexplored cultural phenomenon that may distinguish between Inupiat and Yupik groups in Siberia and Alaska. The correlates of patrilineage and formal social ranking, found more frequently in Yupik groups than in Inupiaq, suggest the existence of a Yupik cultural pattern that has not been carefully examined. In this respect, it seems apparent that Yupik groups that inhabited areas of stable, abundant resources tended to develop increasingly complex social structures more readily than did Inupiaq groups living under the same types of conditions. Complex Yupik sociopolitical development south of Bering Strait can be found on Nunivak Island[5] and in the Koniag culture of Kodiak Island (Clark 1984:192). Examples of increased social complexity on Nunivak are the patrilineage and the inheritance of symbolic objects, songs, and certain designs with special powers. On Kodiak, there were social classes with a nobility, commoners, slaves, and chiefs that passed their positions down within their circle of relatives.

---

5. Lantis (1984:218) notes that the patrilineage was prevalent among traditional Nunivak Island (Yupik) Eskimos, and with accompanying inheritance of objects and songs in the male line. Although in this instance titles and formal "prerogatives" (Lantis 1984:218) were not formally passed down in the patrilineage, a tendency for actual continuity in a lineage from one generation to the next did occur.

CONCLUSION

Future paleoenvironmental studies will undoubtedly give insights into the distributions and availability of whales to human hunters who learned to station themselves in strategic locations in order to intercept the spring and fall migrations. The different patterns of Eskimo subsistence and social organization that operate in what is generally the same type of environment on both sides of Bering Strait suggest that prehistoric environments were not important causal factors in the cultural developments. But is important that paleoenvironmental studies continue, because they provide insight into the relationship between people and the environment, regardless of the form of culture they possessed. And, while it is also important that future studies assume that many cultural elements were shared by the whaling people, there were also significant and fundamental differences in the culture histories of Chukotka, the Bering Sea islands, and northwestern Alaska over the past 2000 years.

The central themes in the issues regarding the relationships between partially contemporaneous Punuk and Birnirk people that have been outlined above are biological affinities of each group and their respective sociocultural interaction spheres. It is important that each of these prehistoric cultures be approached as potentially distinctive biological groups whose successful maritime adaptations supported their existence over the course of several centuries. This approach is advocated here even though there are substantial methodological problems related to this type of inquiry. Examples are seen in the early historic descendants of these cultures displaying a variety of subsistence orientations that included foci on seals, caribou, and salmon as well as walrus and whales, and variable composition of social groupings that often included individuals from territories other than those in which they came to reside.

In spite of our best efforts, archaeological methods and other reconstructive studies may ultimately fall short of providing adequate explanations for the processes of culture change and development that produced the whale-hunting complex facies of Eskimo culture. Nevertheless, those areas in which issues remain to be resolved also define opportunities for gaining insights into this important 2000 year period of human occupation of the Bering Strait region.

REFERENCES

Ackerman, R.
- 1962 Culture Contact in the Bering Sea: Birnirk-Punuk Period. In: Prehistoric Cultural Relations Between Arctic and Temperate Zones of North America, edited by J. M. Campbell, pp. 27-34. *Arctic Institute of North America Technical Paper* No. 11.
- 1984 Prehistory of the Asian Eskimo Zone. In: *Handbook of North American Indians*, Vol. 5, *Arctic*, edited by D. Damas, pp. 106-118. Smithsonian Institution, Washington, D.C.

Anderson, D.
- 1984 Prehistory of North Alaska. In: *Handbook of North American Indians*, Vol. 5, *Arctic*, edited by D. Damas, pp. 80-93. Smithsonian Institution, Washington, D.C.
- 1988 Onion Portage: The Archaeology of a Stratified Site from the Kobuk River, Northwest Alaska. *Anthropological Papers of the University of Alaska* 22(1-2).

Arutyunov, S.
- 1979 Problems of Comparative Studies in Arctic Maritime Cultures Based on Archaeological Data. *Arctic Anthropology* 16(1):27-31.

Beardsley, R., P. Holder, A. D. Krieger, B. J. Meggers, J. B. Rinaldo, and P. Kutsche
    1956   Functional and Evolutionary Implications of Community Patterning. In: Seminars in Archaeology: 1955, pp. 129-157. *Society for American Archaeology Memoir* 11.

Binford, L. R.
    1983   Long-Term Land Use Patterns: Some Implications for Archaeology. In: Lulu Linear Punctated: Essays in Honor of George Irving Quimby, edited by R. C. Dunnell and D. K. Grayson, pp. 27-53. *Anthropological Papers of the University of Michigan* No. 72.

Birdsell, J. B.
    1968   Some Predictions for the Pleistocene-Based Equilibrium Systems Among Recent Hunter-Gatherers. In: *Man the Hunter,* edited by R. Lee and I. DeVore, pp. 229-240. Aldine-Atherton, Chicago.

Bockstoce, J.
    1976   On The Development of Whaling in the Western Thule Culture. *Folk* 18:41-46.
    1979   The Archaeology of Cape Nome, Alaska. *University of Pennsylvania, University Museum Monograph* 38.

Burch, E., Jr.
    1976   The "Nunamiut" Concept and the Standardization of Error. In: Contributions to Anthropology: The Interior Peoples of Northern Alaska, edited by E. Hall, pp. 52-97. *National Museum of Man, Mercury Series, Archaeological Survey of Canada Paper* No. 49.
    1980   Traditional Eskimo Societies in Northwest Alaska. In: Alaska Native Culture and History, edited by Y. Kotani and W. Workman, pp. 253-304. *Senri Ethnological Studies* No. 4. National Museum of Ethnology, Osaka.
    1988   Toward a Sociology of the Prehistoric Inupiat: Problems and Prospects. In: The Late Prehistoric Development of Alaska's Native People, edited by R. Shaw, R. Harritt, and D. Dumond, pp. 1-16. *Aurora: Alaska Anthropological Association Monograph Series* No. 4.

Chang, K. C.
    1962   Typology of Settlement and Community Patterns in Some Circumpolar Societies. *Arctic Anthropology* 1(1):28-41.

Clark, D.
    1984   Pacific Eskimo Historical Ethnography. In: *Handbook of North American Indians*, Vol. 5, *Arctic*, edited by D. Damas, pp. 185-197. Smithsonian Institution, Washington, D.C.

Collins, H. B., Jr.
    1937   Archaeology of St. Lawrence Island, Alaska. *Smithsonian Miscellaneous Collections* 96(1).
    1940   Outline of Eskimo Prehistory. *Smithsonian Miscellaneous Collections* 100:533-592.
    1954   Comment on Time Depths of American Linguistic Groupings, by Morris Swadesh. *American Anthropologist* 56(3):364-372.
    1964   The Arctic and Subarctic. In: *Prehistoric Man in the New World,* edited by J. D. Jennings and E. Norbeck, pp. 85-114. University of Chicago Press, Chicago.

Damas, D.
    1967   The Diversity of Eskimo Societies. In: *Man the Hunter*, edited by R. B. Lee and I. DeVore, pp. 111-117. Aldine-Atherton, Chicago.
    1969a   Introduction: The Study of Cultural Ecology and the Ecology Conference. In: Contributions to Anthropology: Ecological Essays, edited by D. Damas, pp. 1-12. *National Museum of Canada Bulletin* No. 230; *Anthropological Series* No. 86.

1969b Environment, History, and Central Eskimo Society. In: Contributions to Anthropology: Ecological Essays, edited by D. Damas, pp. 40-64. *National Museum of Canada Bulletin* No. 230; *Anthropological Series* No. 86.

Debets, G.
1975 Paleoantropologicheskiye materialy iz dreneberingo-mokikh Uelen i Ekven (Paleoanthropological Materials from the Early Bering Sea Cemeteries of Uelen and Ekven). In: *Problemy etnicheskoy istorii Beringomorya*. S. Arutynov and D. Sergeev, pp. 198-201. Nauka, Moscow.

Donald, L. and D. Mitchell
1975 Some Correlates of Local Group Rank Among the Southern Kwakiutl. *Ethnology* 14(4):325-346.

Dumond, D. E.
1965 On Eskaleutian Linguistics, Archaeology, and Prehistory. *American Anthropologist* 67(5):1231-1257.
1980 A Chronology of Native Alaska Subsistence Systems. In: Alaska Native Culture and History, edited by Y. Kotani and W. Workman, pp. 23-47. *Senri Ethnological Studies* No. 4. National Museum of Ethnology, Osaka.
1982 Trends and Traditions in Alaska Prehistory: The Place of Norton Culture. *Arctic Anthropology* 19(2):39-51.
1987 *The Eskimos and Aleuts*. Revised Edition. Thames and Hudson, London.
1988 Trends and Traditions in Alaskan Prehistory: A New Look at an Old View of the Neo-Eskimo. In: The Late Prehistoric Development of Alaska's Native People, edited by R. Shaw, R. Harritt, and D. Dumond, pp. 17-26. *Aurora: Alaska Anthropological Association Monograph Series* No. 4.

Ford, J.
1959 Eskimo Prehistory in the Vicinity of Point Barrow, Alaska. *Anthropological Papers of the American Museum of Natural History* 47(1).

Fried, M.
1967 *The Evolution of Political Society: An Essay in Political Anthropology*. Random House, New York.

Gerlach C. and O. K. Mason
1992 Calibrated Radiocarbon Dates and Cultural Interaction. *Arctic Anthropology* 29(1):54-81.

Giddings, J. L.
1964 *The Archaeology of Cape Denbigh*. Brown University Press, Providence.

Giddings J. L. and D. D. Anderson
1986 Beach Ridge Archeology of Cape Krusenstern. *National Park Service, Publications in Archaeology* No. 20.

Harritt, R. K.
1992 Relationships Between Whale Hunting, Human Social Organization, and Subsistence Economies in Coastal Areas of Northwest Alaska During Late Prehistoric Times. Paper presented at the International Conference on the Role of the Polar Regions in Global Change, June 11-15, University of Alaska, Fairbanks.
1994a Eskimo Prehistory on the Seward Peninsula, Alaska. *National Park Service Resources Report NPS/ARO/RCR/CRR-93/21*.
1994b Some Correlates of the Development of Prehistoric Whale-Hunting and Climate Change in Bering Strait. Paper presented at the Conference on the Arctic Climate System, Nov. 7-10, Gteborg, Sweden.

Hassan, F.
    1981   *Demographic Archaeology*. Academic Press, New York.

Hirsch, D.
    1954   Glottochronology and Eskimo and Eskimo-Aleut Prehistory. *American Anthropologist* 56(5):825-838.

Hughes, C.
    1984a   Asiatic Eskimo: Introduction. In: *Handbook of North American Indians*, Vol. 5, *Arctic*, edited by D. Damas, pp. 243-246. Smithsonian Institution, Washington, D.C.
    1984b   Siberian Eskimo. In: *Handbook of North American Indians*, Vol. 5, *Arctic*, edited by D. Damas, pp. 247-261. Smithsonian Institution, Washington, D.C.
    1984c   St. Lawrence Island Eskimo. In: *Handbook of North American Indians*, Vol. 5, *Arctic*, edited by D. Damas, pp. 262-277. Smithsonian Institution, Washington, D.C.

Krupnik, I.
    1983   Early Settlements and Demographic History of Asian Eskimos of Southeastern Chukotka (Including St. Lawrence Island). In: *Culture and History of the Bering Sea Region: Papers From an International Symposium*, edited by H. Michael and J. VanStone, pp. 84-111. International Research and Exchanges Board, New York.
    1985   The Male-Female Ratio in Certain Traditional Populations of the Siberian Arctic. *Etudes/Inuit/Studies* 9(1):115-140.

Lantis, M.
    1984   Nunivak Eskimo. In: *Handbook of North American Indians*, Vol. 5, *Arctic*, edited by D. Damas, pp. 209-223. Smithsonian Institution, Washington, D.C.

Larsen, H.
    1972   The Tareormiut and the Nunamiut of Northern Alaska: A Comparison Between Their Economy, Settlement Pattern and Social Structure. In: *Circumpolar Problems*, edited by G. Berg, pp. 119-126. Pergamon Press, Oxford.

Larsen, H. and F. Rainey
    1948   Ipiutak and the Arctic Whale Hunting Culture. *Anthropological Papers of the American Museum of Natural History* 42.

Lee, R. B. and I. DeVore
    1968   Problems in the Study of Hunters and Gatherers. In: *Man the Hunter*, edited by R. B. Lee and I. DeVore, pp. 3-12. Aldine-Atherton, Chicago.

Maschner, H.
    1992   *The Origins of Hunter and Gatherer Sedentism and Political Complexity: A Case Study from the Northern Northwest Coast*. Unpublished Ph.D. dissertation, Department of Anthropology, University of California, Santa Barbara.

Mason, O. and S. C. Gerlach
    n.d.   Chukchi Hot Spots, Paleo-polynyas and Caribou Crashes: Climatic and Ecological Dimensions of North Alaska Prehistory. *Aurora: Alaska Anthropological Association Monograph Series*. In press.

Mathiassen, T.
    1930   Archaeological Collections from the Western Eskimos. *Report of the Fifth Thule Expedition 1921-24* 10(1). Copenhagen.

Nelson, E.
    1983   *The Eskimos About Bering Strait*. Smithsonian Institution Press, Washington, D.C. (1899)

Oswalt, W.
  1967   *Alaskan Eskimos*. Chandler Publishing Company, San Francisco.

Price, T. and J. Brown
  1985   Aspects of Hunter-Gatherer Complexity. In: *Prehistoric Hunter-Gatherers: The Emergence of Cultural Complexity*, edited by T. D. Price and J. A. Brown, pp. 3-20. Academic Press, San Diego.

Rainey, R. and E. Ralph
  1959   Radiocarbon Dating in the Arctic. *American Antiquity* 24(4): 365-374.

Ray, D. J.
  1975   *The Eskimos of Bering Strait, 1650-1898*. University of Washington Press, Seattle.
  1983   *Ethnohistory in the Arctic: The Bering Strait Eskimo*. The Limestone Press, Kingston, ON.

Riches, D.
  1982   *Northern Nomadic Hunters-Gatherers: A Humanistic Approach*. Academic Press, London.

Schledermann, P.
  1980   Polynyas and Prehistoric Settlement Patterns. *Arctic* 33(2): 292-302.

Sheehan, G.
  1985   Whaling as an Organizing Focus in Northwestern Alaskan Eskimo Society. In: *Prehistoric Hunter-Gatherers: The Emergence of Cultural Complexity*, edited by T. D. Price and J. Brown, pp. 123-154. Academic Press, New York.

Spencer, R.
  1959   The North Alaskan Eskimo: A Study in Ecology and Society. *Bureau of American Ethnology Bulletin* 171.
  1984   North Alaska Coast Eskimo. In: *Handbook of North American Indians*, Vol. 5, *Arctic*, edited by D. Damas, pp. 320-337. Smithsonian Institution, Washington, D.C.

Stanford, D.
  1976   The Walakpa Site, Alaska: Its Place in the Birnirk and Thule Cultures. *Smithsonian Contributions to Anthropology* No. 20.

Steward, J.
  1942   The Direct Historical Approach to Archaeology. *American Antiquity* 7(4):337-343.
  1968   Causal Factors in the Evolution of the Pre-farming Societies. In: *Man the Hunter*, edited by R. Lee and I. DeVore, pp. 321-334. Aldine, Chicago.

Turner, C. G., II
  1988   A New View of Alaskan Population Structure at About Historic Contact. In: The Late Prehistoric Development of Alaska's Native People, edited by R. Shaw, R. Harritt, and D. Dumond, pp. 27-36. *Aurora: Alaska Anthropological Association Monograph Series* No. 4.

Utermohle, C.
  1984   *From Barrow Eastward: Cranial Variation of the Eastern Eskimo*. Unpublished Ph.D. dissertation, Department of Anthropology, Arizona State University.
  1988   The Origin of the Inupiat: The Position of the Birnirk Culture in Eskimo Prehistory. In: The Late Prehistoric Development of Alaska's Native People. edited by R. Shaw, R. Harritt, and D. Dumond, pp. 60-73. *Aurora: Alaska Anthropological Association Monograph Series* No. 4.

Wobst, M.
   1974   Boundary Conditions for Paleolithic Social Systems: A Simulation Approach. *American Antiquity* 39(2):147-178.

Woodbury, Anthony C.
   1984   Eskimo and Aleut Languages. In: *Handbook of North American Indians*, Vol. 5, *Arctic*, edited by D. Damas, pp. 49-63. Smithsonian Institution, Washington, D.C.

Worl, R.
   1980   The North Slope Inupiat Whaling Complex. In: Alaska Native Culture and History, edited by Y. Kotani and W. Workman, pp. 305-320. *Senri Ethnological Studies* No. 4. National Museum of Ethnology, Osaka.

Whale Traps on the North Pacific?

Don E. Dumond
Department of Anthropology
University of Oregon
Eugene, OR 97403

Abstract. *The presence on the Alaska Peninsula at Izembek Lagoon of a prehistoric house with rafters of whale mandibles combines with recent studies of the behavior of the Pacific gray whale (Eschrichtius robustus)* to suggest that the lagoon systems on the Bering Sea coast of the Peninsula served in summer as aboriginal whale traps. A survey of that coast shows a consistent tendency for archaeological sites of unusual size to be located on those shallow lagoons. Now and in past centuries, these lagoons have provided potential human subsistence resources not only in the form of seals, waterfowl, and migratory fish, but also as active seasonal feeding grounds for gray whales. Although evidence for whale capture in the bays is inconclusive, it is permissive insofar as the question can be pursued with data available. Gray whale habits also suggest the possibility of similar aboriginal practices at various sections of the Northwest Coast.*

INTRODUCTION

In this paper, I report evidence that is suggestive, if indirect and inconclusive, for the taking of the Pacific gray whale (*Eschrichtius robustus*) by prehistoric natives of southwestern Alaska. Both the species taken and the techniques presumably used represent a departure from common North American whaling practices as they are reported in the ethnographic literature. Nevertheless, the habits of this whale may well have made it an important subsistence resource in certain areas of its range.

The gray whale of the American coasts is known for its return from near-extinction after intensive exploitation by Euroamerican whalers in the 19th century, as well as for its habit of calving in lagoons along the coast of Baja California that almost led to its extinction. The major annual feeding ground of the species, however, is far to the north of Baja California in the near-shore areas of the Bering Sea, and especially in the shallow bays that punctuate its coastline. It is this area that is the focus of attention here.

AN UNEXPECTED WHALE BONE HOUSE

Izembek Lagoon is located on the Bering Sea shore near the tip of the Alaska Peninsula (Fig. 1). Measuring some 35 km in length and a maximum of about 11 km in width, the lagoon depth at mean low tide now is less than 1 m over much of the area, so that the

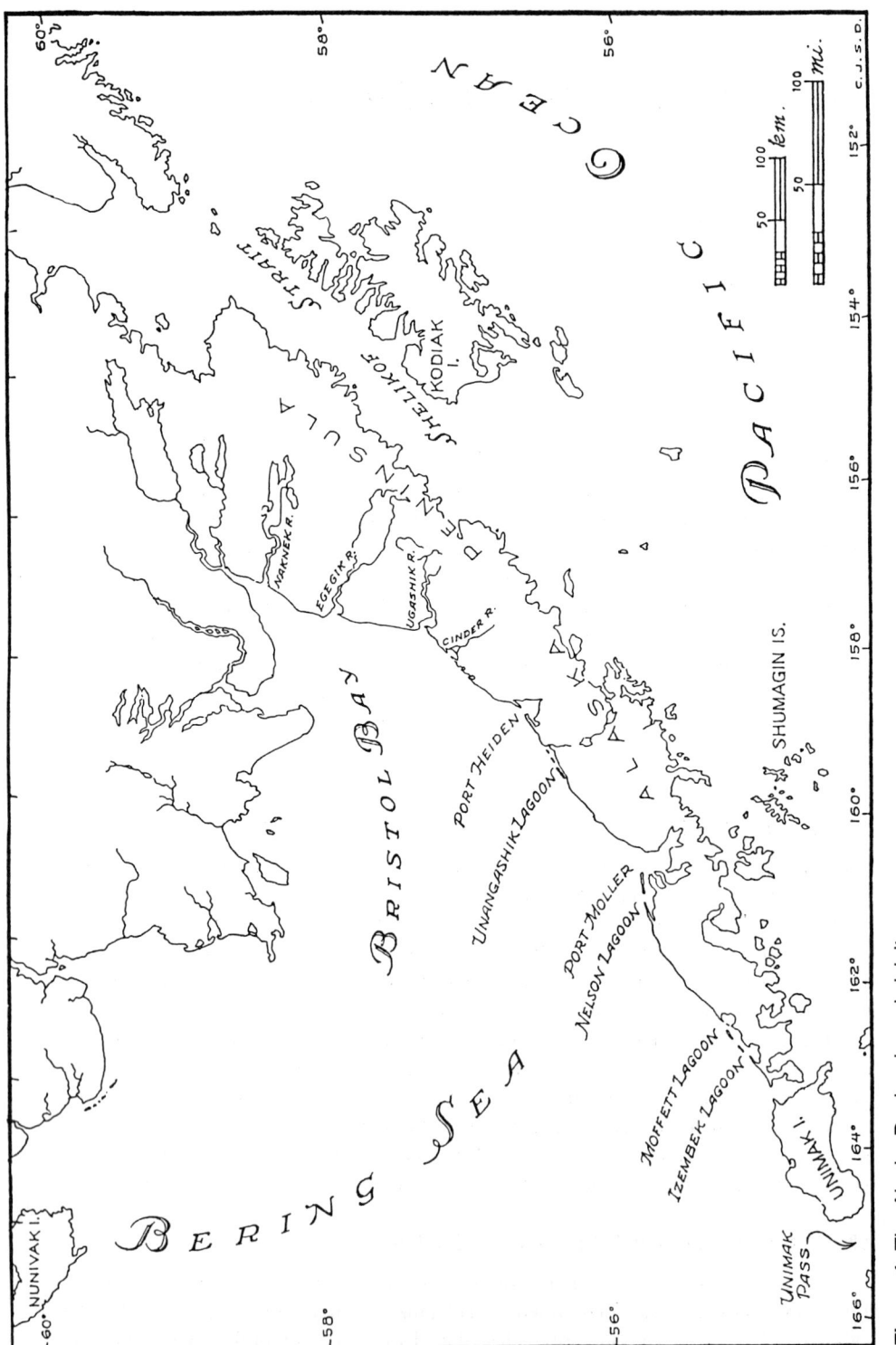

Figure 1. The Alaska Peninsula and vicinity.

usual tide of less than 2 m raises the depth to less than 3 m. Enormous flocks of migratory birds stage in the area in both spring and fall, seals breed in the lagoon in spring and summer, and streams debouching into the lagoon receive modest runs of salmon.

Nearly 20 years ago, Allen McCartney (1974) described the excavation of an aboriginal house located on an Izembek shore rising 5 to 8 m above high tide. Three sites tested on the lagoon consisted of saucer-shaped house depressions. In all but one of these, the absence of major stone and bone construction members suggested seasonal or temporary shelters, although occasional whale bones and beach cobbles were present.

The remaining structure was different (site XCB-003, House 1), for it had been framed with 32 to 34 whale mandibles, supported by boulders. The single radiocarbon determination acceptable to the excavator indicated a date of about AD 945 (1005 B.P. ± 105; SI-916) in uncorrected radiocarbon years.

The whale mandibles were of baleen species — so far as I know not further identified — and were described as "up to 5.7 m long and weighing about 200-400 pounds" (McCartney 1974:65). The absence of bone structural members in any of the few other houses tested, as well as the fact that the mandibles had remained in the one location for a millennium after the house was abandoned, suggested to McCartney that there was no local practice of selectively scavenging whale structural elements from abandoned habitations. He concluded that the bones for the one special house had been hauled some 8 km by skin boat across the shallow lagoon from the Bering Sea coast, on the grounds that the lagoon was too shallow for large whales. The season of occupation was inferred to be summer and fall, although the conclusion was based upon a less than complete identification of faunal remains.

There are some obviously troubling elements in these conclusions, in particular the suggestion that 16 or 17 pairs of whale jaws were painfully transported across the lagoon and up more than 5 m of bank in order to build one house.

LOCAL HABITS OF THE GRAY WHALE

The gray whale is a medium-sized baleen whale now limited to the Pacific and Arctic oceans, the only remaining viable population being that which migrates along the North American coast. Since the Izembek house was excavated, several descriptions have appeared of the migratory and feeding habits of this mammal (e.g., Reeves and Mitchell 1988, with references).

Although long-distance travelers, an outstanding characteristic of the species is its consistent habit of movement near shore along the continental shelf. Calving in bays of the Baja California shore in January and February, pods of the whales then move northward, pass closely along the coasts of Oregon, Washington, and British Columbia between late February and late April, skirt the shores of the Gulf of Alaska to pass Kodiak Island, then cruise through the Aleutian Islands at Unimak Pass, immediately west of Unimak Island (Fig. 1). They turn northeastward in May and June to coast the Alaska Peninsula at least as far as the mouth of the Ugashik River at Ugashik Bay, sometimes as far as the Egegik River, and thence head northward across Bristol Bay, moving west of Nunivak Island and toward Bering Strait as the ice clears (Braham 1984). The return, over much the same route, occurs as the animals leave the Bering Sea through Unimak

Pass in November, to cruise along the southern Northwest Coast again in December and January.

The gray whale is apparently the only large whale to feed almost exclusively upon ocean bottom organisms, plowing the bottom sediments to do so. Virtually lying on its side, a gray whale roots along in the mud while straining the contents through its baleen-lined mouth (Nerini 1984). Preference is for relatively shallow waters, and even the long-distance migration path is restricted to near-shore areas, so that, for instance, most of the animals pass salient points on the Oregon coast at a distance of no more than 5 km from shore (Herzing and Mate 1984).

Little feeding occurs in Baja California, apparently because of a shortage of appropriate food organisms. On the northward migration, some whales have been observed at various points along the way feeding in bays and on beach margins shallow enough to be exposed at low tide (Ezzell 1991). In some areas, such as the outer coast of Vancouver Island, there appear to be small summer resident populations (Darling 1984; Murison et al. 1984). Most feeding, however, occurs in Alaskan and Siberian waters on the floor of the shallow seas covering the continental shelf both south and north of Bering Strait, and it begins in earnest in late spring as the animals enter the Bering Sea and move along the northwestern shore of the Alaska Peninsula (Moore and Ljungblad 1984; Nerini 1984).

In that area in June, the whales seen are almost all within 1 km and very commonly are within 30 m of shore, where they feed in only a few meters of water. They are particularly prone to enter and feed in lagoon systems from Izembek Lagoon northeastward to Ugashik Bay. They are especially partial just now to Nelson Lagoon (Gill and Hall 1983), a body somewhat smaller than Izembek Lagoon, although perhaps a meter deeper, where a higher range of tide may raise the water depth over much of the area to 4 m or somewhat more. A great deal of their feeding, however, occurs as they lie on their sides in water that may be less than 3 m deep (Moore and Ljungblad 1984).

With this feeding behavior in mind, the presence of the whale bone house on the shores of Izembek Lagoon seems to take on new significance.

SEA LEVEL CHANGES ON THE ALASKA PENINSULA

Commonly accepted models of sea level change worldwide deny that the past millennium has brought much alteration to general ocean depth. But study of the lower Alaska Peninsula and Shumagin Island region by Margaret Winslow (1992) has produced a model of the post-Pleistocene rise of sea level *and* the concurrent crustal rebound of the Peninsula that does provide a sea level change in that area. Whereas the end of the Pleistocene experienced a sharp rise in seas that flooded land surfaces still warped downward from the disappearing ice load, sometime after about 8000 years ago there began an exponential decrease in the rate of sea level rise. For the past 4000 years, the eustatic rise has been on the order of 1 to 2 m per millennium, while at the same time the isostatic rebound of the land of the lower Peninsula has been at a rate closer to 6 m per millennium, according to Winslow. The two processes created a practical net rate of rise of land relative to sea over the past 4000 years of around 4 m per millennium.

If correct, this means that only a millennium ago Izembek Lagoon was as much as 4 m deeper than at present. An indication that gross features of Winslow's model are descriptive is provided by evidence of strand lines near the tip of the Peninsula suggesting former still-stands at about 3 m and 16 m above present sea level (Funk 1973), with strand lines comparable to the higher of these visible along the Bering Sea coast of the Peninsula as far northeast as the Ugashik River area (Detterman et al. 1981, 1987). But even if the land has risen much less slowly than Winslow suggests, the situation a millennium ago would almost certainly have been conducive to regular visits by gray whales. A small amount of gray whale activity has been noted within Izembek Lagoon even in its present shallow state, and no more than a single meter of additional depth would evidently make it consistently attractive to the shallow-feeding cetaceans.

THE ARCHAEOLOGY OF OTHER LAGOONS

I had arrived at this speculative point several years ago when I began a helicopter-based survey of much of the Alaska Peninsula (Dumond 1987). It was my intention to examine the remaining lagoon systems of the Alaska Peninsula in the hope of finding evidence of hitherto unrecorded archaeological sites on their banks. This hope was only partly fulfilled. The first flight of the coast made it clear that several of the lagoons that had seemed promising on the map were formed by beach and bar constructions so recent and so low as to be unsuitable for permanent occupation even now, much less under conditions of higher relative sea level. Although two of these lagoons now support, or have recently supported, settlements on their enclosing bars (i.e., Nelson Lagoon, located on the outer bar of the lagoon of the same name, and Ilnik, located on the southern portion of the lagoon formed by the Seal Islands and referred to as Unangashik Lagoon on Figure 1), both of these settlements were established within the past 50 years, upon what was and remains very low ground.

On the other hand, the same flight allowed an identification of those few lagoons bounded by ground elevated enough to have been attractive for settlement several millennia ago, accepting Winslow's model as at least generally descriptive of the relationship of land and sea levels. In addition to Izembek, where high ground all around the southern edge of the lagoon is formed by a heavy glacial outwash deposit, five other embayments also have high ground in the immediate vicinity. Four of these areas of high ground exhibit habitation sites that are known to be at least two millennia in age, and one of them also supports a site of unusual size, although its date is unknown. The four additional embayments are Moffet Lagoon, Port Moller, Unangashik Lagoon, and Ugashik Bay, which will next be described briefly in turn. Together with Izembek, these provide the locations of all of the major coastal sites of the Bering Sea side of the lower Alaska Peninsula.

Immediately northeast of Izembek Lagoon proper, and partly connected to it, lies Moffet Lagoon, which in 1988 was visited on the ground by Bureau of Indian Affairs archaeologists, who examined the large and undated site referred to above (which is registered as XCB-028) and reported more than 800 depressions that presumably represent prehistoric habitations. There are other sites in the immediate vicinity, at least two of which front an ancient (raised) shoreline that is now more than 1 km from the lagoon shore. At one of these (XCB-029), radiocarbon ages in excess of 3000 years were obtained and a whale rib was noted (Cooper and Bartolini 1991; see also Dumond 1987:143-144).

Farther northeast, inside the main Port Moller bay system that is fronted by Nelson Lagoon, the Hot Springs site (XPM-001), with more than 200 visible house depressions has yielded remains dating predominantly between 2000 B.C. and A.D. 500 (Dumond 1987, Table 2.5; Okada and Okada 1974; Okada et al. 1976, 1979, 1984, 1986). Whale bone is common in these excavations, and the original excavator suggested that they had been used as structural members in the houses (Weyer 1930). So far as I am aware, the species of whale has never been identified.

The north end of Unangashik Lagoon, which lies behind the Seal Islands, is formed by glacial outwash a number of meters above sea level, upon which sits the historic site of Unangashik (CHK-015). It was abandoned about 1920 according to local people, and beneath it lies older remains that were dated by charcoal taken in a survey test in 1975 at about A.D. 800 (Dumond n.d.). This is a major site that has never been excavated. Whether whale bone is present is not known.

At Ugashik Bay, still a favorite feeding location for gray whales, beach ridges now about 8 m above high tide contain evidence for what must be literally hundreds of houses (UGA-029) that have never been fully surveyed. Very brief tests of two of the house depressions in 1975 produced dates of about A.D. 900 and A.D. 1000, very few artifacts, and, unfortunately, almost no faunal remains (Henn 1978).

In addition, another lagoon formed at the mouth of Cinder River has a small prehistoric site (XBB-001) located on a beach ridge so low as not to qualify as the kind of high ground I had in mind. The date is not known, but on the basis of a few artifacts in the hands of collectors, I should judge it to be at least earlier than A.D. 1400 (Dumond n.d.).

The only one of the embayments with high ground in the vicinity that has yielded no evidence of a prehistoric site of unusually significant size is Port Heiden. Such remains as underlie the modern village of Heiden, on the northeast side of the bay, are scanty and appear to be either historic or very late prehistoric (Dumond n.d.; Yesner 1983).

Finally, Klingler (1979, 1985; pers. comm., 1993) reports that a 1979 survey of Unimak Island, sponsored by the U.S. Fish and Wildlife Service, located a house with whale bone structure at the head of the relatively small Peterson Lagoon, on the northwest side of the island. The site (UNI-073) is said to include an area of depressions on a 16 m bluff, another on a 6 m terrace, and a third consisting of the whale bone structure which is eroded and evidently slumped toward the lagoon's inlet stream from a hill described as lower than the elevations just mentioned. A detailed description of the situation of the structure has not been given, and a more complete evaluation of the site is not here possible.

All in all, present data seem to promise a significant association of major sites, whale bone, lagoons, and gray whale feeding grounds, although, of course, whales were by no means the only summer resource available in these bird-rich lagoon systems. Yet, there is one fact that does not perfectly fit a hypothesis that occupation on some of the Alaska Peninsula lagoons was at least partly for the purpose of taking gray whales. That fact is that some of the mandibles found by McCartney (1974) at his whale bone house on Izembek Lagoon are too large to have come from gray whales. Specifically, he reports that the largest of the mandibles in the house was more than 5 m in length, whereas the longest that could be expected from even a large gray whale would be scarcely 3 m. Rather, one must suppose the very long mandible reported by McCartney to represent either a blue whale (*Balaenoptera musculus*) or a bowhead (*Balaena mysticetus*; see Table 1). The former is thought to migrate through Unimak Pass in the eastern Aleutians, whereas the latter normally confines its movements considerably farther to the north and west than the Alaska Peninsula, although it is impossible to say where occasional drifting carcasses might wash ashore. But, given the lack of any species identification of the Izembek mandibles as a whole, worry over this point seems at present premature.

ABORIGINAL WHALING

As indicated at the outset, the current ethnographic literature provides no clear account of the taking of gray whales in North America. Northern Eskimo whalers have historically focused on the bowhead, and the specific interests of the more southerly Aleut and Pacific Eskimo whalers, as well as of the whaling Northwest Coast Indians, seem most unclear in the literature (O'Leary 1984).

That the gray whale may not have been a favored prey species by many peoples is suggested by its reputation with Euroamerican whalers as an uncommonly pugnacious

Table 1. Approximate maximum lengths of mandibles of Alaskan whales.*

Species	Maximum Whale Length (m)	Length Ratio, Mandible/Total	Maximum Mandible Length (m)
Blue (Balaenoptera musculus)	30	2/9	7
Fin (Balaenoptera physalus)	22	1/5	4.5
Sei (Balaenoptera borealis)	15	1/5	3
Minke (Balaenoptera acutorostrata)	10	1/5	2
Bowhead (Balaena mysticetus)	20	1/3	6.5
Humpback (Megaptera novaengliae)	14	1/5	3
Right (Eubalaena glacialis)	20	1/4	5
Gray (Eschrichtius robustus)	13	1/5	2.75

*Based on information in *Alaska Geographic* (1978), Martin (1977), Ridgway and Harrison (1985), Scammon (1968), and True (1983). Migration paths of all but the bowhead include passes within the eastern Aleutian Islands leading into portions of the southern Bering Sea, which would place them in reasonable proximity to the tip of the Alaska Peninsula. The bowhead is normally well to the north and west of the Peninsula. Only the gray and minke whales cover virtually all of Bristol Bay (see e.g., *Alaska Geographic* 1978).

beast. It was known by some as the devil fish, by others as the hard head, for its propensity for ramming and sinking small boats (Henderson 1984). For the same reason, evidently, mature gray whales were avoided by many natives of the eastern coast of the Chukchi Peninsula of Siberia. Although some mature grays were reportedly harpooned with regularity by aboriginal Chukchi or Eskimos of the northeastern corner of the Chukchi Peninsula (despite a strong preference there for bowhead whales), their neighbors immediately to the south around Mechigmen Bay carefully avoided mature gray whales but consistently took juveniles, including nursing calves. On the northern Okhotsk Sea, however, mature gray whales were hunted by the Koryak people (Krupnik 1984, 1987).

Given the existence of the lagoon network on the Alaska Peninsula, however, a technique that seems more appropriate is that reported by natives to have been used into the 20th century for beluga or white whale (*Delphinapterus leucas*) a toothed whale no more than a third the length of the gray. On the Naknek River, for instance, pods of beluga entering the river in pursuit of smelt or salmon were reportedly herded upstream by kayak, with paddlers slapping the water, until the whales reached a shallow lagoon at upper tidal limit (see also Friesen and Arnold, this volume). There at the drop of tide they were beached and butchered (Dumond, field notes).

It seems entirely possible that a variation of this technique was practiced against lagoon-feeding gray whales, making use of tide, noise, shallow water, and lances. Such a technique is in keeping with the general Aleut and Koniag practice of whaling with lances not affixed to lines, with struck whales simply allowed to die and drift ashore as luck and ocean currents directed. Examination of this potential use of lagoons could be realized, given no more than the sites and locations I have just mentioned, and has the promise of yielding far more than the answer to the enigmatic whale bone house of Izembek. It could result in that most tantalizing of archaeological ends: the documentation of a human practice for which there is no clear historic or ethnographic record.

Furthermore, as noted above, some gray whales apparently feed along the Northwest Coast during their annual migrations, and modest numbers of them are even known to summer in selected locations on that coast (Darling 1984). Thus it is by no means inconceivable that lagoon-traps for gray whales may be found along the Pacific coast of southeast Alaska, a possibility for which the coastal-inclined among us should be alerted.

Acknowledgments. I thank Edward Mitchell and Steven Klingler for calling my attention to some of the sources cited herein. Brief versions of this material have been presented at the annual meetings of the Northwest Anthropological Conference (1987), the American Anthropological Association (1991), and the Alaska Anthropological Association (1993).

REFERENCES

Alaska Geographic
   1978   Alaska Whales and Whaling. *Alaska Geographic* 5(4).

Braham, Howard W.
   1984   Distribution and Migration of Gray Whales in Alaska. In: *The Gray Whale*, Eschrichtius robustus, edited by M. L. Jones, S. L. Swartz, and S. Leatherwood, pp. 249-266. Academic Press, Orlando.

Cooper, D. Randall and Joseph D. Bartolini
  1991   New Perspectives on Settlement and Ethnicity on the Lower Alaska Peninsula. Paper presented at the 18th Annual Meeting of the Alaska Anthropological Association.

Darling, James D.
  1984   Gray Whales off Vancouver Island, British Columbia. In: *The Gray Whale,* Eschrichtius robustus, edited by M. L. Jones, S. L. Swartz, and S. Leatherwood, pp. 267-287. Academic Press, Orlando.

Detterman, Robert L., T. P. Miller, M. E. Young, and F. H. Wilson
  1981   Quaternary Geologic Map of the Chignik and Sutwik Island Quadrangles, Alaska. *U.S. Geological Survey, Miscellaneous Investigations Series*, Map I-1292.

Detterman, Robert L., F. H. Wilson, M. E. Young, and T. P. Miller
  1987   Quaternary Geological Map of the Ugashik, Bristol Bay, and Western Part of Karluk Quadrangles, Alaska. *U.S. Geological Survey, Miscellaneous Investigations Series* Map I-1801.

Dumond, Don E.
  1987   Prehistoric Human Occupation in Southwestern Alaska: A Study of Resource Distribution and Site Location. *University of Oregon Anthropological Papers* No. 36.
  n.d.   Archaeological Reconnaissance in the Chignik-Port Heiden Region of the Alaska Peninsula. In: *Contributions to the Anthropology of Southcentral and Southwestern Alaska*, edited by R. Jordan, F. de Laguna, and A. Steffian. *Anthropological Papers of the University of Alaska* 24(1,2). In press.

Ezzell, C.
  1991   Hungry Whales Take a Bite Out of the Beach. *Science News* 140:167.

Funk, James M.
  1973   *Late Quaternary Geology of Cold Bay, Alaska, and Vicinity*. Unpublished MS thesis, Department of Geology, University of Connecticut.

Gill, Robert E., Jr., and John D. Hall
  1983   Use of Nearshore and Estuarine Areas of the Southeastern Bering Sea by Gray Whales (*Eschrichtius robustus*). *Arctic* 36(3):275-281.

Henderson, David A.
  1984   Nineteenth Century Gray Whaling: Grounds, Catches and Kills, Practices and Depletion of the Whale Population. In: *The Gray Whale,* Eschrichtius robustus, edited by M. L. Jones, S. L. Swartz, and S. Leatherwood, pp. 159-186. Academic Press, Orlando.

Henn, Winfield
  1978   Archaeology on the Alaska Peninsula: The Ugashik drainage, 1973-1975. *University of Oregon Anthropological Papers* No.14.

Herzing, Denise L. and Bruce R. Mate
  1984   Gray Whale Migrations along the Oregon Coast, 1978-1981. In: *The Gray Whale,* Eschrichtius robustus, edited by M. L. Jones, S. L. Swartz, and S. Leatherwood, pp. 289-307. Academic Press, Orlando.

Klingler, Steven L.
  1979   Description of Archaeological Survey Findings, Izembek Lagoon and Unimak Island, 1979. MS. in the possession of S. L. Klingler.
  1985   Archaeological Survey on Unimak Island, 1979. Paper presented at the 12th Annual Meeting of the Alaska Anthropological Association, Anchorage.

Krupnik, Igor I.
 1984 Gray Whales and the Aborigines of the Pacific Northwest: The History of Aboriginal Whaling. In: *The Gray Whale,* Eschrichtius robustus, edited by M. L. Jones, S. L. Swartz, and S. Leatherwood, pp. 103-120. Academic Press, Orlando.
 1987 The Bowhead vs. the Gray Whale in Chukotkan Aboriginal Whaling. *Arctic* 40(1):16-32.

Martin, Richard Mark
 1977 *Mammals of the Oceans.* G. P. Putnam's Sons, New York.

McCartney, Allen P.
 1974 Prehistoric Cultural Integration along the Alaska Peninsula. *Anthropological Papers of the University of Alaska* 16(1):59-84.

Moore, Sue E. and Donald K. Ljungblad
 1984 Gray Whales in the Beaufort, Chukchi, and Bering Seas: Distribution and Sound Production. In: *The Gray Whale,* Eschrichtius robustus, edited by M. L. Jones, S. L. Swartz, and S. Leatherwood, pp. 543-559. Academic Press, Orlando.

Murison, Laurie E., Debra J. Murie, Karen R. Morin, and Jeannette da Silva Curiel
 1984 Foraging of the Gray Whale along the West Coast of Vancouver Island, British Columbia. In: *The Gray Whale,* Eschrichtius robustus, edited by M. L. Jones, S. L. Swartz, and S. Leatherwood, pp. 451-463. Academic Press, Orlando.

Nerini, Mary
 1984 A Review of Gray Whale Feeding Ecology. In: *The Gray Whale,* Eschrichtius robustus, edited by M. L. Jones, S. L. Swartz, and S. Leatherwood, pp. 423-450. Academic Press, Orlando.

Okada, Hiroaki, and Atsuko Okada
 1974 Preliminary Report of the 1972 Excavations at Port Moller, Alaska. *Arctic Anthropology* 11(sup.):112-124.

Okada, Hiroaki, A. Okada, Y. Kotani, and K. Hattori
 1976 *The Hot Springs Village Site (2): Preliminary Report of the 1974 Excavations at Port Moller, Alaska.* Hiratsuka Printing Co, Hachioji, Tokyo.

Okada, Hiroaki, A. Okada, and Y. Kotani
 1979 *The Hot Springs Village site (3): Preliminary Report of the 1977 Excavations at Port Moller, Alaska.* Institute for the Study of North Eurasian Cultures, Hokkaido University, Sapporo.

Okada, Hiroaki, A. Okada, K. Yajima, and B. Yamaguchi
 1984 Preliminary Report of the 1980 and 1982 Excavations at Port Moller, Alaska. In: *The Qaluyaarmiut (2),* by H. Okada, A. Okada, K. Yajima, O. Miyaoka, M. Oshima, and B. Yamaguchi, pp. 1-52. Department of Behavioral Science, Hokkaido University, Sapporo.

Okada, Hiroaki, A. Okada, K. Yajima, and M. Sugita
 1986 Preliminary Report of the 1984 Excavations at Port Moller. In: *The Qaluyaarmiut (3),* by H. Okada, A. Okada, K. Yajima, O. Miyaoka, M. Oshima, and B. Yamaguchi, pp. 1-34. Department of Behavioral Science, Hokkaido University, Sapporo.

O'Leary, Beth L.
 1984 Aboriginal Whaling from the Aleutian Islands to Washington State. In: *The Gray Whale,* Eschrichtius robustus, edited by M. L. Jones, S. L. Swartz, and S. Leatherwood, pp. 79-102. Academic Press, Orlando.

Reeves, Randall R., and Edward Mitchell
 1988 Current Status of the Gray Whale, *Eschrichtius robustus. The Canadian Field-Naturalist* 102:369-390.

Ridgway, Sam H., and Sir Richard Harrison (editors)
   1985   *The Sirenians and Baleen Whales. Handbook of Marine Mammals*, Vol. 3. Academic Press, London.

Scammon, Charles M.
   1968   *The Marine Mammals of the Northwestern Coast of North America* Dover Publications, New York. (1874)

True, Frederick W.
   1983   *The Whalebone Whales of the Western North Atlantic ... With Some Observations on the Species of the North Pacific*. Smithsonian Institution Press, Washington. (1904)

Weyer, Edward M.
   1930   Archaeological Materials from the Village Site at Hot Springs, Port Moller, Alaska. *Anthropological Papers of the American Museum of Natural History* 31(4).

Winslow, Margaret A.
   1992   Modeling Paleoshorelines in Geologically Active Regions: Applications to the Shumagin Islands, Southwest Alaska. In: *Paleoshorelines and Prehistory: An Investigation of Method*, edited by L. L. Johnson and M. Stright, pp. 151-169. CRC Press, Boca Raton.

Yesner, David R.
   1983   Archaeological Theory, Ethnohistory, and Cultural Resource Management: Some Notes from the Alaska Peninsula. *Contract Abstracts and CRM Archeology* 3(2):109-116.

Prehistoric Use of Cetacean Species in the Northern Gulf of Alaska

Linda Finn Yarborough
USDA Forest Service, Chugach National Forest
3301 C Street, Suite 300
Anchorage, AK 99503

Abstract. *This paper reviews current knowledge about the seasonality, distribution, and population sizes of various species of cetaceans in the North Pacific, and compares it to ethnographic and archaeological evidence for human use of whales, particularly in the Alutiiq area of the Gulf of Alaska and Prince William Sound. While the prehistoric southern Alaskan archaeological Cetacea sp. samples are too small to allow comparisons of mortality rates, or to thus make strong suggestions about hunting or scavenging by human populations, there is clear evidence that large baleen whales, orcas, and porpoises were being utilized by at least 4000-3000 B.P. The development of whaling may be linked to an increase in social complexity in the North Pacific. The most useful models for understanding the development of prehistoric complexity in this area are likely to be those which take environmental change and intensification of resource use into consideration.*

INTRODUCTION

Whale hunting has long been recognized as an important subsistence activity among arctic peoples. It required development of specialized technologies, and held — in some cases still holds — a special cultural significance. Among the Alaskan coastal groups, the Inupiat are well known for their historic and prehistoric use of bowhead whales as a subsistence resource. Both large baleen whales and beluga whales have been important to the Yupik of the Bering Sea, the species available varying with differences in the marine habitat. Whaling was also an important pursuit among the subarctic Aleuts and Eskimos of the North Pacific, who had access to many more species than their northern whaling neighbors.

At the time of historic contact in the 18th century, the southern coastal inhabitants of Alaska were observed to be more than simple hunter-gatherers. They, along with the whaling groups of northwestern Alaska and northern Canada and the people of the Northwest Pacific Coast, had many of the characteristics associated with social complexity: sedentism, elaborate burial, occupational and task specialization, the potential for dietary surplus, redistributive economies, territorial boundaries, long-distance exchange, technological innovation, and warfare (Price and Brown 1985; Sheehan 1985). Although disease quickly decimated populations and swift Westernization obfuscated portions of the aboriginal cultures, early explorers and missionaries observed several levels of social inequality among the Aleut, the Koniag, and the Chugach, enough to warrant categorizing these groups as ranked societies (Townsend 1980). Special note was taken of the importance, influence, and ritual activities of whalers.

Archaeological evidence for the prehistoric use of cetaceans is available in the Gulf of Alaska from Ocean Bay tradition times and continues through the European contact period. Indications of active hunting of whales date from the Kachemak tradition (Thomas Amorosi and Philomena Hausler-Knecht, pers. comm., 1993). While it is often difficult to identify material to the species in archaeological faunal remains, due to fragmentary or deteriorated condition, it is apparent that both large whales and the smaller porpoises have been utilized for at least several thousand years. Their use generally coincides with Neoglacial paleoclimatic events of the late Holocene (Wendland and Bryson 1976).

Numerous models have been proposed to explain the prehistoric development of North Pacific coast technological and social complexity. Although these fall into the categories of stability models, environmental change models, and resource intensification models (Yesner 1992), all of the models consider the subsistence base in its entirety. Because whaling is ethnographically associated with an investment of community time and labor, special knowledge, individual training and leadership, high risk, and prestige, a closer look at the prehistoric evidence of whaling in the context of one or more of these models may provide additional insight into the development of social complexity in the North Pacific.

CURRENT CETACEAN POPULATIONS

Seasonality and distribution studies conducted in the late 1970s and early 1980s indicate that modern populations of mysticetes (baleen whales) are present in the Gulf of Alaska and Prince William Sound during spring, summer, and fall and are rare in winter, while odontocetes (toothed whales) are present year-round, although in smaller numbers in winter than summer. The individual mysticete species whose ranges include the North Pacific are the humpback (*Megaptera novæangliæ*), minke (*Balænoptera acutorostrata*), fin (*Balænoptera physalus*), California gray (*Eschrichtius robustus*), right (*Eubalaena glacialis*), blue (*Balaenoptera musculus*), and sei (*Balaenoptera borealis*) whales. Odontocete species include the sperm (*Physeter macrocephalus*), pilot (*Globicephala macrorhynchus*), killer (*Orcinus orca*), Baird's beaked (*Berardius bairdii*), and beluga (*Delphinapterus leucas*) whales, Dall (*Phocoenoides dalli*) and harbor (*Phocoena phocoena*) porpoises, and Pacific White-sided dolphin (*Lagenorhynchus obliquidens*; Leatherwood et al. 1988).

Current populations of right, blue, fin, humpback, and perhaps sei whales are considered severely depleted from pre-European contact levels (Braham 1984a:2, 1992:46; see Table 1). Hall (1981:xii) suggests that 18th century populations of gray and humpback whales in the North Pacific Ocean and Bering Sea may have been low due to environmental stress associated with late Quaternary ice advances. Neoglacial advances corresponding to worldwide northern hemisphere climate changes have been documented on the Kenai Peninsula as having occurred at about 3600 B.P. and 1400 B.P. (Wiles and

Table 1. Some taxa in the Order Cetacea, found in the Gulf of Alaska area.

	Estimated Population	
	1970-1980	Precontact
Mysticetes		
humpback whale (*Megaptera novæangliæ*)	850-1,200	15,000
minke whale (*Balaenoptera acutorostrata*)	98*	
fin whale (*Balænoptera physalus*)	16,500	
California gray whale (*Eschrichtius robustus*)	15,000-20,000	15,000
right whale (*Eubalaena glacialis*)	200-300	
blue whale (*Balaenoptera musculus*)	1,500	5,000
sei whale (*Balaenoptera borealis*)	16,000	70,000
Odontocetes		
orca (*Orcinus orca*)	286 (NE Pacific)	
	50-75 (Prince William Sound*)	
Pacific white sided dolphin (*Lagenorhynchus obliquidens*)	200-300	
Dall porpoise (*Phocoenoides dalli*)	6,700	
harbor porpoise (*Phocoena phocoena*)	900	
sperm whale *(Physeter macrocephalus)*		290,000

* Estimate for Prince William sound; no overall Northeast Pacific estimate available (from Gambell 1985:180; Hall 1981; Heyning and Dalheim 1988; Brent and Leatherwood 1985)

Calkin 1992). The most recent episode, also known as the Little Ice Age, began about 700 B.P. and ended late in the 19th century (Wendland and Bryson 1974; Wiles and Calkin 1992). These episodes appear to have increased both the amount and duration of sea ice in the Arctic Ocean and Bering Sea, and may have been associated with increased precipitation, an increased amount of sea ice, and reduced sea surface and air temperatures in Prince William Sound as well. It is likely that such changes may have affected biological habitat, migration routes, and/or seasonal location patterns.

Seasonal presence and population distribution in the Gulf of Alaska vary from species to species (Fig. 1). Large baleen whales pass through the North Pacific during migrations west and north to Bering Strait and the Arctic Ocean, exploiting the continental shelf and certain areas of Prince William Sound during periods when primary biomass is particularly high (Hall 1981:72). Historically recorded sightings of at least nine species of whales within Prince William Sound include both mysticetes and odontocetes. Both

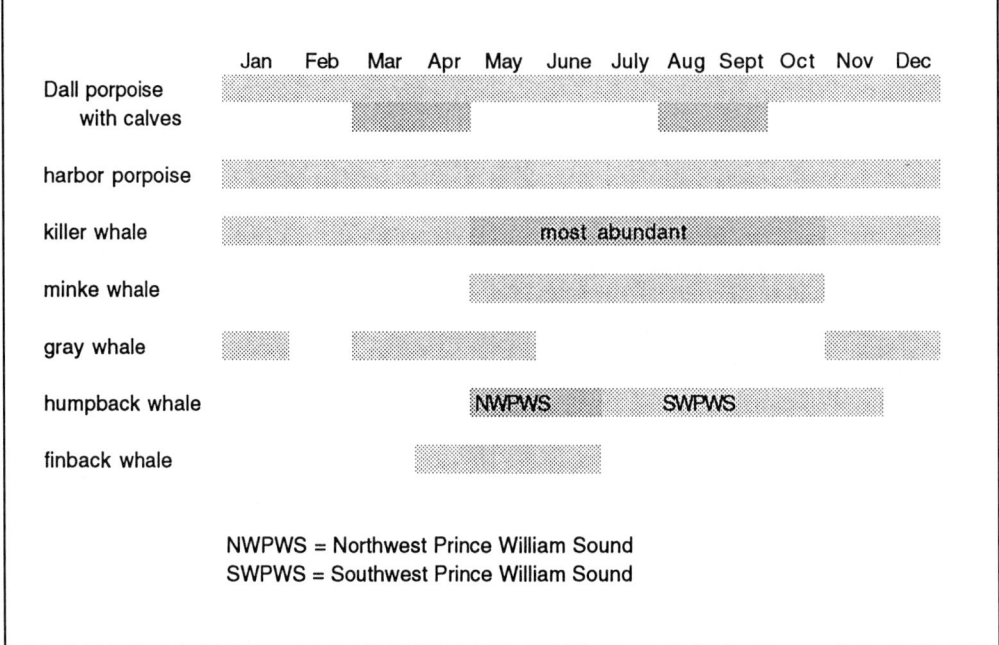

Figure 1. Seasonal occurrences of some cetacean species in the Gulf of Alaska (from Braham 1984b; Hall 1981; Stewart and Leatherwood 1985; Leatherwood et al. 1988; Wolman 1985).

suborders occur in the sound, but only small numbers of odontocetes remain in the area during winter; large whales are very rare at that time of year (Hall 1981:39, 1986). The three mysticete species recorded from within the sound are humpback, minke, and fin whales.

Gray whales migrate north along the northeast Pacific coast in two waves, between March and June. Males, nonparturient females, and immatures compose the first wave, and are followed in May by females with calves (Braham 1984b:261, 263). They migrate within 5 km, and more commonly within 2 km, of shore, and sometimes even occur in the surf zone, apparently as a result of feeding behavior (Braham 1984b). This behavior suggests the possibility that these whales could be taken by subsistence hunters without going far out into the open ocean. Their route takes them along the south edge of Montague and Hinchinbrook islands. While they have not been recorded from Prince William Sound, it is possible that they occasionally enter these waters (Braham 1984b; Hall 1981). It has been suggested that most gray whales continue their migration along the south side of Kodiak Island rather than through Shelikof Strait (Braham 1984b:257). The southward migration occurs between October and January (Braham 1984b:263-264). Females migrate first, particularly those who are pregnant, and adults tend to move south before immature individuals (Wolman 1985:73).

Other baleen whales are found in the Gulf of Alaska during spring and summer, apparently migrating north from more temperate zones where they reside year-round (Stewart and Leatherwood 1985:101). Fin whales are common among the outer islands of Prince William Sound and along the southern Alaska Peninsula (Leatherwood et al. 1988:24). There may be instances of minkes remaining in the northeast Pacific year-round (Stewart and Leatherwood 1985:112). These small whales generally occur in small groups of only two to three, which tend to be age- and sex-dominated, and are easy to approach. Sometimes they approach boats themselves (Stewart and Leatherwood 1985:117-118). Feeding occurs near the surface, after driving fish up from deeper waters (Stewart and Leatherwood 1985:120).

Odontocetes found in Prince William Sound include both belugas and orcas (killer whales). The southern limit of the beluga is the Subarctic; its northward range includes the Arctic Ocean. While individuals winter along the edge of pack ice and in arctic polynyas (Brodie 1985:124), seasonal movements bring this species into coastal and freshwater areas such as Cook Inlet and Prince William Sound in summer. They are found in relatively shallow water (Brodie 1985:132), which may have made them fairly easy prey for prehistoric human hunters.

The movements of orcas, many of which remain in the Gulf of Alaska year-round, appear to be closely related to the movements and availability of their prey species. Fin fish seem to be the preferred prey, with a switch made to marine mammal prey species "when fish are less abundant or not available" (Braham and Dahlheim 1982:643). Orca sightings are especially frequent in Prince William Sound, the northern Gulf of Alaska, and along the southeast side of Kodiak Island, where they occur most often within 20 km of shore. Sightings of Dall's porpoise and orcas seem to often coincide with each other (Braham and Dahlheim 1982:644).

HISTORIC OBSERVATIONS

The circumstance of late initial European contact with the coastal whaling people of northern Alaska and Canada allowed documentation of many of their cultural traits and some understanding of the multiple factors that comprised their late prehistoric complexity. In contrast, the decimation of southern Alaska Native coastal cultures following Russian contact in the early 18th century led to less detailed ethnographic information concerning social organization and resource utilization. However, Russian-American Company policies, which forced employees to rely at least partly on local resources, encouraged local people to preserve some knowledge of late prehistoric subsistence patterns. Observations by Gedeon (1989), Lisianski (1968), and Davydov (1977) in the early 19th century provide some indications of how whaling was being practiced in early historic times. In the case of traditional whaling information obtained by ethnographers in the 20th century, Black (1987:11) has suggested that informants may have combined knowledge of techniques and practices that were used at various times in the past.

The subsistence base of the natives of the northeastern Pacific at the time of European contact included a wide variety of sea mammals, land mammals, fish, birds, intertidal zone resources, and plants, according to early historic records and archaeological evidence. Whales were observed to contribute to the diet of North Pacific peoples from the time of early Russian contact with the Aleutian Islanders. Veniaminov (1984:43, 356) noted during his stay in Alaska from 1824 to 1834 that the Aleuts recognized 12 species of whales, including both baleen and toothed whales (Table 2). The species appear to include Pacific right whale, bowhead, possibly blue whale, beluga, orca, and two kinds

Table 2. Whale species recognized by the Aleut.

Aleut Name	Russian Name	Probable Species Identification
Kuulamax	kulema	Pacific right or possibly bowhead
Chikaaxlux		possibly bowhead or gray whale
Aalamax	polosatik	probably humpback or possibly blue whale
Mangidax	(all baleen whales	possibly gray or humpback
Umguulix	with a fin, possibly	
Agamagchix	different sizes of the same species as Aalamax)	possibly minke or little piked whale
Chiiduxn	plavun, cachalot	beaked or sperm whale
Agdaaxchix		noted to be "the same" as sperm, but without teeth
Xaadax		beluga
Aglux	kosatka	orca
Alladax	svinka	harbor porpoise
Kdaang or Kdan	svinka	porpoise or dolphin with white spots

(from Bergsland 1980; Veniaminov 1984:34, 356; Wolman 1985:68)

of porpoise. According to Wrangell (1980), the Koniag recognized five species of whales. These included *polosatik*, a baleen whale with a fin which may be the blue whale; *aliama*, a baleen whale smaller than the *polosatik* which may correspond to fin whale; *kulema*, the Pacific right whale; *cachelot*, the sperm whale; and *uchulukhpag* (Wrangell 1980:28). Holmberg (1985:47-48) believed that only one species, *kulema* or Pacific right whale, was hunted during his visit in 1850-1851. However, his command of Russian and Alutiiq was apparently poor, and what he believed were different names for variously aged right whales were actually generic names for whales, or, in one case, an indication of size (Lydia Black, pers. comm., 1993).

Seasonality and Areas of Whale Hunting

Davydov (1977:115) sighted whales at the north end of Kodiak Island on March 31, although he did not mention the species. Baleen whales "of the striped type" were present in the archipelago in June, when Davydov observed one being butchered on Spruce Island (Davydov 1977:122). He was told that porpoise were hunted mainly by people living on the north shore of Kodiak. Holmberg (1985:48) recorded that a large number of whales were present in the Kodiak area in June, feeding on small fish and jellyfish, and that they moved into bays in the following two months.

At the time of Birket-Smith's (1953:33) Alaskan visit, he was informed that the people living in the old village of Chenega still hunted large whales, which he supposed might mean sperm or humpback whales, as well as "little finners, white whales, blackfish and porpoises." Traditionally, whaling was said to have occurred year-round, "but especially in winter." This would seem to indicate porpoise and/or orca, as other species are rare in Prince William Sound in winter.

Preparations for Whale Hunting

Of the necessary preparations for a successful whale hunt, perhaps the most important was the transmission of knowledge from one generation to the next. In the Gulf of Alaska, this occurred only among select individuals. Training was apparently done in secret, and whaling weapons were kept hidden away from villages. Birket-Smith's (1953) informants recalled that whalers were not supposed to associate with menstruating women. In addition, young men refrained from sexual intercourse for three days prior to hunting, although this restriction did not apply to old men. Prior to a hunt, whalers were said to travel to special locations where they drew pictures of various animals to aid in spiritual or "magical" preparation for a successful hunt. Birket-Smith (1953:34) did not record any general food taboos for whalers, although he intimated that there might be short-term taboos immediately prior to a whale hunt. Both whalers and their wives were expected to dress neatly during the course of the hunt, and wives had to stay at home while their husbands were whaling. Certain rituals were also observed after a whale was obtained. One Chenega resident indicated that a drink of fresh water was given to the dead whale, a practice similar to that followed by arctic whalers, while another reported that it was the whalers who received the drink after a hunt (Birket-Smith 1953:34-36).

Arctic Eskimos held spring ceremonies directly related to whaling prior to the beginning of the whale hunting season (Murdoch 1988:142). While no ceremonies particularly oriented to taking cetaceans were recorded by early travelers to Gulf coast villages, it may be that some of the festivals noted and attended by Lisianski, Gedeon, and Davydov (1977:108) were similar ceremonies whose purpose was unrecognized.

Poison

The use of poison in southern Alaskan whaling has been discussed by numerous researchers and is here only briefly summarized. In the Koniag and Chugach areas, poison was reportedly used on whaling lances. Both Davydov (1977) and Gedeon (1989) were told that this poison was prepared from the fat of corpses, especially those of successful whalers or other important people. As a result of his research into this question, Heizer (1943:437) maintained that the stories among the Koniag and Chugach of poison made from human fat were myths developed by the whalers to keep knowledge of the true poison — aconite — from the common, uninitiated people. In 1790, Sauer (1802) noted that Kodiak Islanders prepared this poison from the roots of a species of *Aconitum*. Birket-Smith's 20th century Chugach informants denied the use of aconite poison on whaling lance blades, but allowed that poison from human fat might have been used. They also indicated a second use for poison during whaling, telling a story of pouring poison made from human fat into the water between the whale and the open ocean, across which the whale would not travel (Birket-Smith 1953:34). A similar story was recorded by Hrdlicka (1944:126) for the Kodiak area. Birket-Smith believed that the preparation of poison was accompanied by magical songs with percussion instruments (rattles). Black (1987:20) has suggested that the use of poison spread from the Gulf of Alaska area to the Aleutians, although aconite poison is documented in some areas of the Asian Pacific as well (Heizer 1943; Bisset 1976).

It has been suggested that if poison was used, it probably was not itself a cause of death, but may have resulted in loss of balance, infection, and/or paralysis which could lead to drowning (Bisset 1976:117; Black 1987:26). Studies of other mammals have shown that the local effect of aconite poison in a wound is a burning, prickling sensation. As the poison spreads through the body, it causes nausea, choking, vomiting, slowed breathing, lowered blood pressure, heart fibrillations, giddiness, and weakness due to loss of muscle power (Bisset 1976:115). Even a small amount introduced into a sensitive area with conglomerations of blood vessels, such as the flippers, thoracic region, or the base of the brain, would spread rapidly through the body and could seriously disable a whale (Bisset 1976:117). Death would not necessarily have been instantaneous, but the introduction of a toxic substance could have contributed significantly to a successful hunt.

Watercraft

Although all coastal Alaskan Eskimos used both large open boats and smaller kayaks or *bidarkas*, with the possible exception of the people of St. Lawrence Island (Nelson 1983:218), the method of pursuing cetaceans from an open boat in historic times was practiced almost exclusively by hunters of the Bering Strait and Arctic Ocean. *Umiaks* were better suited than kayaks to carry the large whaling harpoons and mass of equipment which had been developed to actively pursue and retrieve large whales. However, the Inuit of Labrador used kayaks in addition to *umiaks* (Taylor 1979:294). The smaller craft were apparently faster and more maneuverable than *umiaks* and, therefore, useful in lancing wounded large whales and in hunting immature whales.

During the 18th century, Veniaminov (1984) observed that the Unalaska natives attacked whales from single kayaks. Davydov (1977) noted the same technique among the Koniag in the first years of the 19th century. Although this same method was still used during the mid-1800s (Wrangell (1980:27), Holmberg (1985:48) noted that while the Koniag pursued whales from single kayaks, they hunted in groups for safety. Birket-Smith (1953:33-37) recorded three accounts of whale hunting from Chugach elders in Prince William Sound in the 1930s. Two of the methods described were very similar to those practiced on Kodiak, with one or two men attacking whales from kayaks. A third method, which involved using an open boat, was similar to the technique traditionally used by arctic whale hunters. Open skin boats (*umiaq*) similar to those used historically in Greenland were recorded by Cook among the Chugach at the time of contact; these were large enough to hold as many as 20 or 30 people (Birket-Smith 1953:49). However, because the "traditional" pursuit of large whales had not occurred for several generations at the time that Birket-Smith interviewed his informants, Birket-Smith himself questioned the origin of the information, suspecting that it had been somehow passed along from Bering Sea Eskimos. However, he did not totally rule out the possibility that this method may have been used in the Gulf of Alaska (1953:36).

Hunting Technology and Technique

Large whales hunted by northern Eskimos were generally taken with the use of a heavy harpoon, to which several floats were attached. In contrast, cetaceans were hunted in the northern Gulf of Alaska with a long spear or harpoon propelled from a throwing board, and the use of floats was not observed. The use of special harpoons with detachable heads is especially well documented for the Kodiak area of the Gulf of Alaska (Davydov 1977; Holmberg 1985; Gedeon 1989; Wrangell 1980). Two whaling lances from Kodiak, purchased by Holmberg for the Danish National Museum, each have barbed polished slate blades hafted in long cylindrical bone sockets which are in turn bound to wooden shafts. Their points, made to detach or break off in the whale, are only loosely set in their wooden foreshafts. One lance is about 154 cm long, and the other measures about 142 cm long. These lances are each fletched with three half feathers, and are designed to be hurled from a throwing board (Birket Smith 1941:138). Four "harpoons" at the museum described as used for hunting small whales are very similar to sea otter hunting darts. Their points are only about 5-6 cm long, and their total length varies from about 120 to 131 cm. On the other hand, Birket-Smith (1941:136) believed that two harpoons with barbed toggle-type bone heads were probably used for hunting at least small whales.

Two of Birket-Smith's three Chugach informants described the traditional use of 3 m-long harpoons with detachable barbed slate points for the Prince William Sound area as well (Birket-Smith 1953:35-36). A third described using a sea lion or seal toggle harpoon with a detachable point for small whales, but a lance with a fixed blade for large whales (Birket-Smith 1953:33). Additional auxiliary gear described by two informants included sea lion stomach floats and sealskin drag and towing lines for bringing dead whales into shore. Although mention was made of butchering the whale in the water, "to save the trouble of towing it home" (Birket-Smith 1953:36), there was no suggestion of special flensing knives or waterproof clothing such as that noted by Taylor (1979:296-298) for eastern arctic whalers.

Whalers attempted to hit the whale under its flipper (Davydov 1977:224; Holmberg 1985), but if the ventral area was missed, the hunter would attempt to strike the back fin or tail (Gedeon 1989:68). It was common knowledge that (a) a good hit made below the fin would cause the whale to surface on the third day, (b) a hit towards the tail would delay the whale surfacing for 5 or 6 days, and (c) a whale struck under the tail fin would not surface for 8 or 9 days (Gedeon 1989:68). No attempt apparently was made to retrieve the animal, but it was hoped that it would wash up on shore a few days later. This method left the hunter dependent on wind and ocean currents, with the possibility that the whale might wash up somewhere else. Instances were recorded in the early 1800s of whales killed in the vicinity of Kodiak being washed up on Unalaska (Holmberg 1985:48). A hunter hoped that his marked spearhead would be recognized and that his village would be contacted for a portion of the whale. However, there was no guarantee that this would happen.

Non-Food Uses

According to ethnographic accounts, whales were hunted in the Gulf of Alaska not only for their food value, but also for the raw manufacturing material provided by their baleen, sinew, and bone. Baleen was used to make runners for the bottoms of sledges and for repair of wooden implements (Birket-Smith 1953:50, 60). Davydov (1977:202) recorded that split and scraped whalebone (baleen) was used in the Kodiak area in the frame of *baidaras* or *umiaqs*. Thread was made of the sinew of large whales and porpoises (Birket-Smith 1953:77). Although whale bones were often used as structural supports for houses on the Bering Sea and Arctic Ocean coasts, there is currently no evidence for such applications in Prince William Sound or Cook Inlet or in the Kodiak archipelago.

Risk

Whale hunting in general involved the risk of "time and effort," with little return if a whale was not both struck and captured (Ackerman 1988:69). Whalers additionally ran the risk of loss of life and property if they were too close to the animal as it thrashed about after being harpooned (Gedeon 1989; Holmberg 1985; Wrangell 1980). In order to lessen the first risk, whale hunters needed to have sufficient resources on hand to supply the group in the event that a struck whale was not recovered. The second aspect of risk was reduced by choosing to hunt only in particular weather conditions — fairly still days were said to be necessary — and by individuals participating in a group hunt in single-person kayaks. However, the very real possibility of death gave the hunters more prestige in the eyes of the uninitiated (Holmberg 1985:48), while the successful retrieval of a whale increased the hunters' status. The difficulty and danger of whale hunting may also have been lessened by the choice of particular age animals as targets. Holmberg (1985:48) noted that "only the young and one-years-olds are hunted." Gedeon (1989:67) wrote that yearling whales were chosen "because their meat and blubber are tastier and more tender." This age bias in selection of prey is similar to that noted for northern whalers (McCartney and Savelle 1985:45; Savelle and McCartney 1991:206; see McCartney, this volume), and corresponds to prehistoric whale mortality profiles in the eastern Canadian Arctic (Savelle and McCartney 1991:214). The risk and relatively small energy investment, in terms of technological preparation and energy expended in hunting whales with the intention of merely wounding them, may have been worth the caloric return when one actually was harpooned and then washed up on shore.

ZOOARCHAEOLOGICAL STUDIES

Cetaceans may have contributed substantially to subsistence and dietary surplus along the southcentral Alaskan coast on at least a seasonal basis. A beluga whale might weight as much as 1100 lbs (500 kg), while even a small baleen whale might weigh between 6.5-13 tons (6000-12,000 kg; Leatherwood et al. 1988; McCartney and Savelle 1985:41). A newborn gray whale weighs about 500 kg, and an adult can weigh over 8800 kg (Wolman 1985:69). Dall porpoise can grow to 200 kg, while harbor porpoise may attain an adult weight of 90 kg (Leatherwood et al. 1988). While porpoise remains increase over time in the faunal assemblages of sites in the Kodiak archipelago and in Cook Inlet, evidence of the use of larger whales is sparse. This may be due to the sheer size of baleen whales, such that only those portions which were to be utilized would have been brought to a residential site. McCartney and Savelle (1985:41-42) have noted that very large portions of whales could have been used, consumed, or decomposed without leaving any archaeological "signature."

The earliest whaling in the Bering Sea has long been attributed to the Old Whaling culture complex of 3400-3300 B.P., and associated with climatic cooling, increasing populations, and an expanding subsistence resource base (Ackerman 1988:66; Wendland and Bryson 1974). However, whaling does not seem to have been an important part of the succeeding Choris and Norton cultures. Ackerman (1988:67) has theorized that this apparent early whaling culture was the result of a temporary expansion of Siberian sea mammal hunters. After reanalysis of faunal data, Mason and Gerlach (this volume) suggest that the use of whales by the inhabitants of Cape Krusenstern is unclear, and the archaeological remains may reflect scavenging rather than active whaling. In a proposal similar to the traditional view of northwestern Alaskan whaling, Black (1987) has proposed that use of whales as a subsistence resource in the Aleutians was not continuous over time and space, and that both hunting knowledge and technology were lost and then reintroduced several times. This interpretation does not apply to the Aluutiq area to the east. Although present evidence is sparse, it appears that cetaceans were used more or less continuously for several thousand years in the North Pacific.

The archaeological data for the northern Gulf of Alaska suggest that cetacean use may have been part of a diachronic trend towards subsistence intensification. Currently the earliest evidence of the use of cetacean species in the gulf is from Kodiak Island, where deteriorated pieces of large whale bone have been recovered from the Ocean Bay tradition site at Rice Ridge, dated to about 6500-6200 B.P. The earliest indication of *active* whaling comes from the later Kachemak tradition site at Crag Point, in the form of a large whale rib with a projectile point imbedded in it (Thomas Amorosi, pers. comm., 1993). The use of large whales and two species of porpoise continued through the late prehistoric Koniag period, with evidence from sites on both the Shelikof Strait side and the open Gulf of Alaska side of the Kodiak archipelago in the form of whole, fragmented, and modified bones (Clark 1974: 46-50; Richard Knecht and Amy Steffian, pers. comm., 1994). Two occurrences of associated whale and human remains are recorded from Koniag period sites on the south side of the archipelago: in a pit-like structure at Three Saints Bay, and in a burial cairn at Rolling Bay (Clark 1974:47).

Porpoise remains were a significant portion of the Kachemak tradition faunal assemblage at the Chugachik Island site (2700-1500 B.P.) in Kachemak Bay, on the west side of the Kenai Peninsula. The proportion of remains shows an increase over time, and porpoise bones are second only to those of harbor seal in numbers (Lobdell 1980). Lobdell's (1980) and Yesner's (1992) analyses of faunal assemblages at the Fox Farm site, Yukon Island, Kachemak Bay, indicate that porpoise was an increasingly important resource during the second half of the first millennium A.D. In the later occupation levels, over one-third of the faunal assemblage is porpoise (Yesner 1992:173). Large whales are represented by a few specimens in the later levels.

Cetacean species remains are known from archaeological sites in Prince William Sound beginning at about 2250 B.P. (Yarborough 1993). Bones of both porpoise and large whales were recovered from Uqciuvit, in northwestern Prince William Sound, during excavations in 1988 (Yarborough and Yarborough 1991). The porpoise specimens consisted of vertebrae and one auditory bulla, which appear to be from harbor rather than Dall porpoise. All of the large whale remains are deteriorated and morphologically unidentifiable to species. The 13 phocoenoid specimens are all from late prehistoric Chugach phase deposits throughout the site, as are 10 of the fragmentary cetacean specimens from larger whales. In addition, one layer in one square identified as Late Palugvik phase yielded 26 large whale specimens.

De Laguna's (1956) 1933 excavations at Palugvik, in southeastern Prince William Sound, resulted in recovery of both odontocetes and mysticetes from various stratigraphic layers throughout the midden. Large baleen whale remains include three vertebral epiphysis fragments that are large enough to belong to gray, pilot, right, or fin whales (Carl Kinze, pers. comm., 1993). One epiphysis from an early Palugvik phase context has a deep cut mark on it. One skull fragment and two fragments of bone from large whales were also identified during Ulrik Møhl's initial study of the faunal assemblage in the 1940s. During my recent studies of previously unidentified bone from the site, two orca teeth were identified, as well as an two additional bone fragments and a portion of a vertebral epiphysis from a large baleen whale in a Late Palugvik context. Although 26 porpoise elements were originally reported from the site, my reexamination of Møhl's notes and the collection indicates that 42 elements were actually identified. An additional six porpoise specimens have been identified during the continuing analysis. All appear to be harbor porpoise rather than Dall porpoise specimens. The majority of the porpoise remains occur in layers 5 through 11, which correspond to the Early Palugvik phase and possibly early Late Palugvik phase, with the greatest concentration in layers 5 through 8.

MODELS

The prehistoric inhabitants of the northern Gulf of Alaska are categorized as having a modified maritime adaptation (Fitzhugh 1975:344; McCartney 1988:33). In other words, based on currently available faunal analyses, they used both terrestrial and marine resources, with an emphasis on the latter. According to Binford's (1980) definition, the northern Gulf of Alaska was occupied, at least in late prehistoric and protohistoric times, by "logistically mobile" hunter-gatherer-fishers who sent out task groups from a residential base. The ethnographic and recently obtained archaeological evidence points to these residents' social complexity, at least later in time, for which a number of models have been suggested.

Yesner (1992:176) suggests that past models fall into three different categories — maritime stability models, environmental change models, and subsistence intensification models — which attempt, in part, to account for the changes seen over time in faunal assemblages in the Kachemak area of lower Cook Inlet. The maritime stability model is one which has been more or less assumed by many researchers for the North Pacific, resulting in explanations for subsistence change that exclude population increases or either catastrophic or gradual environmental changes (Jordan 1988, 1994). Schalk's (1981) model for the distribution of complex societies, which he characterized as a result of the "distributional structure of the environment," fits this particular category. He hypothesized that these societies developed due to the "clumped" nature of resources in more northern regions, rather than as a result of increasing population size, circumscription, or general regional resource availability. Environmental change models have been developed for the most part on local bases, such as Kotani's (1980) and Yesner's (1982) respective explanations for changes in molluscan fauna at the Hot Springs site on the Alaska Peninsula and the Chaluka site on southwest Umnak Island. Subsistence intensification models, such as that proposed by Haggerty et al. (1991) for the northern Gulf of

Alaska, assume increasing population growth and circumscription, with increasing stress on subsistence resources over time.

The development of whaling may have been associated with both environmental change and subsistence intensification. Whaling in both the Arctic and Subarctic appears to have developed during just the past 5000 years, the portion of the Holocene during which at least three major Neoglacial advances and retreats have affected precipitation, temperature, glaciation, and sea ice. Prehistoric adjustments to changing climate have been well documented for inhabitants of the Arctic Ocean and Bering Sea coasts (Ackerman 1988) and have been recently suggested for the inhabitants of Prince William Sound as well (Yarborough and Yarborough 1991).

When Neoglacial climate changes are combined with the possibilities of increasing populations and potential human pressure on resources, a more realistic picture of human interaction with subsistence resources may emerge. Sheehan (1985) has discussed the relatively large community size necessary to technologically prepare arctic whalers for a successful hunt. Successful whaling villages in the Arctic had developed large enough populations to send out as many as a dozen 6-8 person crews, in part because of the low return of whales captured as opposed to whales struck (Ackerman 1988:69). Krupnik (1988) has noted the community versatility necessary for the subsistence success of recent Asiatic Eskimo whaling groups, as well as the necessity for territorial economic buffer zones, surplus storage, and flexibility in hunting the most accessible prey in order to ensure coping with environmental changes. His study of modern Eskimos and previous archaeological studies of arctic coastal sites (Fitzhugh 1975; Maxwell 1985; McCartney 1980; Savelle 1987) suggest that such changing strategies, together with environmental factors, have probably been common throughout time. Cognizance of the risks associated with whaling and development of prestige for those participating in the chase may have gone hand-in-hand with changes in levels of social organization necessary to prepare for and execute a successful whale hunt. This may, in turn, have required growth to a minimum level of population and evolution of a particular degree of societal complexity for the inclusion of whaling as a regular part of the subsistence round.

DISCUSSION

Current whale populations indicate the variety of this resource that would have been available prehistorically to aboriginal populations in the northeastern Pacific. Some of the ethnographically recorded techniques for whaling suggest that they were developed with certain great whales in mind, in particular the gray, humpback, and right whales. These whales expose their ventral area during feeding (Winn and Reichley 1985:258; Wolman 1985:80). Late prehistoric and early historic whaling techniques and technology in the northern Gulf of Alaska were not isolated, but appear to be a portion of a continuum of knowledge stretching along the coast from the North Pacific to the eastern Canadian Arctic. While the prehistoric evidence for use of cetaceans is not as voluminous along the south-central Alaskan coast as it is from sites along the Bering Sea coast or the Arctic Ocean, the evidence is present and appears to be fairly continuous for several millennia. This use increases in late prehistoric times in at least the Cook Inlet and Kodiak areas. Despite the lack of substantial faunal samples, the Kachemak tradition evidence from

Kodiak indicates that Gulf of Alaska inhabitants had the capability to not only scavenge but also hunt whales during much of that period.

Ethnographic examples and archaeological data on the size and complexity of community organization in arctic whaling villages, the technological requirements and risks involved with whale hunting, and the coincidence of the development of whaling with recent periods of neoglaciation make the concept of a combined environmental change/subsistence intensification model look promising in understanding the development of social complexity and the place of whaling in southern Eskimo culture. However, in developing such models, a variety of labor and social input levels should be considered in light of the differences in whaling techniques and technology documented ethnographically for the North Pacific, the Bering Sea, and the Arctic Ocean. The greater diversity of cetacean species in the northern Gulf of Alaska must also be considered. The "hunt and hope" method of hunting large whales as recorded for southern Eskimos may indicate that large whales were not as critical a resource for people of the Gulf of Alaska as for those of the Bering, Chukchi, or Beaufort seas. This may have arisen from an ecological system that furnished such a variety or abundance of alternative resources that southern Eskimos could afford to have whaling methods and technology that furnished prestige to its initiates, but did not ensure great success. Consideration should also be made of the possibility that the development of whaling was involved with the growth of an elite social system (Hayden 1981, 1992) as much as 4000 years ago in the Gulf of Alaska, culminating in the society documented for the Alutiiq at European contact.

Acknowledgments. I would like to express my appreciation to Herb Maschner, Henry Bunn, Allen McCartney, T. Douglas Price, and James Stoltman for their encouragement of my zooarchaeological studies, and to the researchers at the Zoological Museum of the University of Copenhagen for their assistance in my studies of the 1933 Palugvik faunal assemblage. In particular I thank Kim Aaris-Sørensen for his permission to study the collection and assistance in translating Ulrik Møhl's faunal identification notes, and Carl Kinze and Jeppe Møhl for their advice on speciation of cetacean specimens. Thanks also go to Howard Braham and Jim Thomason of the National Marine Mammal Laboratory for allowing me to examine their Dall and harbor porpoise specimens. I am also grateful to Thomas Amorosi of Hunter College for allowing me to note his work in progress on cetacean fauna from several sites on Kodiak Island. Richard Knecht kindly kept me informed of his current research on Kodiak and Afognak Island archaeological sites. My thanks go to Amy Steffian of Northern Land Use Research, Inc. for her critical review and insightful comments. Lydia Black of the University of Alaska Fairbanks helpfully advised me on Holmberg's ethnographic notes. My continuing studies of the Palugvik fauna have been funded, in part, by grants from Sigma Xi, the Kenaitze Indian Tribe, and the USDA Forest Service, Chugach National Forest.

REFERENCES

Ackerman, Robert E.
  1988 Settlements and Sea Mammal Hunting in the Bering Chukchi Sea Region. *Arctic Anthropology* 25(1):52-79.

Bergsland, Knut
  1980 *Atkan-Aleut-English Dictionary*. National Bilingual Materials Development Center, University of Alaska Anchorage.

Binford, Lewis R.
  1980 Willow Smoke and Dogs' Tails: Hunter-Gatherer Settlement Systems and Archaeological Site Formation. *American Antiquity* 45(1):4-20.

Birket-Smith, Kaj
  1941 Early Collections from the Pacific Eskimo; Ethnographical Studies. *Nationalmuseets Skrifter Etnografisk Række* 1:121-163. Copenhagen.
  1953 The Chugach Eskimo. *Nationalmuseets Skrifter Etnografisk Række* 6. Copenhagen.

Bisset, N. G.
  1976 Hunting Poisons of the North Pacific Region. *Lloydia, The Journal of Natural Products* 39(2-3):89-124.

Black, Lydia
  1987 Whaling in the Aleutians. *Etudes/Inuit/Studies* 11(2):7-50.

Braham, Howard W.
  1984a The Status of Endangered Whales: An Overview. *Marine Fisheries Review* 46(4):2-6.
  1984b Distribution and Migration of Gray Whales in Alaska. In: *The Gray Whale*, edited by M. L. Jones, S. L. Swartz, and S. Leatherwood, pp. 249-266. Academic Press, Orlando.
  1992 Scientific Investigations of the National Marine Mammal Laboratory, 1990. *Polar Record* 28(164):43-46.

Braham, Howard W. and Marilyn E. Dahlheim
  1982 Killer Whales in Alaska Documented in the Platforms of Opportunity Program. *Report of the International Whaling Commission* 32:643-646.

Brodie, Paul F.
  1985 The White Whale *Delphinapterus leucas* (Pallas, 1776). In: *Handbook of Marine Mammals, The Sirenians and Baleen Whales*, Vol. 3, edited by S. H. Ridgway and Sir R. Harrison, pp. 119-143. Academic Press, London.

Clark, Donald W.
  1974 Koniag Prehistory. *Tübinger Monographien zur Urgeschichte* No. 1. Universitat Tubingen, Stuttgart.

Davydov, Gavriil I.
  1977 *Two Voyages to Russian America, 1802-1807*. Translated by Colin Bearne, edited by R. A. Pierce. Limestone Press, Kingston, ON.

Fitzhugh, William W.
  1975 A Comparative Approach to Northern Maritime Adaptations. In: *Prehistoric Maritime Adaptations of the Circumpolar Zone*, edited by W. Fitzhugh, pp. 339-386. Mouton, The Hague.
  1983 Introduction to: *The Eskimo About Bering Strait*, Edward William Nelson, pp. 5-49. Smithsonian Institution Press, Washington, D.C.

Gambell, Ray
   1985   Fin Whales *Balaenoptera physalus* (Linnaeus, 1758). In: *Handbook of Marine Mammals, The Sirenians and Baleen Whales*, Vol. 3, edited by S. H. Ridgway and Sir R. Harrison, pp. 171-192. Academic Press, London.

Gedeon, Hieromonk
   1989   *The Round the World Voyage of Hieromonk Gedeon 1803-1809*. Translated by Lydia T. Black, edited by R. A. Pierce. Limestone Press, Kingston, ON.

Haggarty, James C., Christopher B. Wooley, Jon M. Erlandson, and Aron Crowell
   1991   *The 1990 Exxon Cultural Resource Program: Site Protection and Maritime Cultural Ecology in Prince William Sound and the Gulf of Alaska*. Exxon Shipping Company and Exxon Company, USA, Anchorage.

Hall, John D.
   1981   *Aspects of the Natural History of Cetaceans of Prince William Sound, Alaska*. Unpublished Ph.D. dissertation, University of California, Santa Cruz.
   1986   Notes on the Distribution and Feeding Behavior of Killer Whales in Prince William Sound, Alaska. In: *Behavioral Biology of Killer Whales*, edited by B. C. Kirkevold and J. S. Lockard, pp. 69-84. Alan R. Liss, Inc., New York.

Hayden, Brian M.
   1981   Subsistence and Ecological Adaptations of Modern Hunter/Gatherers. In: *Omnivorous Primates*, edited by R. S. O. Harding and G. Teleki, pp. 344-421. Columbia University Press, New York.
   1990   Nimrods, Piscators, Pluckers, and Planters: The Emergence of Food Production. *Journal of Anthropological Archaeology* 9(1):31-69.

Hazard, Katherine W. and Lloyd F. Lowry
   1984   Benthic Prey in a Bowhead Whale from the Northern Bering Sea. *Arctic* 37(2):166-168.

Heyning, John E. and Marilyn E. Dahlheim
   1988   *Orcinus orca*. *Mammalian Species* No. 304:1-9. The American Society of Mammalogists.

Heizer, Robert F.
   1943   Aconite Poison Whaling in Asia and America: An Aleutian Transfer to the New World. *Bureau of American Ethnology Bulletin* 133:415-468.

Holmberg, H. J.
   1985   Holmberg's Ethnographic Sketches. Edited by M. W. Falk, translated by F. Jaensch. *Rasmussen Library Historical Translation Series* 1. University of Alaska Press, Fairbanks.

Hrdlicka, Ales
   1944   *The Anthropology of Kodiak Island*. Wistar Institute of Anatomy and Biology, Philadelphia.

Jordan, Richard H.
   1988   Kodiak Island's Kachemak Tradition: Violence and Village Life in a Land of Plenty. Paper presented at the 15th Annual Meeting of the Alaska Anthropological Association. Fairbanks.
   1993   Qasqilutng: Feasting and Ceremonialism among the Traditional Koniag of Kodiak Island, Alaska. In: *Anthropology of the North Pacific Rim*, edited by W. Fitzhugh and V. Chaussonnet, pp. 147-173. Smithsonian Institution Press, Washington, D.C.

Kotani, Yoshinobu
  1980 Paleoecology of the Alaska Peninsula as Seen from the Hot Springs Site, Port Moller. In: Alaska Native Culture and History, edited by Y. Kotani and W. B. Workman, pp. 113-121. *Senri Ethnological Studies* 4. National Museum of Ethnography, Osaka.

Krupnik, Igor I.
  1988 Asiatic Eskimos and Marine Resources: A Case of Ecological Pulsations or Equilibrium? *Arctic Anthropology* 25(1):94-106.

Laguna, Frederica de
  1956 Chugach Prehistory: The Archaeology of Prince William Sound, Alaska. *University of Washington Publications in Anthropology* No. 13. University of Washington, Seattle.
  1975 *The Archaeology of Cook Inlet, Alaska* (2nd edition). Alaska Historical Society, Anchorage.

Leatherwood, Stephan, Randall R. Reeves, William R. Perrin, and William E. Evans
  1988 *Whales, Dolphins and Porpoises of the Eastern North Pacific and Adjacent Arctic Waters*. Dover Publications, New York.

Lisianski, Urey
  1968 *A Voyage Round the World, 1803-1806*. N. Israel, Amsterdam, and Da Capo Press, New York.

Lobdell, John E.
  1980 *Prehistoric Human Populations and Resource Utilization in Kachemak Bay, Gulf of Alaska*. Unpublished Ph.D. dissertation, Department of Anthropology, University of Tennessee.

McCartney, Allen P.
  1980 The Nature of Thule Eskimo Whale Use. *Arctic* 33(3):517-541.
  1988 Maritime Adaptations in Southern Alaska. In: *Proceedings of the International Symposium on Maritime Adaptations in the North Pacific*, edited by H. Okada, pp. 19-55. Abashiri, Hokkaido, Japan.

McCartney, Allen P. and James M. Savelle
  1985 Thule Eskimo Whaling in the Central Canadian Arctic. *Arctic Anthropology* 22(2):37-58.

Maxwell, Moreau
  1985 *Prehistory of the Eastern Arctic*. Academic Press, Orlando.

Murdoch, John
  1988 *Ethnological Results of the Point Barrow Expedition*. Smithsonian Institution Press, Washington, D.C. (1892)

Nelson, Edward
  1983 *The Eskimo About Bering Strait*. Smithsonian Institution Press, Washington, D.C. (1899)

Price, T. Douglas and J. Brown
  1985 Aspects of Hunter-Gatherer Complexity. In: *Prehistoric Hunter-Gatherers: The Emergence of Cultural Complexity*, edited by T. D. Price and J. Brown, pp. 3-20. Academic Press, Orlando.

Sauer, Martin
  1802 *An Account of a Geographical and Astronomical Expedition to the Northern Parts of Russia, by Commodore Joseph Billings, in the Years 1785 to 1794*. London.

Savelle, James M.
  1987 Collectors and Foragers: Subsistence-Settlement System Change in the Central Canadian Arctic A.D. 1000-1960. *British Archaeological Reports, International Series* 358.

Savelle, James M. and Allen P. McCartney
- 1988 Geographical and Temporal Variation in Thule Eskimo Subsistence Economies: A Model. *Research in Economic Anthropology* 10:21-72.
- 1991 Thule Eskimo Bowhead Whale Procurement. In: *Human Predators and Prey Mortality*, edited by M. C. Stiner, pp. 202-216. Westview Press, Boulder.

Schalk, Randall F.
- 1981 Land Use and Organizational Complexity Among Foragers of Northwestern North America. In: Affluent Foragers: Pacific Coasts East and West, edited by S. Koyama and D. H. Thomas. *Senri Ethnological Studies* 9. National Museum of Ethnology, Osaka.

Sheehan, Glenn W.
- 1985 Whaling as an Organizing Focus in Northwestern Alaskan Eskimo Society. In: *Prehistoric Hunter-Gatherers: The Emergence of Cultural Complexity*, edited by T. D. Price and J. Brown, pp. 123-154. Academic Press, Orlando.

Stewart, Brent S. and Stephen Leatherwood
- 1985 Minke Whale. In: *Handbook of Marine Mammals, The Sirenians and Baleen Whales*, Vol. 3, edited by S. H. Ridgway and Sir R. Harrison, pp. 91-136. Academic Press, London.

Taylor, J. Garth
- 1979 Inuit Whaling Technology in Eastern Canada and Greenland. In: Thule Eskimo Culture: An Anthropological Retrospective, edited by Allen P. McCartney, pp. 292-300. *National Museum of Man, Mercury Series, Archaeological Survey of Canada Paper* No. 88.

Tikhmenev, Petr A.
- 1978 *A History of the Russian-American Company*. Translated and edited by R. A. Pierce and A. S. Donnelly. University of Washington Press, Seattle.

Townsend, Joan B.
- 1980 Ranked Societies of the Alaskan Pacific Rim. In: Alaska Native Culture and History, edited by Y. Kotani and W. B. Workman, pp. 123-156. *Senri Ethnological Studies* No. 4. National Museum of Ethnology, Osaka.

Veniaminov, Ivan
- 1984 *Notes on the Islands of the Unalashka District*. Translated by Lydia T. Black and R. H. Geoghegan, edited by R. A. Pierce. Limestone Press, Kingston, ON.

Wendland Wayne M. and Reid A. Bryson
- 1974 Dating Climatic Episodes of the Holocene. *Quaternary Research* 4:9-24.

Wiles, Gregory C. and Parker E. Calkin
- 1992 Late Holocene, High Resolution Glacial Chronologies and Climate, Kenai Mountains, Alaska. Preprint, in review, *Geological Society of America Bulletin*.

Winn, Howard E. and Nancy E. Reichley
- 1985 Humpback Whale *Megaptera novæangliæ* (Borowski, 1781). In: *Handbook of Marine Mammals, The Sirenians and Baleen Whales*, Vol. 3., edited by S. H. Ridgway and Sir R. Harrison, pp. 241-273. Academic Press, London.

Wolman, Allen A.
- 1985 Gray Whale *Eschrichtius robustus* (Lilljeborg, 1861). In: *Handbook of Marine Mammals, The Sirenians and Baleen Whales*, Vol. 3, edited by S. H. Ridgway and Sir R. Harrison, pp. 67-89. Academic Press, London.

Wrangell, Ferdinand Petrovich
- 1980 *Russian America Statistical and Ethnographic Information*. Translated by Mary Sadouski, edited by R. A. Pierce. Limestone Press, Kingston, ON.

Yarborough, Linda Finn
- 1993 Preliminary Report on Prehistoric Ecological Adaptations at Palugvik, Prince William Sound, Alaska. MS. on file at Sigma Xi, The Research Foundation, Research Triangle Park, NC.

Yarborough, Michael R. and Linda Finn Yarborough
- 1991 Uqciuvit: A Multicomponent Site in Northwestern Prince William Sound, Alaska. MS. on file, USDA Forest Service, Chugach National Forest, Anchorage.

Yesner, David R.
- 1982 Analysis of Faunal Remains from the 1981 Salvage Excavations at the Chaluka Site, Umnak Island, Eastern Aleutians. MS. on file, U.S. Public Health Service, Anchorage.
- 1989 Osteological Remains from Larsen Bay, Kodiak Island, Alaska. *Arctic Anthropology* 26(2):96-106.
- 1992 Evolution of Subsistence in the Kachemak Tradition: Evaluating the North Pacific Maritime Stability Model. *Arctic Anthropology* 29(2): 167-181.

Whale Size Selection by Precontact Hunters of the North American Western Arctic and Subarctic

Allen P. McCartney
Department of Anthropology
University of Arkansas
Fayetteville, AR 72701

Abstract. *Baleen whales were probably important as food and fuel resources for western arctic and subarctic native societies for the past two millennia. However, no systematic study of archaeological whale remains has been conducted in this region. Measurement of archaeological bowhead bones in the Central Canadian Arctic indicates that precontact hunters there consistently selected yearling and other immature animals. A review of archaeological and ethnohistoric records suggests that small animals may have been selected in the western region for the same reasons: optimal productivity that comes from taking immature whales which are less dangerous to hunt, easier to tow ashore and butcher, and, thus, more likely to be processed prior to spoilage. Regular hunting of larger whales by Eskimo hunters appears to have been made possible only with the introduction of bomb darts and guns and block and tackle during the late 19th century. Measurements of archaeological whale bones and derivation of original animal sizes would provide evidence for any precontact prey patterning that may have existed in the past.*

INTRODUCTION

The 1948 publication of *Ipiutak and the Arctic Whale Hunting Culture* by Helge Larsen and Froelich Rainey marked approximately 20 years of systematic western Eskimo archaeology and the conclusion that "whale hunting seems to have been the most important economic factor" underlying some precontact Eskimo societies (Larsen and Rainey 1948:39). While archaeologists since then have established regional culture histories for the Gulf of Alaska, Bering Sea, Chukchi Sea, and Beaufort Sea coasts, including early to mid-Holocene cultures, almost no additional work has refined our knowledge of whaling and whale use by Okvik-Old Bering Sea, Norton, Birnirk, Punuk, and Thule Eskimos of the past 2000 years who created the so-called "arctic whale hunting culture."

Whaling, as a specialized subsistence activity, is important to an understanding of precontact adaptations to arctic and subarctic coastal zones. Questions such as when and where whaling originated, how it spread, what economic and social significance it had for the people who practiced it, and what methods of hunting or scavenging were used are obvious ones to pose. We have few answers to any of these questions, in spite of the fact that western arctic and subarctic Eskimos and Aleuts are commonly characterized as having been native whalers, par excellence, within the archaeological and ethnohistoric literature.

These questions are drawn together here by addressing animal size selection expressed in precontact whaling. Whale hunting, in contrast to whale scavenging, implies that some size selection occurred as animals were encountered. Methods used to procure, butcher, store, and use whale blubber, meat, organs, bones, and baleen and the amounts of these products that were obtained were directly related to the size of animals taken. Most northern observers have been so impressed by the huge size of baleen whales (or their carcasses or bones), compared to other sea and land mammals, that the questions of "how huge?" and "what kind of whale?" have gone largely unanswered. Often, maximum lengths of adult whales are given in species descriptions (such as 20 m for bowheads), which can lead to the erroneous impression that natives commonly took animals of that size. While whale size has been a focus of cetacean biological studies since the 1970s, it was not among the early chroniclers of Alaska's native peoples, nor has it been among archaeologists until recently. Larsen and Rainey presented almost no information about the practice or implications of prehistoric whaling in their monograph. They devote one page to contemporary bowhead whaling at Point Hope (Larsen and Rainey 1948:28), although Rainey (1940, 1947) published a note and a longer report about 1940s whaling there.

For the past 15 years, James Savelle and I have studied sizes and numbers of bowhead bones found at Thule Eskimo winter sites of c. A.D. 1000-1500 in the Central Canadian Arctic (see McCartney and Savelle 1993 for a summary). We find that Thule Eskimos selected yearlings and slightly older animals, to the almost total exclusion of calves or adults, from the Davis Strait stock. Because Canadian Thule Eskimos originally derived from antecedent western arctic populations who used bowheads of the Bering Sea stock, a logical question is whether those earlier Okvik-Old Bering Sea, Norton, Birnirk, and Punuk populations also selected small bowheads or whether the Central Canadian Arctic pattern was unique.

That question has simultaneously and independently been investigated by Igor Krupnik and his Russian colleagues, who studied bowhead, gray, and humpback bones found in ancient and modern contexts along the Chukchi Peninsula shore in the late 1970s and 1980s. A recently published paper by Krupnik (1993b) particularly addresses the issue of whale sizes represented at Chukchi Peninsula sites and settlements. Calves and other juvenile animals are commonly found at many of these sites. Because the papers by Krupnik and colleagues have characterized whale bone remains found on the Northeast Asian shore, this paper will address comparable archaeological, ethnohistoric, and recent information about baleen whale size preferences for Alaska and the Western Canadian Arctic. As one might expect, native groups on both sides of the Bering and Chukchi seas hunted whales in a similar manner, expressing their common cultural base. The conclusion of this paper, following Krupnik, is that while size data are equivocal because of their scarcity, there seems to be a pattern of small baleen whale use in the Western Arctic and Subarctic during the precontact or aboriginal period.

Finally, readers interested in the biology, behavior, and human predation of bowheads, one of the major western arctic whales, are referred to a new, state-of-the-species volume edited by Burns, Montague, and Cowles (1993), which will serve as a thorough compendium for many years to come.

CANADIAN THULE ESKIMO WHALING

In 1975-1976, I directed a survey and excavation project that focused on Thule winter house sites in the Central Canadian Arctic. In 1978, we began systematically counting and measuring bowhead bones associated with such sites on southeastern Somerset Island, using a measurement schedule suggested by Edward D. Mitchell (Arctic Biological Station, Ste. Anne de Bellevue, Quebec; McCartney and Savelle 1993). This schedule includes 16 cranial, 12 mandibular, five scapular, and five cervical vertebral measurements. These measurements were used in multiple regression models to estimate live animal length. The models were based on 14 skeletons of recently killed North Alaskan bowheads, collected by Floyd Durham for the Los Angeles County Museum (Burns 1993:746-747). Craig Gerlach, John George, and Robert Suydam (1993) recently and independently developed models for estimating bowhead length based on scapular and mandibular lengths that give similar results.

Since 1980, Savelle and I have expanded the measured bowhead sample at Thule sites, in terms of both bone numbers and locales. Sites on northern Somerset, Prince of Wales, Bathurst, Cornwallis, Devon, and Baffin islands have been studied (Savelle and McCartney 1994). To date, approximately 10,500 bowhead bones have been counted, and 354 crania, 784 mandibles, 213 scapulae, and 120 cervical vertebrae have been used to estimate bowhead lengths. We consistently find that the estimated live animal length used by Thule whalers averages approximately 8.5 m, with a sharp clustering between 7-10 m (Fig. 1a). This average size is that of yearling bowheads. We have collected additional measurements from the bones of Holocene-aged stranded bowheads found on uplifted beaches of Brodeur Peninsula and Borden Peninsula, northwestern Baffin Island, and on Somerset Island that date as old as 10,000 years B.P. (see Dyke and Morris 1990). This stranded sample clearly reflects the contemporary living bowhead population profile, with estimated lengths ranging between 6.5-18.5 m (see Fig. 1c for modern Alaskan bowhead size range), and stands in marked contrast to the hunted Thule sample (Savelle and McCartney 1991, 1994). The small whale selection pattern is also expressed by modern northwestern Alaskan hunters, who usually take 7-10 m bowheads (Fig. 1b).

A basic tenet of our research is that bowheads were used as a food source as well as a raw material source for heating oil, baleen and bones for implements, and bones for house construction. Based on our central Canadian Arctic studies, the following conclusions may be drawn:

(a) Thule Eskimos selected primarily yearling and, secondarily, other juvenile bowheads for use;
(b) hunting live animals rather than scavenging stranded carcasses was the predominant procurement pattern;

Figure 1.
Comparison of bowhead whale lengths.

(A) Estimated lengths derived from bones associated with Thule Eskimo features of the Central Canadian Arctic.

(B) Actual lengths of bowheads taken by Alaska Eskimos between 1973-1989 (from Philo et al. 1993:Fig. 8.1).

(C) Estimated lengths of live Beaufort Sea bowheads derived photogrammetrically from surfaced animals (from Koski et al. 1988:Fig. 9).

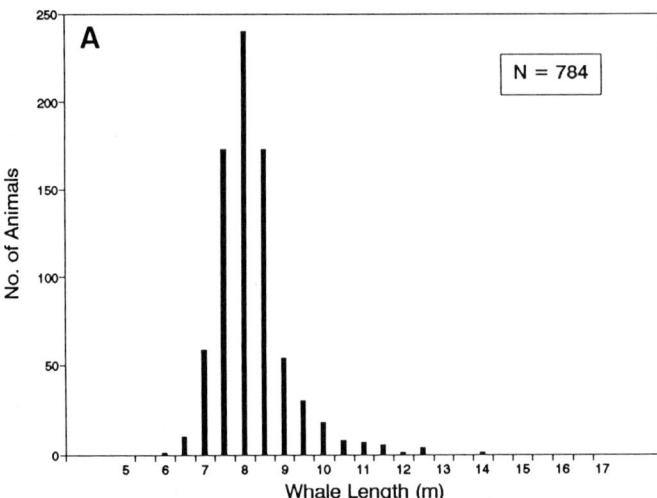
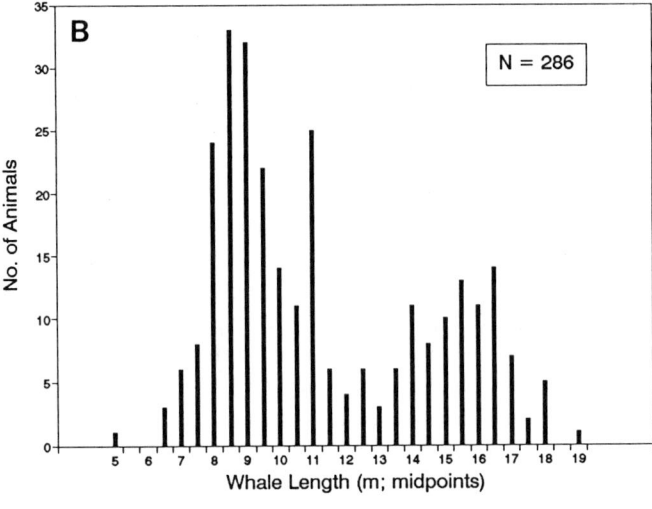
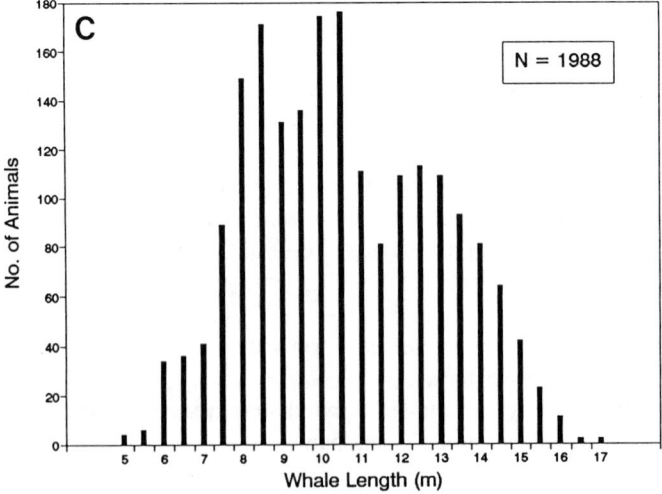

(c) Thule settlement patterns consistent with whale hunting, butchering, and caching have been recorded along central Canadian Arctic shores;
(d) bowheads were the primary subsistence prey where open summer waters were present, whereas caribou and seals were dominant prey where year-round sea ice restricted whale migrations; and,
(e) several selective stages were involved in whale use, from choosing small whales during hunts, to saving bones for future house construction, to removing certain bones from abandoned houses for other secondary use.

This pattern of bowhead age-selective culling, while unique to northern latitudes, is generally comparable to age-selective predation (in contrast to scavenging, herd stampeding, and other confrontational methods) of megafauna such as mammoths (Saunders 1980; Yesner, this volume).

WESTERN ARCTIC AND SUBARCTIC WHALING

Precontact natives variously hunted baleen whales and used their stranded carcasses throughout the Western Arctic and Subarctic. These include bowheads along the central and western parts of Bering Sea and northward throughout the Chukchi Sea and the southern Beaufort Sea, gray whales in the Gulf of Alaska, the eastern and western Bering Sea, and throughout the Chukchi Sea, and humpback, sei, and fin whales along the Gulf of Alaska and Aleutian chain (*Alaska Whales and Whaling* 1978). These whales were prey animals to the degree that they swam within range of native hunters, be that through Aleutian Island passes, through Bering Strait, or along the narrow leads of the North Alaskan shore during the spring. In the latter region, bowheads usually swim within 7 km of shore as they migrate (Marquette et al. 1982). Baleen whales tend to migrate in specific age/size groups or "runs" (e.g., Maher and Wilimovsky 1963; Rice and Wolman 1971; Nerini et al. 1984; Moore and Reeves 1993), and immature bowheads, for example, are known to swim relatively close to shore (e.g., Durham 1979, n.d.; Marquette et al. 1982). Although these migrating whales have large ranges, they were not uniformly taken by coastal hunters. Water depth, ice patterns, feeding habits, and migration routes are important determining factors as to whether they are available on a seasonal basis at any particular locale.

Baleen whales were pursued using different technologies. Both Yupik and Inupiat Eskimos of western and northern Alaska used large *umiaks*, harpoons with floats, and lances, while eastern Aleuts and Pacific or Allutiq Eskimos of the Gulf of Alaska used a quite different kayak and darting procedure. These native hunting procedures are well documented and are not reviewed here (see e.g., Heizer 1943; Hughes 1960; Black 1987; Stoker and Krupnik 1993). In northern coastal areas where baleen whales were absent, such as Norton Sound and Kotzebue Sound, beluga whales were hunted in number and were the largest alternative sea mammals to baleen whales (Ray 1975). Whale carcasses grounded on shore by high tides and storms were commonly used for oil and dog food, even if their flesh and organs were too decayed for human consumption.

For our purposes, whaling may be temporally divided between the pre-Euroamerican period (c. A.D. 1-1750), the early contact period (c. A.D. 1750-1870), and the late contact period (c. A.D. 1870-1940). Alternatively, Krupnik et al. (1983) divide Northeast Asian whaling into Aboriginal (ancient times-c. 1870), Traditional (c. 1870-1930s), Transitional (1930s-early 1960s), and Modern (early 1960s-present) periods. Marquette and Bockstoce (1980) use Aboriginal (ancient times-1880s), Commercial (1880s-1910), and Subsistence (1910-present) phases for western and northern Alaskan bowhead whaling. The three-fold division used here stems as much from our ability to understand native whaling from the archaeological and ethnohistoric records as it does from the whaling methods employed. For instance, our only knowledge of precontact whaling is from the archaeological record, although native oral traditions in some cases may reflect this period as well as more recent ones. The mid-18th century brought direct Russian and other European contact with southwestern Alaskan Eskimos and Aleuts, and began a period of cultural conflict, significant cultural change, and early documentation of native lifestyles, including subsistence patterns. The 1870s ushered in American control of the territory, extensive impacts on the baleen whale stocks, and continuous contact between Yankee whalers and north Alaskan whaling villages.

The expansion of the Yankee whaling fleet into the Bering Sea in the 1840s and into the Chukchi Sea after 1848 had the greatest impact on native whaling there (Bockstoce 1986). Yankee whalers directly and indirectly impacted coastal native settlements in many ways. They began the wholesale slaughter of baleen whales through pelagic "fishing" and processing, thereby greatly reducing all of the stocks by the end of the century. In addition, they transferred 19th century commercial whaling technology to Eskimos, the effects of which are still seen today in North Alaska subsistence whaling (see below). Finally, they disrupted the social fabric and economic organization of Alaskan and Asian coastal Eskimos through the slaughter of other subsistence animals such as walrus, the introduction of foreign trade goods (including liquor), and the creation of a cash/wage economy in which Eskimos sold baleen and oil in a world market system (see Bailey and Hendee 1926; Bockstoce 1986; Stoker and Krupnik 1993). For example, during the heyday of Yankee whaling, "many Eskimo made a small fortune out of the sale of baleen" when a "very large whale might produce up to $10,000. worth of this 'whalebone'" (Rainey 1947:281; see also Bockstoce and Burns 1993).

The technology transfer that most affected native whalers included the dart gun and the shoulder gun, for killing whales with explosive charges, and large sets of block and tackle and manila lines with which to haul bowhead carcasses onto the sea ice for processing (VanStone 1958; Bockstoce 1986). These transfers, made during the 1870s-1880s, made it possible to kill, land, and butcher larger whales than were generally possible with aboriginal, nonexplosive harpoons, lances, and walrus-skin ropes. While small whales are still preferred by North Alaskan whalers, the range of bowheads taken in recent years demonstrates the fact that medium-sized to very large animals may successfully be taken with the 19th century-style guns and tackle (Figs. 1b & 2).

As mentioned above, one effect that the Yankee commercial whalers had on natives was the decimation of stocks through relentless high seas hunting. For example, Yankee whalers could attack bowheads in late winter and early spring in the southern Bering Sea, where whales wintered south of the pack ice, and continue hunting them in the Chukchi and Beaufort seas in the summer when the ice cleared. By 1860, over 8000 bowheads had been killed in the Western Arctic by Yankee whalers from an estimated maximum pre-1848 population of 23,000 animals (Woodby and Botkin 1993). By 1867, when the U.S. purchased Alaska, the bowhead population had been reduced to half that of the

Figure 2. An 8-9 m (26-30 ft) bowhead shown hauled up on the ice at Barrow, Alaska (c. 1992; photo by Karen Brewster).

precontact period. A second effect of pelagic whaling was the creation of wariness on the part of whales to sounds of whale ships and small whaleboats. Recent studies have further demonstrated that bowheads identify and react to underwater noise (Richardson and Malme 1993).

The diminishing whale population and the "spooking" of animals through intensive pursuit was perceived relatively early in the commercial whaling period in North Alaska. The journals of Rochfort Maguire, who visited Barrow in 1852-1854, contain the observation that the native whaling "success ... has been very partial" (Maguire 1988:2:376). He cites his comrade, Doctor Simpson, in stating that the lack of whales may be accounted for "by the number of ships that visit this sea every summer decreasing the number of whales and limiting this peoples' supply without their knowing how to account for it...except by saying (as I believe they do) that our Ship has brought them bad luck." Others besides Maguire noted the increased difficulty in approaching bowheads. American whalemen noted that the "whales appear very shy" and that bowheads clearly discriminated between natural sounds and those from American whaleboats (Bockstoce 1986:101; Bockstoce and Burns 1993:573).

With regard to documenting these 19th century changes, it is important to consider that commercial whaling contact generated observations about baleen whales and native whaling, while at the same time it caused great changes in both. To the degree that Yankee whaling impacts on whales and indigenous native subsistence were dramatic during the early years (1840s-1860s), we must consider the fact that 19th century descriptions of native whaling do not reflect the pre-1840 period, when whales were more abundant and aboriginal techniques were still in use.

EVIDENCE FOR WHALE SIZE SELECTION IN THE WESTERN ARCTIC AND SUBARCTIC

The evidence for any size selection of larger or smaller whales exhibited by native hunters of the Gulf of Alaska, Bering Sea, Chukchi Sea, and Beaufort Sea coasts is relatively rare. Most historic accounts fail to specify any size preferences, if they existed, and little information about whale sizes is known from archaeological sites, with the exception of some Chukchi Peninsula sites (see below). The following are examples found in the ethnohistoric and archaeological literature. These are arranged in the following regional order: southwestern Alaska, Bering Sea-St. Lawrence Island, northwestern Alaska, and Beaufort Sea.

Southwestern Alaska

Ethnohistoric and Recent Evidence

Several authors remark on small whale selection in the Kodiak-Aleutian area during the 19th century. For example, Father Gideon, who visited Kodiak during the first decade of the 19th century, reports that Koniag hunters "go out singly, in one-hatch *baidarkas*, and choose yearling whales because their meat and fat are tastier and tender" (Black 1977:102). Holmberg (1985:47-48) describes Kodiak whaling of the 1850s by differentiating between various sizes of balaenoptera (probably fin) whales. These include 10 fathom-long old whales, 8 fathom middle-aged ones, 6 fathom yearlings, and 3.5-4 fathom "younger whales," of which "only the young and one-year-olds are hunted." Bisset (1976:106-107) cites Pinart's 1870s publications for eastern Aleutian-Kodiak whaling that mention smaller humpback and fin whales being taken. Bisset (1976:109) also cites Lisiansky's record of his 1804 visit to Kodiak, in which he mentions only small whales being hunted. Dall (1870:404) refers to Kodiak whale hunters, who "only attempted to kill the smaller specimens."

In the adjacent eastern Aleutian Islands, Laughlin (1963:76) refers to Chamisso's study of whales when the latter spent several months on Unalaska in 1817. Based on Chamisso's work, Laughlin states that: "It seems clear that the Aleut hunters preferred small whales to large ones, and young rather than old. The Humpback whale was extensively hunted, especially the calves." Veniaminov (1984:277) describes Unalaska Aleuts of the 1824-1834 period getting between 10-30 whales a year, "but, generally, the whales here are only of the small kind so that it is easily possible to load a whole whale into one baidara." Elliott (1886:152) reports that Aleuts of the early American period selected smaller whales: "Carefully looking the whales over, the hunter finally recognizes that yearling, or the calf, which he wishes to strike; for it is not his desire to attack an old bull or angry cow-whale."

The Kodiak-eastern Aleutian area is well known for "poison" whaling, wherein the darts used on whales were coated with an aconitum-based material or with the body fluids of dead whalers (Heizer 1943; Bisset 1776; Black 1987). It is not known whether these materials served as magical "poisons" only, or whether they had toxic effects on whales.

Given the amount of material that could coat a dart head and the large size of even small whales, aconitum-based toxins may have only affected flipper and fluke muscle groups when struck in those spots, thereby disabling the animal's swimming balance (Bisset 1976). Another possibility is that whales were killed by sharp, deeply buried dart points that lacerated internal organs or veins/arteries as the animals swam. Traditionally, these whales would float, possibly ashore, in three days, at which time the carcass was butchered. If it was not aconite poison and subsequent drowning that finally killed the animal in three days, then it was internal bleeding that killed it. Putrefaction gases accumulated in sufficient quantity to float the animal to the surface and, depending upon the currents, to shore.

The fact that whales were not towed ashore using this technique might suggest a greater latitude of animals sizes chosen. However, small animals may have been more effectively killed with the relatively small darts used on them than large whales. Thus, the available technology may have been important, as in northern Alaska with harpoon and float whaling, in limiting sizes of whales chosen to be attacked.

Archaeological Evidence
Although whale bones are frequently reported at southwestern Alaskan archaeological sites, almost no data are presented about species or size, and scale drawings of these are equally scarce. Archaeological whale bones are mentioned for Prince William Sound (Yarborough, this volume), for Kodiak (Hrdlicka 1944), and for the Aleutian Islands (Jochelson 1925; Hrdlicka 1945). Four late prehistoric-early historic whale crania (possibly humpback) of medium size and other bones were recently found at a Reese Bay, Unalaska, longhouse. At Izembek Lagoon, near the southwestern tip of the Alaska Peninsula, an oval house with original superstructure of large whale bones was excavated in 1971 (McCartney 1974; Dumond, this volume). This house dates to c. A.D. 915. However, the large size of the 30+ mandibles (>5 m long) and the fact that they are not from humpback or gray whales suggest they are probably from blue whales that were collected from stranded rather than from hunted whales.

Bering Strait-St. Lawrence Island

Ethnohistoric and Recent Evidence
This region is famous for migrating whales and the Eskimo whalers who exploited them. Western Seward Peninsula, the Diomede Islands, and St. Lawrence Island were all well-known 19th century whaling locales (Hughes 1960; Ray 1975). Photographs of early 20th century whales taken at Wales show Eskimos butchering small (9 m) baleen whales (Bernardi 1981). Nelson (1899:259) describes 1880s whale bone houses and storerooms on St. Lawrence Island, but with bone sizes unspecified. Collins (1929:Fig. 136; 1935:Plate 10) also illustrates historic house ruins on Punuk Island that include some small to large whale mandibles.

On the Siberian side, Krupnik (1993a:232-233; 1993b) reports the selection of gray whale calves as prey from prehistoric times until the 1940s or 1950s.

Archaeological Evidence
Excavations on St. Lawrence Island have revealed many whale bones of various sizes. Presumably, most of these are bowhead bones, but some "summer" whales (fin, gray, and humpback; Hughes 1960:111) were taken when available. Geist and Rainey (1936) report

mandibles, ribs, and vertebrae found in Thule to recent period meat caches. The illustrated mandibles appear to measure approximately 4-4.5 m long and, thus, are adult-sized specimens. Collins (1937a:68ff, 189, 258, 285-286, Plates 1-11) reports small to large whale bones in Old Bering Sea and Punuk houses and caches, but includes no measurements or scale drawings from which size estimates can be made. He also excavated small whale mandibles in old Eskimo midden sites at Wales (Collins 1937b:Fig. 57). Krupnik (Stoker and Krupnik 1993:589, Fig. 15.2) also refers to "very small skulls and mandibles, representing bowhead calves and yearlings" that he observed at Wales. More recently, Hofmann-Wyss (1987) investigated early Punuk burial cairns which incorporated whale bones. Of eight more or less whole mandibles shown in her scale drawings, three are calf size (160-185 cm) and five are yearling size (220-240 cm), assuming that these are from bowheads.

As noted in the introduction, Krupnik and his Russian colleagues have published major reports about both precontact and historic/recent whaling along the Chukchi Peninsula coast, including details on sizes of animals represented (Krupnik 1984, 1987, 1993a,b; Krupnik et al. 1983; Arutiunov et al. 1982; Bogoslovskaya et al. 1982; Stoker and Krupnik 1993). Gray whale calf bones and juvenile scapulae and 1.5 m long mandibles are found associated with the Ekven Old Bering Sea cemetery (Arutiunov and Sergeev 1975), and later with Birnirk period graves. In addition, Punuk graves have bowhead bones associated with them, while other old sites, such as those on Arakamchechen Island, have mostly 6-8 month old calf bones of gray whales found at them (skull widths of 65-90 cm). Krupnik (1984) estimates that as many as 2000-3000 small gray whale skulls are located on the eastern Chukchi Peninsula shore. Farther to the east, in the Okhotsk Sea, Vasel'evsky found bones of 200 small to medium whales within an Ancient Koryak site on Kony Peninsula (Krupnik 1984:116). Mandibles at this site are 1.2-1.5 m long, while skull widths are less than 1 m; clearly these derive from young animals, probably gray whales.

Northwestern Alaska

Ethnohistoric and Recent Evidence
Moving to North Alaska, there are a number of references to suggest that small bowheads are preferred over large ones. McVay (1973:28) notes that modern Inupiat whalers "markedly prefer small whales over large because their muk-tuk ... and meat are more tender and taste better. Another reason is that it is hard to haul the big ones upon the ice." In contrast, the largest bowheads are characterized by North Alaskan Eskimos as having "tough meat, little oil," and being "too hard to catch" (McVay 1973:28). As an example from the early 1850s, Maguire (1988:1:370) reports a bowhead with 40 inches long baleen taken at Barrow, which is interpreted by John George, a modern cetacean biologist, as being an 8 m (25 ft) animal.

In the Chukchi Sea-Beaufort Sea area, there has long been an identification of a small and separate whale category called *inito, ingutuk, inyutok,* or *ingutok,* a native classification category reported at the end of the 19th century by American whalers such as James Allen and Charles Brower and up through this century by such writers as Bailey and Hendee (1926), Rainey (1947), Foote (1964), and others. Krupnik (1993a) reports this designation used among Siberian Eskimos as well.

Regardless of what the *ingutok* category is taxonomically (see Braham et al. 1980), it is usually reported as a small animal (Bailey and Hendee 1926; Foote 1964:56ff). At Point Hope and other North Alaskan communities, it has been the preferred whale to hunt. Rainey (1947:261) clearly states the preference for these small animals at Point Hope in 1940: "Even the smallest bowhead whales, the *ingutuk* (young ones), preferred by the Eskimo, are too heavy to be drawn up onto the ice." In the last century "[t]hese were so numerous that hunters pursued only the young ones (*ingutuk*) and merely threw ice at the older and larger whales to drive them away." Bailey and Hendee (1926:26) remark that Charles Brower, the famous Barrow whaler, told them that the *ingutok* were "usually easier to approach in the water than the bowheads." Further, they claim that Eskimos preferred the jaws and ribs from the *ingutok* over those of bowheads for making sled runners because they were "solid and very hard."

In 1962, Don Foote (1964:71) counted calf, *ingutok*, and adult whale mandibles at Point Hope in old house ruins, graves, present-day houses, and other features, and found a pattern that mimics the Central Canadian Arctic Thule profile: five calf, 5622 *ingutok*, and 87 adult mandibles. Photographs of Point Hope taken between the 1880s and 1920s show many small bowhead mandibles used in these features (Burch 1981:Plates 7-12). Many of the bowhead mandibles eroding from the Old Tigara midden are from yearlings or other small animals (*Point Hope Beach Erosion* 1972:Plates 14-17). On the other hand, some old Point Hope houses, perhaps of the protohistoric period, contain large mandibles as well (R. Newell, pers. comm., 1994).

Archaeological Evidence

Because of the association of North Alaskan Eskimo culture and whaling, this is an appropriate region to look for size information. The so-called Old Whaling culture, dating to c. 1700 B.C., has for years been considered the earliest reflection of northwest Alaskan whaling (Giddings 1967:223ff; Giddings and Anderson 1986:231-267). However, Mason and Gerlach (this volume) challenge the interpretation of ancient whaling among this group.

Ford (1959) excavated at the Birnirk site, dating to c. A.D. 500-900, near Barrow during the early 1930s. He found winter houses made of timbers and bowhead bones, and refers to both small and large crania, mandibles, and scapulae in his excavations (Ford 1959:38-49). I measured the whale bones included in his scaled feature maps, and identified five mandibles of approximately 2.4 m long in Mound A (Table 1). These compare closely with 2.4 m long mandibles of recent Barrow bowheads and 2.5 m long mandibles of yearling bowheads found in the Central Canadian Arctic sample (see also the scatter plot of mandible length to animal length in Gerlach et al. 1993:57). Further, six crania from Mound A and one from Mound B at the Birnirk site measure about 1.25 m wide, and these compare well with 1.2-1.4 m widths from the Thule sample (Table 1). Finally, Ford (1959:72) illustrates one bowhead mandible that measures approximately 1.5 m long from the Utkiavik site, also near Barrow, which is almost certainly from a bowhead calf.

Bee and Hall (1956:167) made several measurements on 34 mandibles and 20 crania "from abandoned dwellings and meat cellars at Point Barrow." They also report cranial widths "of the largest seven skulls at Birnirk Mounds," in addition to one scapula width (1.3 m; location not specified). These various measurements are compared in Table 1.

Table 1. Comparison of old and recent North Alaskan bowhead bone samples.

Site	Source	Old Sample (cm)	Recent Sample (cm)*	Total Est. Whale Length (cm)**
Barrow, Birnirk, Mound A	Ford (1959)	Mandible length N = 5 (200, 215, 230, 245, 290) $\bar{X}$ = 236	N of animals = 8 N of bones = 15 R = 225-263 $\bar{X}$ = 238	N of animals = 8 R = 760-880 $\bar{X}$ = 823
		Cranial width N = 6 (122, 122, 122, 122, 137, 152) $\bar{X}$ = 129.5	N = 9 R = 112.8-140.6 $\bar{X}$ = 126	N = 9 R = 760-880 $\bar{X}$ = 828
Mound B	Ford (1959)	Cranial width N = 1 104	N = 1 93.6	N = 1 610
Utkiavik	Ford (1959)	Mandible length N = 1 153	N = 1 165	N = 1 610
Barrow, Abandoned houses/ meat cellars, Point Barrow	Bee and Hall (1956)	Mandible ramus height *** Small: N = 27 R = 20.1-29.5	N of animals = 7 N of bones = 14 R = 25.4-29.8 $\bar{X}$ = 27.4	N of animals = 7 R = 760-880 $\bar{X}$ = 824
		Medium: N = 5 R = 32-33	N of animals = 2 R = 29.3-30.8 $\bar{X}$ = 29.9	N of animals = 2 N of bones = 4 R = 910-940 $\bar{X}$ = 925
		Large: N = 3 (36.6, 45, 53) $\bar{X}$ = 44.9	N of animals = 2 N of bones = 3 R = 31.1-36.6 $\bar{X}$ = 33	N of animals = 2 R = 1040-1070 $\bar{X}$ = 1055
		Total: N = 34 R = 20.1-53 $\bar{X}$ = 27.7	N of animals = 11 N of bones = 21 R = 25.4-36.6 $\bar{X}$ = 28.7	N of animals = 11 R = 760-1070 $\bar{X}$ = 885
		Width of occipital condyles N = 20 R = 30.5-37.5 $\bar{X}$ = 33.1	N = 12 R = 27.4-33.2 $\bar{X}$ = 29.5	N = 12 R = 760-1070 $\bar{X}$ = 864

Table 1 *continued*. Comparison of old and recent North Alaskan bowhead bone samples.

Site	Source	Old Sample (cm)	Recent Sample (cm)*	Total Est. Whale Length (cm)**
Birnirk	Bee and Hall (1956)	Cranial width of "largest" of 7 skulls N = 1 156	N = 3 (largest specimens) R = 140-150.6 $\bar{X}$ = 143.7	N = 3 R = 910-1070 $\bar{X}$ = 973
Barrow, Utqiagvik, Mounds 8 & 37	Dekin (pers. comm., 1981)	Mandible condylar head height N = 12 R = 23.5-32.5 $\bar{X}$ = 27.8	N of animals = 7 N of bones = 14 R = 25.4-29.8 $\bar{X}$ = 27.4	N of animals = 7 R = 760-880 $\bar{X}$ = 824
		Mandible condylar head width N = 13 R = 22-28 $\bar{X}$ = 24	N of animals = 7 N of bones = 14 R = 21.9-24.6 $\bar{X}$ = 22.9	N of animals = 7 R = 760-880 $\bar{X}$ = 824
		Scapula width N = 2 R = 45.4-52.2 $\bar{X}$ = 48.8	N of animals = 4 N of bones = 7 R = 45.2-50.9 $\bar{X}$ = 48.4	N of animals = 4 R = 760-880 $\bar{X}$ = 830
Barrow, Utqiagvik, Mound 44	Dekin (pers. comm., 1983)	Mandible coronoid circumference Calf: N = 3 (39.3, 42.7, 44.2) R = 39.3-44.2 $\bar{X}$ = 42.1	N = 1 38.5	N = 1 610
		Yearling: N = 2 R = 50.7-53.9 $\bar{X}$ = 52.3	N of animals = 7 N of bones = 14 R = 52.5-64 $\bar{X}$ = 56.4	N of animals = 7 R = 760-880 $\bar{X}$ = 824
		Scapula length N = 12 R = 46.2-55.9 $\bar{X}$ = 50.4	N of animals = 4 N of bones = 7 R = 45.2-50.9 $\bar{X}$ = 48.4	N of animals = 4 R = 760-880 $\bar{X}$ = 830

* Floyd Durham bowhead series collected at Barrow during the 1960s-early 1970s and measured by McCartney at the Los Angeles County Museum in 1980.

** Estimated live whale lengths based on regression models derived from the Durham series.

*** Bee and Hall claim to treat 34 ramus heights but present measurements for 35; these figures are shown as the authors present them.

Most of the mandible measurements correspond to those of recent animals measuring, on average, 8.25 m long (yearling size). Whereas Bee and Hall (1956:167) conclude that these measurements make it "clear that Eskimos took principally young whales," it is not clear what period is being characterized, since "abandoned dwellings and meat cellars" could date from this century to many hundreds of years old.

Finally, recent excavations at the Utkiavik site (Barrow) by the SUNY-Binghamton Utquiagvik Archaeology Project (1981-1983) also produced small whale bones. Thirteen mandibles and two scapulae from Houses 8 and 37 (Cargill 1990: 266) and 12 scapulae and five mandibles from House 44 are all from 7-9 m long animals and, thus, are from yearlings or 2-3 year-old bowheads (Table 1).

The general precontact pattern reflected in this limited sample from Barrow region sites seems to be that of small/young bowhead selection. It is also the impression of trained cetacean biologists who work for the North Slope Borough Department of Wildlife Management that skulls found around old Barrow area sites are of 7-9 m animals (J. George, pers. comm., 1994). The few larger bones found in these sites either reflect less frequent hunting of larger bowheads or scavenging of bones from large bowhead carcasses stranded near Barrow.

Beaufort Sea

Ethnohistoric and Recent Evidence

Bowheads migrate eastward past Point Barrow as far as Amundsen Gulf during the late spring and summer, as the winter sea ice decays or recedes there. A return migration to the west occurs during the late summer and fall as the sea ice closes (Moore and Reeves 1993). Open water along the Beaufort Sea coast is controlled by a combination of temperature, winds, and currents. Because migrating bowheads in the Beaufort Sea are not as restricted to shore leads as they are in Northwest Alaska, Eskimos here had less access to them. Thus, 19th century Mackenzie Eskimos took far fewer bowheads than did the Northwest Alaskan Eskimos, and the Avvagmiut of the Cape Bathurst area took only one or two animals a year (McGhee 1974:15-18).

Herschel Island, immediately west of the Mackenzie Delta, was famous as a late 19th century Yankee whaling station (Bockstoce 1986:255ff), and whalers also overwintered along the Amundsen Gulf shore (Stefansson 1913:497). Stefansson points out that prior to the Yankee whaling period, Eskimos pursued bowheads from a number of Beaufort Sea localities as far east as Cape Parry. Jenness (1922:46-47, 105) observes that whereas the Copper Eskimo, whom he studied during the early 20th century, subsisted mainly on caribou, seals, and fish, earlier whale hunting Eskimos occupied whale bone house sites at Cape Kellett, southwestern Banks Island, at c. 1800. These Eskimos were related to those of Cape Bathurst, located on the southern Amundsen Gulf coast. Perhaps colder temperatures of the 19th century were responsible for less summer open water in Amundsen Gulf, which would have led to natives shifting from bowheads to other food animals.

Neither Stefansson, Jenness, nor other 19th century or early 20th century commentators address bowhead size selection by Beaufort Sea-Amundsen Gulf Eskimos. The only size reference that I am aware of is from Thomas Simpson (1843:116), who observed "some bones of an enormous whale, probably stranded here, of which the skull measured eight feet in breadth" on the western end of Herschel Island. This cranium would be from a large adult bowhead. While a prehistoric site is located at the island's western end and the bones could possibly be from a hunted animal, I tend to accept Simpson's interpretation that the bones were from a stranded bowhead. Reeves and Mitchell (1985) review the history of 19th century accounts that contain bowhead whaling references. They refer to an 1865 stranded bowhead just west of Cape Bathurst that was utilized by several Eskimo families.

Archaeological Evidence

Two types of Thule period houses are located around Amundsen Gulf, Dolphin and Union Strait, and Coronation Gulf. These are square to rectangular timber houses and circular whale bone houses. The former follow the western Thule house style of Alaska (see Anderson 1986) and were being used up through the 19th century (Yorga 1980:58-60), whereas the latter follow the house style typical of the Central and Eastern Arctic (see MacNeish 1956:61-66 for Yukon coast houses). A few whale bones are often found in the square timber houses, but these were not the primary construction elements. The following is a summary of precontact houses for the Beaufort Sea-Amundsen Gulf region, but only limited information about whale bone sizes is available from the literature.

Yorga (1980) tested timber houses dating to the early second millennium A.D. at the Washout site on Herschel Island. He illustrates no whale bones in the house ruins, despite the fact that bowheads were common to this area during historic times and baleen was found in the excavations. McGhee (1972:21ff) found very few bowhead bones at the Memorana site on western Victoria Island or at the Bloody Falls site on the Coppermine River near Coronation Gulf. He did, however, excavate a few artifacts made of whale bones at these sites. A similar pattern of few bowhead bones associated with timber houses was noted by Arnold (1986) at Nelson River on southern Banks Island (see also Manning 1956). What appears to be an immature bowhead cranium is illustrated on the surface at another whale bone house site (OhRh-2; Arnold 1986:Fig. 5). Morrison (1983:51ff) excavated oval stone, timber, and whale bone houses at the Clachan site, western Coronation Gulf, which are intermediate between western square timber houses and eastern round whale bone houses. Whale bones were "very rare" and thought to be from stranded carcasses. No size estimates of bowheads represented at these western Canadian Arctic sites are given by the excavators.

Manning (1956:24ff) reports several Thule whale bone house sites located at Cape Kellett, Banks Island, one of which was tested in 1952-1953. Dendrochronology dates of the mid-15th century A.D. were established for this site. He also illustrates a whale bone house ruin near Nelson River, in the vicinity of the timber house that Arnold later investigated (see above; Manning 1956:26, Fig. 16). Jenness (S. Jenness 1990:95-97, Fig. 5) investigated several timber and whale bone house sites at Barter Island, and photographed one house ruin containing at least three bowhead crania. Although no scale is provided in the photograph, the crania appear to be from small bowheads. These houses date somewhere between the early 20th century and several hundred years ago.

OPTIMAL WHALING PREDATION: USING THE SMALLEST OF THE LARGEST ANIMALS

Krupnik (1993b:7-9) suggested that hunters' preying on megafauna calves or other juveniles, be they baleen whales or Pleistocene mammoths, was highly efficient and thus adaptive. These animals could be hunted with less risk than could adults, animal loss was minimized, and a lesser effort was required to obtain the greatest food/raw material package, compared to hunting smaller non-cetacean or non-proboscid animals. I concur with these observations, and summarize the optimal foraging qualities of hunting the smallest of the largest whales during the aboriginal or precontact period as follows:

(a) they were less experienced or wary of predators and, thus, were less dangerous to attack; they were easier to kill with a hand-thrust harpoon in a vital spot;
(b) they were easier and quicker to tow ashore, using fewer boats; carcasses were less likely to be lost to offshore winds or currents;
(c) they were easier to haul onto the ice for butchering, and thinner ice would serve as a butchering platform than was necessary for larger, heavier animals;
(d) they were quicker to butcher and, therefore, there was greater likelihood that unspoiled meat would be taken from the carcass;
(e) they had a higher proportion of edible weight in relation to their total weight; their better tasting and more tender meat and *muk-tuk* made more of their mass acceptable as food; and,
(f) they were easier to handle in terms of transporting meat, blubber, and organs to onshore communities and caches for storage.

Efficiency in processing — related to animal size — is directly related to the edible product of a large whale. Durham (n.d.) describes bowheads taken at Barrow which were impossible to land and butcher because of their very large size. McVay (1973:28) reports that a 20 m bowhead caught at Wainwright took the entire village of about 300 people four days to butcher. Eric Loring (pers. comm., 1993) relates that an 18 m bowhead killed at Wainwright in the late 1980s required Eskimos about 18 hrs of very hard work to tow the carcass to the ice edge. Further, it took an entire day to haul it onto the ice, and another 16 hrs to butcher it. By then, the meat had spoiled, although the *muk-tuk* was still usable. In contrast, a 7 m bowhead taken at Wainwright on another occasion required less than an hour to haul it onto the ice and about 2 hrs to butcher. John George (pers. comm., 1994), a cetacean biologist, has timed whale processing at Barrow, and notes that 20-50 workers completely butchered 7.9, 8.5, and 8.8 m bowheads in 2 hrs, while 15.6 and 16.8 m whales took at least 48 hrs to process.

A related study, carried out by Floyd Durham (n.d.) among Barrow Inupiat during the early 1960s, addressed the utilization rate of different whales in terms of tonnage. In comparing fresh whales, those small ones that could be landed and processed quickly, with "stinkers" or those, often large, animals that started decomposing between 24-48 hrs, he estimated that about 50% of the small whales were edible or usable, whereas only 25% or less of the "stinkers" were usable. In comparing fresh whales, less wastage is culled from small animals than from large ones. John George (pers. comm., 1994) estimates that about 70% of small whales are edible, whereas for larger whales more skin and meat is discarded because it is too tough to eat.

All of these weight and time considerations are tied to whale size, which is usually represented in cetacean biology literature as length in meters. What is deceptive is the fact that length and body size or weight are not linearly related from subadult through adult sizes. Thus, length distributions such as those in Figure 1 do not express the vast weight differences that a few meters can mean to whalers towing and butchering animals. This important point is demonstrated in figures from estimated bowhead weights calculated by George, Philo, and Carroll (1990). For example, the calculated weight of a 7.5 m bowhead is 5097 kg, that for an animal only one-fifth larger (9 m) is 12,126 kg, or over twice as much, and that for an animal only one-half larger (11 m) is 14,750 kg, or three times as much. A fully adult 17 m bowhead is estimated to weigh approximately 80,000-85,000 kg (J. George, pers. comm., 1994). Thus, an adult of this size (and bowheads reach 20 m or longer) weighs eight times what a yearling of 8.0-8.5 m would weigh. Under the best of conditions, it would be a major effort to tow, land, and butcher an adult animal before it spoiled, and under adverse wind and ice conditions, it would be virtually useless to attempt to land and butcher an adult whale using pre-1850s native technology.

For these reasons, young, small baleen whales may be viewed as optimal prey of Eskimo-Aleut hunters either in comparison to larger whales or to nonwhale prey. Optimization is expressed as the most effective fit of search time (cost to locate prey animals) and pursuit time (cost to capture and process them; Winterhalder and Smith 1981; Savelle 1987; Savelle and McCartney 1988). As expressed in the whaling sequence above, pursuit costs go well beyond harpooning a whale, a relatively quick and easy if often dangerous procedure, to its killing, towing, landing, butchering, and storing. These costs are often too high for adult whales, with the result that those animals do not become part of the Eskimo diet.

Compared to all potential prey animals found along the western arctic and subarctic coasts, these small whales are also optimal prey due to their relatively large size and thus dietary productivity. As predictable, if seasonal, prey that are channeled along coastlines and close to hunter settlements, whales require relatively low search and pursuit costs per yield when compared to their equivalent yield in the form of 15-25 belugas or walrus or over a hundred caribou or small seals.

DISCUSSION

Deviation From the Ideal Pattern

While immature baleen whales would have been the ideal prey, local circumstances would have sometimes necessitated some deviation from this pattern. From the western arctic and subarctic literature, we may conclude that any village might expect to land only a few whales per year, and that the meat, blubber, and other materials from this bulk would translate into a successful winter diet. Records from northwestern Alaska show that it was common for a village such as Point Hope to acquire either one or no whales in some years, requiring hunters to pursue alternative resources, such as belugas, walrus, seals, and caribou to make up for the whale shortfall. In modern times (1987-1989), bowheads, on average, have made up 34% of subsistence food pounds, caribou 29%, walrus 11%, and whitefish 7.3% among 54 species of mammals, fish, birds, and other indigenous food resources (Braund 1990).

Climatically-driven deviation in coastal ice patterns (for the northern half of Alaska and the Beaufort Sea coast) can radically alter open lead whaling. For instance, the shore leads can be 2-12 km off Point Hope or Barrow (Rainey 1947; Foote 1964). Secondly, timing the spring hunt with the whale migration close to a coastal village is critical, because if the animals are missed then, there are far fewer chances to kill them later, when open water permits the whales to swim far offshore and out of the whalers' reach.

Given that the migration might not exhibit its regularity in "runs" of whales, that the shore leads were distant from shore and difficult to reach, or that the shorefast ice was dangerous to traverse, a whaling crew might attempt to kill any whale that came within its reach, regardless of size (see Savelle and McCartney 1994). In other words, extraordinary risks might be taken to kill large adult animals if those were the only ones available. Further, *umialiqs* (whaling captains) would always have to balance baleen whaling against a substitute effort to kill smaller, if less prestigious, sea mammals, if it appeared that the preferred whaling pattern was being upset by weather and sea ice conditions.

The Archaeological Record

If the Central Canadian Arctic Thule whale hunting pattern, described above, seems a clearer demonstration of small whale selection than that of the Chukchi Sea region, it is partly because of the archaeological record found on isostatically rebounding beach ridges in the Canadian Arctic. Bones salvaged from hunted carcasses and bones from naturally stranded carcasses are typically uplifted for archaeologists to locate and measure today. Holocene sea levels of northern and western Alaska, on the other hand, have been more stable, because the coasts were not covered with Pleistocene glaciers as they were in the Canadian Arctic. Whale bones would be more likely to be washed away or ice-scoured where uplifting does not occur.

Stranding of bowheads in the Western Arctic occurs for the same reasons that it does in the Central and Eastern Arctic, that is primarily from death due to ice entrapment and, possibly, from killer whale attacks (Mitchell and Reeves 1982). Carcasses of other baleen whales in southern Alaska might be stranded when they die of either killer whale attacks, old age, or other natural causes. The larger whale bones cited above may derive from stranded rather than hunted whales, if the hypothesis that hunted whales were primarily small animals is eventually demonstrated. As pointed out above, the Izembek Lagoon house superstructure is thought to have been made of mandibles from stranded whales, since blue whales were never hunted by southwestern Alaskan natives, given their offshore habits and huge size.

As with the Canadian Thule example, whale bones in western arctic and subarctic features are susceptible to post-occupation disturbance through reuse. Removing bones from the original features for reuse elsewhere or reducing them into tools or other objects limits our opportunity to interpret them as parts of constructed features or to measure them.

Obviously, what is needed in the Western Arctic and Subarctic are systematic studies in order to determine the kinds, numbers, and sizes of whales used in the past. These include studies of archaeological whale bones, oral traditions relating to aboriginal whaling, ethnohistoric literature that may contain details about early contact period whaling, and behavioral/ecological interpretations by cetacean biologists with regard to where, when, and how baleen whales might have been successfully taken by native hunters. Patterns of past cetacean predation may well be found if they are sought using these approaches.

Reconstruction of Whaling Societies

Finally, and importantly, evidence of whales used by prehistoric and early contact native peoples is not the ultimate goal of our inquiries. Rather, these data are needed to evaluate the total subsistence patterns of coastal societies and the significance of whales within that pattern. All of the important sociocultural dimensions that we might normally attempt to understand for a group and that are based on differing kinds of subsistence — the settlement pattern, economy, social structure, political organization, and religion — will be impacted by whale use (Lantis 1938). More specifically, sedentism, ranking, trade, warfare, and ceremonialism are related facets of adaptations to large and predictable food resources such as those afforded by whales. It is not an overstatement to say that the accurate reconstruction of coastal native lifeways depends largely upon our understanding of whaling in those societies, simply because of the bulk of food and raw materials provided by those animals.

CONCLUSIONS

(a) Previous studies of native whaling in the Central Canadian Arctic and along the Siberian coast indicate that small whales (calves, yearlings, and subadults, depending upon the area) were selected by prehistoric and early historic native whalers, presumably because of the relative ease in killing and using them.

(b) From an optimal foraging perspective, adult baleen whales are not ideal prey targets when attacked with an aboriginal technology. They are dangerous to hunt and are more difficult and time-consuming to tow ashore, land, and butcher. In contrast, small, immature baleen whales, while significantly larger than any other prey animal available, are by comparison to adult whales easier to capture and process, and, therefore, are thought to have been optimal prey animals where and when they were available.

(c) It is suggested that larger baleen whales would be pursued in situations when smaller animals were not available for selection. Large archaeological whale bones may reflect less than ideal hunting circumstances, or may simply reflect the use of bones from larger stranded carcasses.

(d) Aboriginal whaling technology, involving either harpoons, floats, and *umiaks* or darts and kayaks, limited natives to hunting small whales; only when commercial whaling gear (primarily dart guns and shoulder guns with explosive charges and block and tackle) was transferred to native whalers during the 1870s and 1880s could they begin to effectively hunt large whales.

(e) Although there are examples of small baleen whale use during the precontact and early contact periods in Alaska and the Western Canadian Arctic, as reflected in small bones found at archaeological sites and in ethnohistoric references, no large-scale, systematic study has been carried out to determine the actual whaling patterns in these regions. Unfortunately, the archaeological record at many Alaskan locales is today threatened by coastal erosion, which is rapidly destroying sites before whale bones in them can be studied.

(f) Almost no attempts have been made to utilize the archaeological record in order to reconstruct the "arctic whale hunting cultures" of the Western Arctic and Subarctic. By identifying, counting, and measuring whale bones associated with features and entire settlements, it is possible to establish the potential importance of whales for food as well as whale bones for framing winter houses and other features. Subsistence scheduling, specialty hunting strategies, and storage may then be keyed to the seasonal migration patterns of different whale species. More importantly, settlement size, manipulation or trade of wealth, transformation of surplus or wealth into social ranking, and ceremonialism may subsequently be better understood, because baleen whaling affects all aspects of a coastal native society.

Acknowledgments. I wish to thank the following persons for providing useful information and suggestions for this paper. John George (Department of Wildlife Management, North Slope Borough) kindly provided me with many ideas and data regarding bowhead biology and sizes for North Alaska. I have incorporated several important points that he made into this paper. Besides George, James Savelle, Igor Krupnik, and Roger Harritt kindly read and commented on a draft of this paper. I also wish to thank Max Friesen for information about several Western Canadian Arctic references to which I did not have ready access. The Canadian Arctic bowhead measurement program referred to in this paper was largely supported by the Canadian Museum of Civilization, the Social Science and Humanities Research Council of Canada, the Department of Indian and Northern Affairs, and the Polar Continental Shelf Project (Department of Energy, Mines and Resources).

REFERENCES

Alaska Whales and Whaling
   1978   *Alaska Geographic* 5(4).

Anderson, Douglas D.
   1986   Ancestors of the Inupiat: Development of the Northern Maritime Tradition. In: Beach Ridge Archaeology of Cape Krusenstern: Eskimo and Pre-Eskimo Settlements Around Kotzebue Sound, Alaska, J. L. Giddings and D. D. Anderson, pp. 58-106. *National Park Service, Publications in Archeology* No. 20.

Arnold, Charles D.
   1986   Thule Pioneers. *Prince of Wales Northern Heritage Centre Occasional Paper* No. 2.

Arutiunov, S. A. and D. A. Sergeev
   1975   *Problemy etnicheskoi istorii Beringomor'ia (Ekvenskii mogil'nik).* (Questions in the Ethnic History of the Bering Sea Area: The Ekven Cemetery). Nauka, Moscow. In Russian.

Arutiunov, S. A., I. I. Krupnik, and M. C. Chlenov
   1982   *Kitovaya alleia: Drevnosti ostorvov proliva Seniavina.* (Whale Alley. Alley: Antiquities of the Senyavin Strait Islands). Nauka, Moscow. In Russian.

Bailey, Alfred M. and Russell W. Hendee
   1926   Notes on the Mammals of Northwestern Alaska. *Journal of Mammalogy* 7(1):9-28.

Bee, James W. and E. Raymond Hall
    1956   Mammals of Northern Alaska on the Arctic Slope. *University of Kansas Museum of Natural History, Miscellaneous Publication* No. 8.

Bernardi, Suzanne R.
    1981   Story of a Whale Hunt. *The Alaska Journal* 11:134-143.

Bisset, N. G.
    1976   Hunting Poisons of the North Pacific Region. *Lloydia* 39(2):87-124.

Black, Lydia T.
    1977   The Konyag (The Inhabitants of the Island of Kodiak) by Iosaf [Bolotov](1794-1799) and By Gideon (1804-1807). *Arctic Anthropology* 14(2):79-108.
    1987   Whaling in the Aleutians. *Etudes/Inuit/Studies* 11(2):7-50.

Bockstoce, John R.
    1986   *Whales, Ice, and Men: The History of Whaling in the Western Arctic.* University of Washington Press, Seattle.

Bockstoce, John R. and John J. Burns
    1993   Commercial Whaling in the North Pacific Sector. In: The Bowhead Whale, edited by J. J. Burns, J. J. Montague, and C. J. Cowles, pp. 563-577. *Society for Marine Mammalogy, Special Publication* No. 2.

Bogoslovskaya, L. S., L. M. Votrogov, and I. I. Krupnik
    1982   The Bowhead Whale Off Chukotka: Migrations and Aboriginal Whaling. *Report of the International Whaling Commission* 32:391-399.

Braham, Howard W., Floyd E. Durham, Gordon H. Jarrell, and Stephen Leatherwood
    1980   Ingutuk: A Morphological Variant of the Bowhead Whale, *Balaena mysticetus. Marine Fisheries Review* 42(9):70-73.

Braund, Stephen R. and Associates
    1990   Quantification of Subsistence Harvest (Marine and Terrestrial Mammals, Birds, and Fish) in Barrow and Wainwright, Alaska, 1987-89. In: *Fifth Conference on the Biology of the Bowhead Whale*, Balaena mysticetus: *Extended Abstracts and Panel Discussion*, edited by T. F. Albert, pp. 146-149. North Slope Borough, Barrow.

Burch, Ernest S., Jr.
    1981   *The Traditional Eskimo Hunters of Point Hope, Alaska: 1800-1875.* North Slope Borough, Barrow.

Burns, John J.
    1993   Epilogue. In: The Bowhead Whale, edited by J. J. Burns, J. J. Montague, and C. J. Cowles, pp. 745-764. *Society for Marine Mammalogy, Special Publication* No. 2.

Burns, John J., J. Jerome Montague, and Cleveland J. Cowles (editors)
    1993   The Bowhead Whale. *Society for Marine Mammalogy, Special Publication* No. 2.

Cargill, Jody
    1990   Analysis of Faunal Remains. In: *The 1981 Excavations at the Utqiagvik Archaeological Site, Barrow, Alaska*, Vol. 1, edited by E. S. Hall, Jr. and L. Fullerton, pp. 263-279. North Slope Borough Commission on Inupiat History, Language, and Culture, Barrow.

Collins, Henry B., Jr.
    1929   The Ancient Eskimo Culture of Northwestern Alaska. *Explorations and Field-Work of the Smithsonian Institution in 1928*, pp. 141-150.
    1935   Archeology of the Bering Sea Region. *Annual Report of the Smithsonian Institution for 1933*, pp. 453-468.

1937a Archeology of St. Lawrence Island, Alaska. *Smithsonian Miscellaneous Collections* 96(1).

1937b Archeological Excavations at Bering Strait. *Explorations and Field-Work of the Smithsonian Institution in 1933*, pp. 63-68.

Dall, William H.

1870 *Alaska and Its Resources*. Lee and Shepard, Boston.

Durham, Floyd E.

1979 The Catch of Bowhead Whales (*Balaena mysticetus*) by Eskimos, With Emphasis on the Western Arctic. *Natural History Museum of Los Angeles County, Contributions in Science* 314:1-14.

n.d. Recent Trends in Bowhead Whaling by Eskimos in the Western Arctic With Emphasis on Utilization. Whale Protection Fund, Center for Environmental Protection, Inc., Washington, D.C.

Dyke, Arthur S. and Thomas F. Morris

1990 Postglacial History of the Bowhead Whale and of Driftwood Penetration; Implications for Paleoclimate, Central Canadian Arctic. *Geological Survey of Canada Paper* 89-24.

Elliott, Henry W.

1886 *Our Arctic Province: Alaska and the Seal Islands*. Charles Scribner's Sons, New York.

Foote, Don Charles

1964 Observations of the Bowhead Whale at Point Hope, Alaska. MS., pp. 77.

Ford, James A.

1959 Eskimo Prehistory in the Vicinity of Point Barrow, Alaska. *Anthropological Papers of the American Museum of Natural History* 47(1).

Geist, Otto and Froelich G. Rainey

1936 Archaeological Excavations at Kukulik, St. Lawrence Island, Alaska. *University of Alaska Miscellaneous Publications* 2. U.S. Government Printing Office, Washington, D.C.

Gerlach, Craig, John C. George, and Robert Suydam

1993 Bowhead Whale (*Balaena mysticetus*) Length Estimations Based on Scapula Measurements. *Arctic* 46(1):55-59.

George, John C., L. Michael Philo, and Geoffrey M. Carroll

1990 Observations on Weights of Subsistence Harvested Bowhead Whales. In: *Fifth Conference on the Biology of the Bowhead Whale*, Balaena mysticetus: *Extended Abstracts and Panel Discussion*, edited by T. F. Albert, pp. 89-93. North Slope Borough, Barrow.

Giddings, J. Louis

1967 *Ancient Men of the Arctic*. Alfred A. Knopf, New York.

Giddings, J. Louis and Douglas D. Anderson

1986 An Interval of Unique Early Coastal Dwellers. In: Beach Ridge Archeology of Cape Krusenstern: Eskimo and Pre-Eskimo Settlements Around Kotzebue Sound, Alaska, J. L. Giddings and D. D. Anderson, pp. 231-267. *National Park Service, Publications in Archeology* No. 20.

Heizer, Robert F.

1943 Aconite Poison Whaling in Asia and America: An Aleutian Transfer to the New World. *Bureau of American Ethnology Bulletin* 133:415-468.

Hofmann-Wyss, Anna B.

1987 *Prähistorische Eskimogräber an der Dovelavik Bay und bei Kitnepaluk im Westen der St. Lorenz Insel, Alaska*. St. Lorenz Insel-Studien. Verlag Paul Haupt, Bern.

Holmberg, Heinrich J.
   1985   Holmberg's Ethnographic Sketches. *Rasmuson Library Historical Translation Series* 1.
Hrdlicka, Ales
   1944   *The Anthropology of Kodiak Island*. The Wistar Institute of Anatomy and Biology, Philadelphia.
   1945   *The Aleutian and Commander Islands and Their Inhabitants*. The Wistar Institute of Anatomy and Biology, Philadelphia.
Hughes, Charles C.
   1960   *An Eskimo Village in the Modern World*. Cornell University Press, Ithaca.
Jenness, Diamond
   1922   The Life of the Copper Eskimos. *Report of the Canadian Arctic Expedition, 1913-18*, Vol. 12.
Jenness, Stuart E.
   1990   Diamond Jenness's Archaeological Investigations on Barter Island, Alaska. *Polar Record* 26(157):91-102.
Jochelson, Waldemar
   1925   Archaeological Investigations in the Aleutian Islands. *Carnegie Institution of Washington Publication* 367.
Koski, William R., Gary W. Miller, and Rolph A. Davis
   1988   The Potential Effects of Tanker Traffic on the Bowhead Whale in the Beaufort Sea. *Indian and Northern Affairs Canada, Environmental Studies* No. 58.
Krupnik, Igor I.
   1984   Gray Whales and the Aborigines of the Pacific Northwest: The History of Aboriginal Whaling. In: *The Gray Whale*, Eschrichtius robustus, edited by M. J. Jones, S. L. Swartz, and S. Leatherwood, pp. 103-120. Academic Press, Orlando.
   1987   The Bowhead vs. the Gray Whale in Chkotkan Aboriginal Whaling. *Arctic* 40(1):16-32.
   1993a *Arctic Adaptations: Native Whalers and Reindeer Herders of Northern Eurasia*. University Press of New England, Hanover, NH.
   1993b Prehistoric Eskimo Whaling in the Arctic: Slaughter of Calves or Fortuitous Ecology? *Arctic Anthropology* 30(1):1-12.
Krupnik, Igor I., L. S. Bogoslovskaya, L. M. Votrogov
   1983   Gray Whaling Off the Chukotka Peninsula: Past and Present Status. *Report of the International Whaling Commission* 33:557-562.
Lantis, Margaret
   1938   The Alaskan Whale Cult and Its Affinities. *American Anthropologist* 40(3):438-464.
Larsen, Helge and Froelich Rainey
   1948   Ipiutak and the Arctic Whale Hunting Culture. *Anthropological Papers of the American Museum of Natural History* 42.
Laughlin, William S.
   1963   The Earliest Aleuts. *Anthropological Papers of the University of Alaska* 10(2):73-91.
McCartney, Allen P.
   1974   Prehistoric Cultural Integration Along the Alaska Peninsula. *Anthropological Papers of the University of Alaska* 16(1):59-84.
McCartney, Allen P. and James M. Savelle
   1993   Bowhead Whale Bones and Thule Eskimo Subsistence-Settlement Patterns in the Central Canadian Arctic. *Polar Record* 29(168):1-12.

McGhee, Robert
   1972   Copper Eskimo Prehistory. *National Museum of Man, Publications in Archaeology* No. 2.
   1974   Beluga Hunters: An Archaeological Reconstruction of the History and Culture of the Mackenzie Delta Kittegaryumiut. *Memorial University of Newfoundland, Newfoundland Social and Economic Studies* No. 13.

McVay, Scott
   1973   Stalking the Arctic Whale. *American Scientist* 61:24-37.

MacNeish, Richard S.
   1956   Archaeological Reconnaissance of the Delta of the Mackenzie River and Yukon Coast. *National Museum of Canada Bulletin* 142:46-81.

Maguire, Rochfort
   1988   *The Journal of Rochfort Maguire, 1852-1854*, 2 Vols., edited by John R. Bockstoce. The Hakluyt Society, London.

Maher, William J. and Norman J. Wilimovsky
   1963   Annual Catch of Bowhead Whales by Eskimos at Point Barrow, Alaska, 1928-1960. *Journal of Mammalogy* 44(1):16-20.

Manning, Thomas H.
   1956   Narrative of a Second Defence Research Board Expedition to Banks Island, With Notes on the Country and Its History. *Arctic* 9(1-2):3-77.

Marquette, Willman M. and John R. Bockstoce
   1980   Historical Shore-Based Catch of Bowhead Whales in the Bering, Chukchi, and Beaufort Seas. *Marine Fisheries Review* 42(9):5-19.

Marquette, Willman M., Howard W. Braham, Mary K. Nerini, and Robert V. Miller
   1982   Bowhead Whale Studies, Autumn 1980-Spring 1981: Harvest, Biology, and Distribution. *Report of the International Whaling Commission* 32:357-370.

Mitchell, Edward D. and Randall R. Reeves
   1982   Factors Affecting Abundance of Bowhead Whales *Balaena mysticetus* in the Eastern Arctic of North America, 1915-1980. *Biological Conservation* 22:59-78.

Moore, Sue E. and Randall R. Reeves
   1993   Distribution and Movement. In: The Bowhead Whale, edited by J. J. Burns, J. J. Montague, and C. J. Cowles, pp. 313-386. *Society for Marine Mammalogy, Special Publication* No. 2.

Morrison, David A.
   1983   Thule Culture in Western Coronation Gulf, N.W.T. *National Museum of Man, Mercury Series, Archaeological Survey of Canada Paper* No. 116.

Nelson, Edward W.
   1899   The Eskimo About Bering Strait. *18th Annual Report of the Bureau of American Ethnology for the Years 1896-1897*, pp. 3-518. Washington, D.C.

Nerini, Mary K., Howard W. Braham, Willman M. Marquette, and David J. Rugh
   1984   Life History of the Bowhead Whale, *Balaena mysticetus* (Mammalia: Cetacea). *Journal of Zoology* 204:443-468.

Philo, L. Michael, Emmett B. Shotts, Jr., and John C. George
   1993   Morbidity and Mortality. In: The Bowhead Whale, edited by J. J. Burns, J. J. Montague, and C. J. Cowles, pp. 275-312. *Society for Marine Mammalogy, Special Publication* No. 2.

Point Hope Beach Erosion, Point Hope, Alaska
   1972   Survey Report, Alaska District, Corps of Engineers, Anchorage, Alaska.

Rainey, Froelich G.
   1940   Eskimo Method of Capturing Bowhead Whales. *Journal of Mammalogy* 21(3):362.
   1947   The Whale Hunters of Tigara. *Anthropological Papers of the American Museum of Natural History* 41(2).

Ray, Dorothy J.
   1975   *The Eskimo of Bering Strait, 1650-1898.* University of Washington Press, Seattle.

Reeves, Randall R. and Edward D. Mitchell
   1985   Shore-Based Bowhead Whaling in the Eastern Beaufort Sea and Amundsen Gulf. *Report of the International Whaling Commission* 35:387-404.

Rice, Dale W. and Allen A. Wolman
   1971   The Life History and Ecology of the Gray Whale (*Eschrichtius robustus*). *American Society of Mammalogists, Special Publication* No. 3.

Richardson, W. John and Charles I. Malme
   1993   Man-Made Noise and Behavioral Responses. In: The Bowhead Whale, edited by J. J. Burns, J. J. Montague, and C. J. Cowles, pp. 631-700. *Society for Marine Mammalogy, Special Publication* No. 2.

Saunders, Jeffrey J.
   1980   A Model for Man-Mammoth Relationships in Late Pleistocene North America. In: The Ice-Free Corridor and Peopling of the New World, edited by N. W. Rutter and C. E. Schweger. *Canadian Journal of Anthropology* 1(1):87-98.

Savelle, James M.
   1987   Collectors and Foragers: Subsistence-Settlement System Change in the Central Canadian Arctic, AD 1000-1960. *British Archaeological Reports, International Series* No. 358.

Savelle, James M. and Allen P. McCartney
   1988   Geographical and Temporal Variation in Thule Eskimo Subsistence Economies: A Model. *Research in Economic Anthropology* 10:21-72.
   1991   Thule Eskimo Subsistence and Bowhead Whale Procurement. In: *Human Predators and Prey Mortality*, edited by M. Stiner, pp. 201-216. Westview Press, Boulder.
   1994   Thule Inuit Bowhead Whaling: A Biometrical Analysis. In: Threads of Arctic Prehistory: Papers in Honour of William E. Taylor, Jr., edited by D. Morrison and J.-L. Pilon, pp. 281-310. *Canadian Museum of Civilization, Mercury Series, Archaeological Survey of Canada Paper* No. 149.

Simpson, Thomas
   1843   *Narrative of the Discoveries on the North Coast of America; Effected by the Officers of the Hudson's Bay Company During the Years 1836-39.* Richard Bentley, London.

Stefansson, Vilhjalmur
   1913   *My Life With the Eskimo.* Macmillan Company, New York.

Stoker, Sam W. and Igor I. Krupnik
   1993   Subsistence Whaling. In: The Bowhead Whale, edited by J. J. Burns, J. J. Montague, and C. J. Cowles, pp. 579-629. *Society for Marine Mammalogy, Special Publication* No. 2.

VanStone, James W.
   1958   Commercial Whaling in the Arctic Ocean. *Pacific Northwest Quarterly* 41(1):1-10.

Veniaminov, Ivan
   1984   *Notes on the Islands of the Unalashka District.* Transls. L. T. Black and R. H. Geohegan. Limestone Press, Kingston, ON.

Winterhalder, Bruce and Eric A. Smith (editors)
 1981 *Hunter-Gatherer Foraging Strategies.* University of Chicago Press, Chicago.

Woodby, Douglas A. and Daniel B. Botkin
 1993 Stock Sizes Prior to Commercial Whaling. In: The Bowhead Whale, edited by J. J. Burns, J. J. Montague, and C. J. Cowles, pp. 387-407. *Society for Marine Mammalogy, Special Publication* No. 2.

Yorga, Brian W. D.
 1980 Washout: A Western Thule Site on Herschel Island, Yukon Territory. *National Museum of Man, Mercury Series, Archaeological Survey of Canada Paper* No. 98.

Prehistoric Beluga Whale Hunting at Gupuk, Mackenzie Delta, Northwest Territories, Canada

T. Max Friesen
Department of Anthropology
McGill University
855 Sherbrooke St. West
Montreal, Quebec
Canada H3A 2T7

Charles D. Arnold
Prince of Wales Northern Heritage Centre
Yellowknife, N.W.T.
Canada X1A 2L9

Abstract. *Protohistoric and early historic Inuit societies of the Mackenzie River Delta relied on beluga whales* (Delphinapterus leucas) *for a large proportion of their diet. The ethnohistoric record from Kittigazuit, the largest Mackenzie Inuit site occupied during the historic period, indicates that beluga whales were hunted by driving entire whale pods into shallow waters where they were harpooned and lanced. This report presents a reconstruction of the earlier, prehistoric Mackenzie Inuit beluga hunt on the basis of archaeological data. Several lines of evidence are advanced in support of the hypothesis that the prehistoric beluga hunt at the Gupuk site, located on Richards Island in the Mackenzie Delta, was performed with the same large-scale drive methods recorded for Kittigazuit. The most important data set consists of beluga whale age determinations based on growth layers observed in beluga mandibles recovered from Gupuk. The resulting mortality profile closely resembles catastrophic mortality, which is consistent with large-scale drive hunting at Gupuk.*

INTRODUCTION

In general accounts of indigenous subsistence economies in the Western Arctic, the hunting of beluga whales (*Delphinapterus leucas*) is often overshadowed by that of the larger baleen whales, represented most spectacularly by bowhead and gray whales. However, a review of the ethnohistoric record indicates that beluga whales were hunted by a broad range of peoples occupying the coastlines of Siberia, Alaska, and northwestern Canada. Therefore, belugas deserve close attention when attempts are made to reconstruct historic and prehistoric subsistence patterns in this region.

The analysis of beluga whale remains from archaeological contexts entails a number of methodological difficulties. Most importantly, their large body size, coupled with the fact that meat and blubber are easily removed from their carcasses, results in bones often being abandoned at butchery sites (Savelle and Friesen n.d.). Therefore, beluga bones will frequently be underrepresented at the home bases to which meat and blubber were transported. As a result, zooarchaeological analyses of sites yielding high frequencies of beluga bones assume an increased significance, because the relatively strong behavioral inferences derived from large samples can increase the degree of confidence with which smaller samples are interpreted.

This report presents analysis of beluga whale hunting methods employed at Gupuk (NiTs-1), one of several prehistoric Mackenzie Inuit sites which have yielded large numbers of beluga bones (Fig. 1). The Mackenzie Inuit of the Mackenzie River Delta, Northwest Territories, displayed what was probably the greatest reliance on small toothed whales of any indigenous society in the Arctic and, perhaps, the world. Ideal hunting conditions were created each summer, when thousands of belugas entered the shallow East Channel of the Mackenzie River. Ethnohistoric data, oral histories, and archaeological research suggest that large Inuit populations congregated there annually for the beluga hunt (McGhee 1974). Several lines of evidence, including whaling-associated artifact types, modified beluga vertebrae, and a mortality profile constructed on the basis of mandible thin sections, are examined here in order to reconstruct the methods used to hunt belugas at Gupuk. It is hoped that the methods and data discussed will prove useful to the investigation of sites which are associated with fewer beluga bones or less complete ethnographic information.

ETHNOHISTORIC BACKGROUND

The Mackenzie Inuit

At the time of first contact with Europeans, the Mackenzie Inuit, or Siglit as they referred to themselves, were distributed along the Beaufort Sea coast from the Yukon-Alaska border to at least as far east as Cape Bathurst (McGhee 1974; Morrison 1990). The major concentration of Siglit was at the mouth of the East Channel of the Mackenzie River. Residents of this area had access to a wide variety of food resources, including arctic and subarctic terrestrial fauna, river and lake fish, waterfowl, and marine mammals. The major advantage of this area, however, lay in the abundance of beluga whales that enter the estuary in the summer months.

The belugas of the Mackenzie Delta belong to the Bering Sea population, which probably exceeds 25,000 individuals and inhabits the Bering, Chukchi, East Siberian, and Beaufort seas. Of this population, the largest stock, numbering at least 11,500 individuals, migrates to the eastern Beaufort Sea during the summer (Seaman et al. 1985). From this Beaufort stock, up to 7000 may enter the Mackenzie River estuary between late June and mid-August (Fraker 1980). Of particular importance to Inuit hunters is the fact that up to 2500 beluga whales may congregate at the same time in Kugmallit Bay (Fraker et al. 1978), near the Mackenzie Inuit settlements of Kittigazuit and Gupuk. Beluga whales probably do not congregate in estuaries to feed on fish, as was previously assumed. Rather, it is more likely that the warm waters provide a safe haven for birth and early growth of calves, requiring a minimum expenditure of energy to maintain body heat (Fraker et al. 1979; Sergeant 1973).

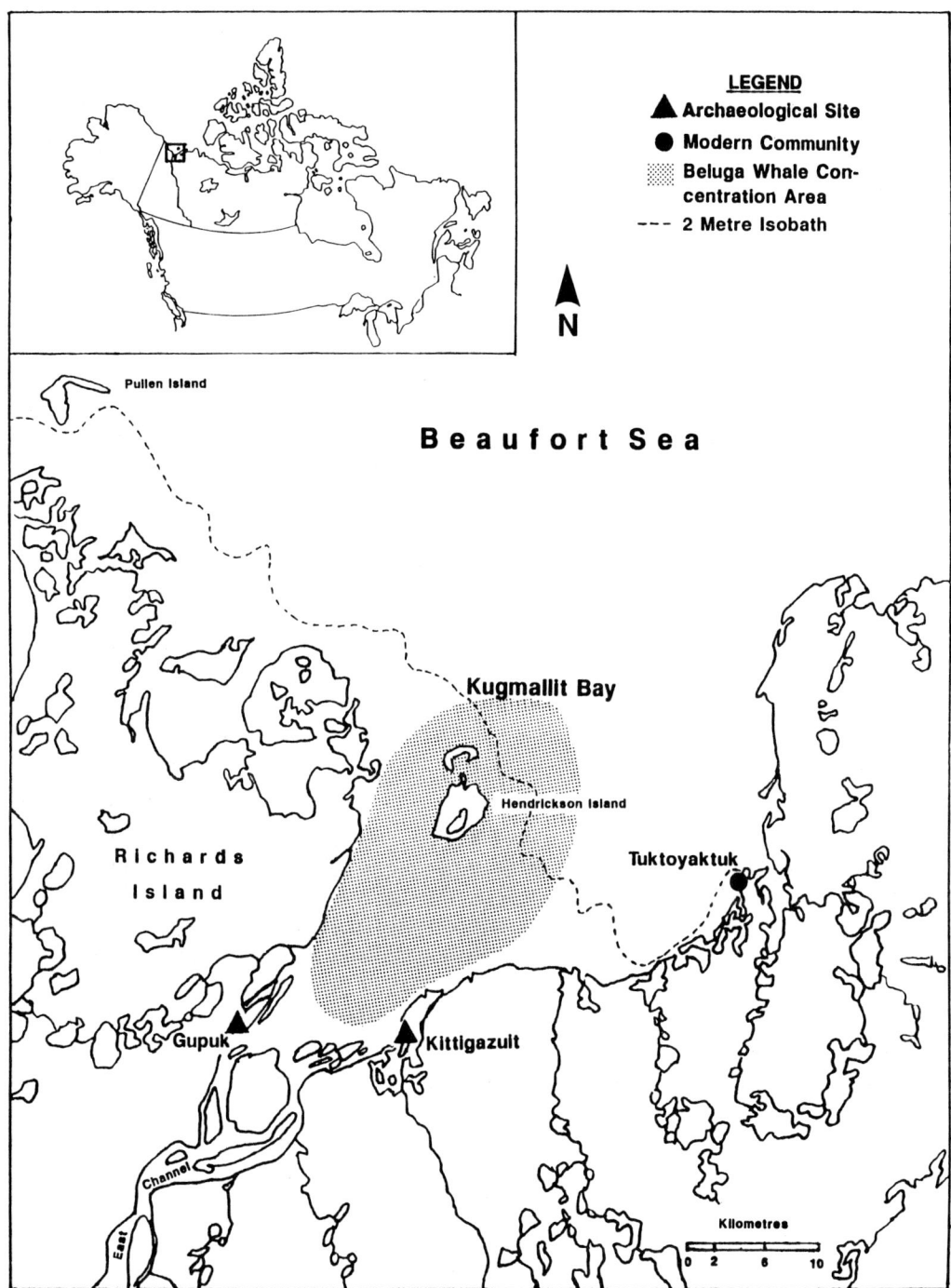

Figure 1. Map of the Kugmallit Bay region of the outer Mackenzie Delta, indicating sites mentioned in the text. Note the extreme shallowness of the Bay as indicated by the 2 m isobath. The shaded zone indicates the summer beluga whale concentration area.

At Kittigazuit, beluga pods were hunted using drive techniques that are described below. After the belugas were towed to camp, meat and blubber were consumed or prepared for storage either by caching in pits, drying, or cutting into small squares and storing in oil-filled bags (Whittaker 1937:176-177). In addition to their use as food, beluga whales were important to the Mackenzie Inuit for their skins, which were used for boat covers, dog harnesses, harpoon lines, boot soles, and tent covers, and for their stomachs, which were used for harpoon floats, bags, and windows (Stefansson 1919; Whittaker 1937).

Aboriginal Beluga Hunting Methods

We can reconstruct the Siglit beluga hunt from the descriptions of three primary observers: Nuligak, a Siglit hunter who lived at the village of Kittigazuit during the early 20th century (Nuligak 1966:15-17), the Earl of Lonsdale, who traveled to the Mackenzie Delta in 1888 (Krech 1989:62-63), and C. E. Whittaker, an Anglican missionary who spent several summers at Kittigazuit in the 1890s (Whittaker 1937:173-179). At the summer whaling camps, observers kept a constant watch from high ground, signaling the presence of belugas with shouts (Krech 1989:63) or arm movements (Whittaker 1937:174). Hunters would then put out to sea in their kayaks, preceded by a temporary hunt leader (Nuligak 1966:16; Whittaker 1937:173). A group of kayakers, numbering between 25 (Whittaker 1937:174) and 100 (Stefansson 1919:172), would then form a line, spaced about 40 m apart (Whittaker 1937:174). They would advance on the whales, splashing the water with their paddles, and shouting in order to drive the whales into shallow water where they were harpooned and lanced.

After the hunt, a blow pipe was inserted below the skin, and the wound was filled with air in order to float the carcasses (Krech 1989:63; Whittaker 1937:176). Whales were towed back to camp behind the kayaks (Nuligak 1966:17; Whittaker 1989:176), or by women in *umiaks* (Krech 1989:63). The harpoons carried ownership marks, allowing individual hunters to claim whales which they had killed (Krech 1989:63; Nuligak 1966:17).

Drive hunting methods similar to those observed at Kittigazuit have been used by a number of other beluga-hunting peoples, including Inuit of eastern Hudson Bay (Saladin d'Anglure 1984:489), northern Alaska (Spencer 1959:34), and Baffin Island (Boas 1888:93). This latter group is reported to have thrown stones into the water in order to drive belugas into shallow bays, where they were harpooned. However, at least four additional beluga hunting methods which predate rifle technology have been recorded. First, special large-mesh nets were used to capture belugas in northwestern Alaska (Nelson 1899:131). Second, hunting of belugas stranded at openings in the sea ice, and thereby cut off from open water, has occurred in Greenland (Birket-Smith 1924:334) and eastern Hudson Bay (Saladin d'Anglure 1984:489). Third, hunting of individual whales from kayaks or the ice edge was probably relatively widespread throughout the North American Arctic, having been recorded in Labrador (Taylor 1984:516), Greenland (Birket-Smith 1924:334; Kleivan 1984:606), Baffin Island (Boas 1888:93), and Alaska (Nelson 1899:137). Fourth, the Dena'ina around Cook Inlet in southern Alaska practiced the unusual method of building wooden platforms in shallow water, from which individual beluga were harpooned (Wrangell 1980:57).

Because of this diverse range of hunting methods reported in the ethnographic record, a specific method cannot be assumed for any given prehistoric site without corroborative evidence. The remainder of this paper outlines the argument for drive hunting at the prehistoric Mackenzie Inuit site of Gupuk.

ARCHAEOLOGICAL INVESTIGATIONS AT GUPUK

Count de Sainville (1984), a French explorer who spent the period 1889-1894 in the lower Mackenzie River area, published a map with a location marked "*vieux village*" on the west side of the East Channel, directly across from Kittigazuit. De Sainville's description can be interpreted as indicating that this settlement had been abandoned by the late 19th century, probably because the continual accumulation of silt made the location unsuitable for beluga hunting (Stefansson 1919:170). In 1954, MacNeish (1956:48) found an archaeological site at about the same map location as de Sainville's "*vieux village*," and, on the basis of information obtained locally, he identified it as the settlement known as Gupuk in the ethnohistoric record. Limited testing was undertaken at the site by Gordon (1972) in 1972, and extensive excavations were conducted by the Prince of Wales Northern Heritage Centre in 1986, 1988, and 1989 (Arnold 1988, 1994).

From the river's edge, the terrain at Gupuk rises approximately 30 m to a series of hills composed of fine sands and gravels which parallel the direction of the river. On the shore side, these hills are quite steep, although erosional fans and spurs moderate the slope in places. Archaeological remains were found mainly on these fans and spurs. Large areas of the site have been lost due to the effects of high water levels and ice scouring during the spring breakup. The 19 house remains still visible at the site are probably a small fraction of the number of dwellings which may have been present in de Sainville's time.

The southernmost zone of the site, designated Area 1, contained one house depression (House 1), which is the source of the beluga bone sample discussed in this paper (Fig. 2). In addition, 16 ground caches and several graves were recorded in this area. Upon excavation, the house was determined to be of the cruciform style which is common in the region (Arnold and Hart 1992). This house type has three interior alcoves, which are raised above floor level, and a long entrance passage which leads toward the water. Much of the floor of the structure, as well as the interior benches, walls, and roof, was constructed of driftwood.

House 1 was excavated by trowel, and all deposits in culture-bearing zones were screened through 6 mm (1/4 in) mesh. Artifacts recovered in situ were recorded in place, and faunal materials were bagged by 10 cm level within each 2 x 2 m excavation unit. Beluga bones were collected separately, and were analyzed as a single large sample. Forty square meters were excavated in Area 1, with some of the units extending down more than a meter before sterile deposits were encountered. Three dates have been obtained for House 1: 730 ± 80 B.P. (AECV-1001C), on unidentified small terrestrial mammal bone; 360 ± 80 B.P. (AECV 1002C), on caribou bone; and 650 ± 140 B.P. (RIDDL-550), an accelerator date on a bone tool. As it lies between the other two dates, the accelerator date on the bone tool is provisionally accepted as indicating the age of House 1. The complete absence of Euroamerican artifacts confirms a prehistoric date for the house.

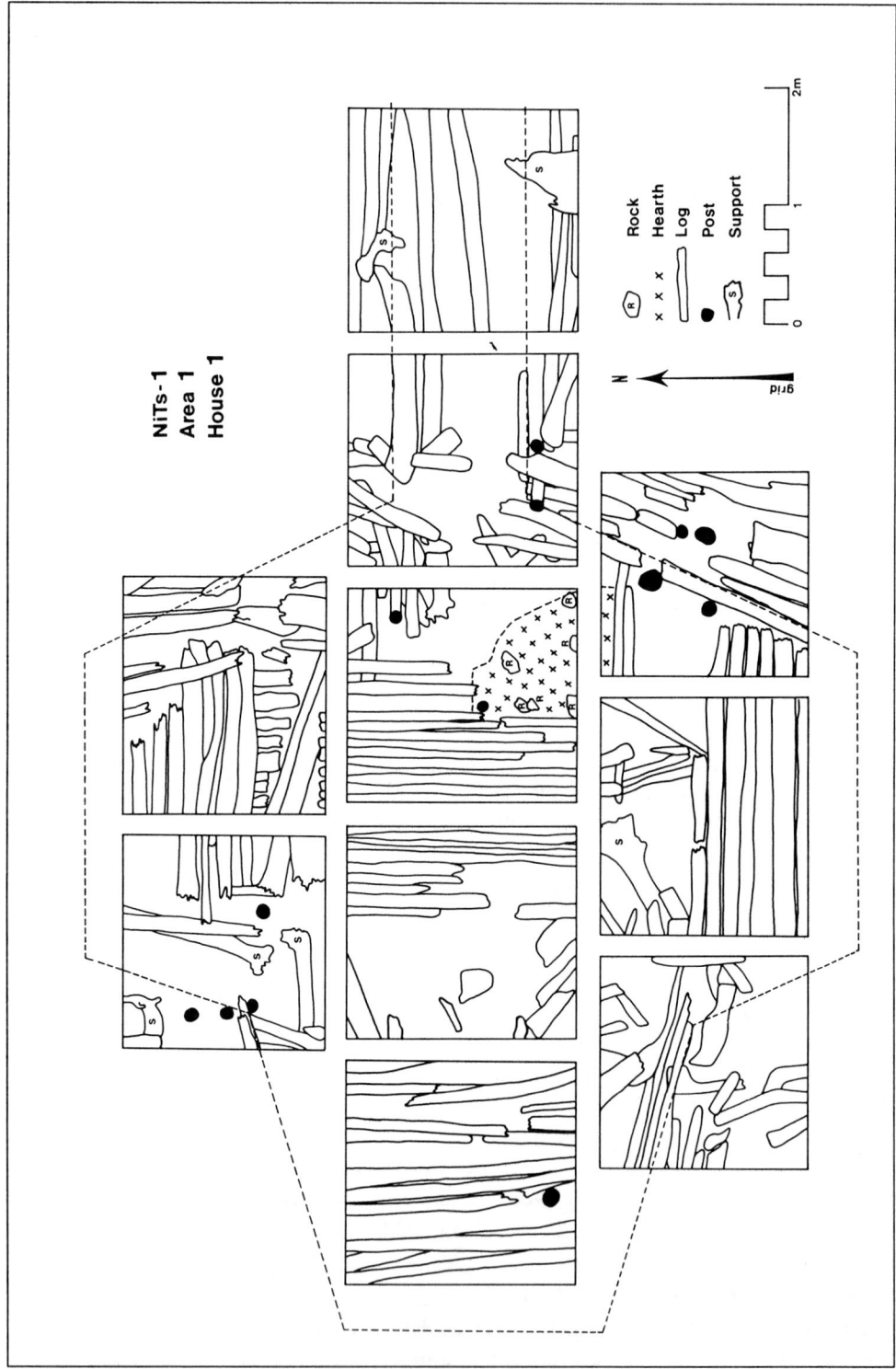

Figure 2. Plan of House 1, a cruciform dwelling from Gupuk Area 1.

THE GUPUK BELUGA HUNT

Based on the ethnohistoric record, the prehistoric beluga hunt at Gupuk can be inferred to have resembled the historically documented hunt at Kittigazuit (Stefansson 1919:24). Several categories of archaeological data recovered from Gupuk reinforce this interpretation. First, the two sites are situated less than 15 km apart. Both are located near extensive shallows that are suitable for stranding belugas, and near the known beluga concentration area in Kugmallit Bay (see Fig. 1). Second, artifact assemblages from the two sites are similar, indicating a close cultural connection (Arnold 1994). Among the artifacts at Gupuk are several which were most likely used for beluga procurement. These include blow pipes used for inflating beluga carcasses prior to towing them to camp, and a type of harpoon head which is larger than the harpoon heads generally used for seal hunting (Fig. 3). Third, beluga whales comprise the most frequently occurring species in the large faunal sample from Gupuk House 1 (Friesen and Arnold n.d.), suggesting their procurement through some intensive hunting technique. Fourth, two beluga vertebrae recovered from Mackenzie Inuit sites on Richards Island retain the broken tips of ground slate harpoon end blades embedded in the bone (Fig. 4). One of these vertebrae was recovered from a midden in Area 2 of the Gupuk site; the second was recovered from the Pond site (NiTs-2), a prehistoric Mackenzie Inuit site located approximately 1 km from Gupuk. These specimens provide direct evidence that harpoons, as opposed to nets or other implements, were used in the hunt. Both specimens are mid-thoracic vertebrae, and the positions of the harpoon end blades indicate that the belugas were struck from above and behind, which is consistent with the drive technique described above.

These varied lines of evidence for drive hunting are largely circumstantial, and additional objective evidence is needed in order to differentiate drive hunting from the many other beluga hunting techniques which have been recorded throughout the North American Arctic. The most important distinction which must be made is between drive hunting and selective hunting of individual animals. This latter method is practiced by the Inuvialuit (modern-day Inuit of the Mackenzie Delta region), who selectively hunt older male animals (Bruemmer 1987), in part because of a preference for larger whales and because power boats and rifles have made selective hunting more practical during the recent period.

The Use of Mandibular Layering to Establish Beluga Mortality

The best test of hunting technique is through the construction of mortality profiles, which can, in certain cases, be reconstructed from archaeological faunal assemblages (Klein 1982; Stiner 1991). Although actual mortality profiles can be extremely variable (see e.g., Wilkinson 1976), three hypothesized ideal types can be differentiated and used for archaeological interpretation. Catastrophic mortality, where all individuals in a population are killed in a single event, is indicated by age sets appearing in the same frequencies as they do in living populations (Voorhies 1969). Living populations are characterized by highest numbers in the neonatal age set, and fewer individuals in each successive age set. Attritional mortality, in which the most vulnerable individuals in a population are hunted, is represented by larger frequencies of young and old individuals (Klein 1982). Finally, selective hunting, where a preferred age class or size is hunted, is indicated by

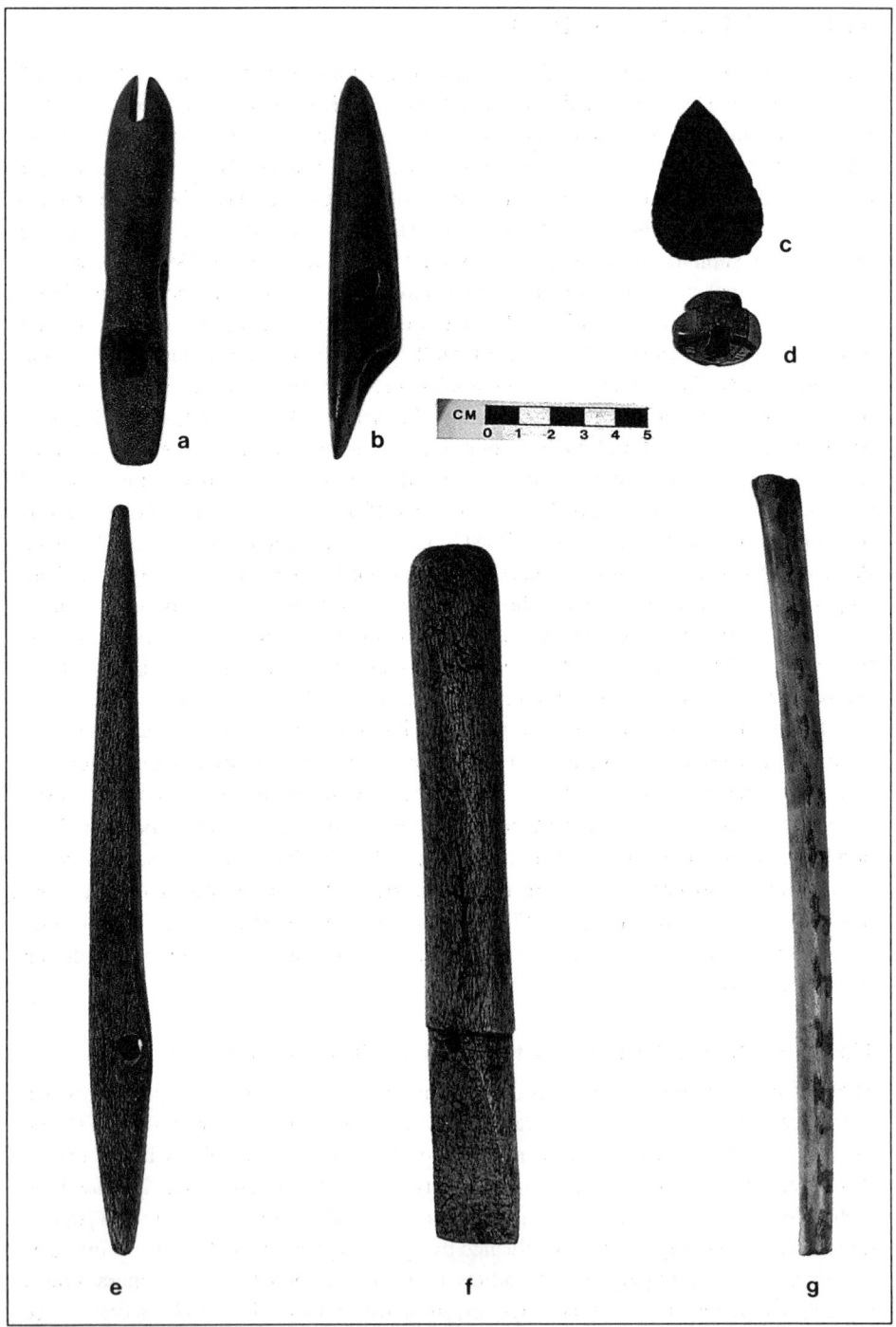

Figure 3. Selected artifacts from Gupuk relating to the beluga hunt: a-b, harpoon heads; c, slate harpoon end blade; d, float inflation nozzle; e, harpoon foreshaft; f, harpoon socket piece; g, bone tube probably used to inflate beluga carcass.

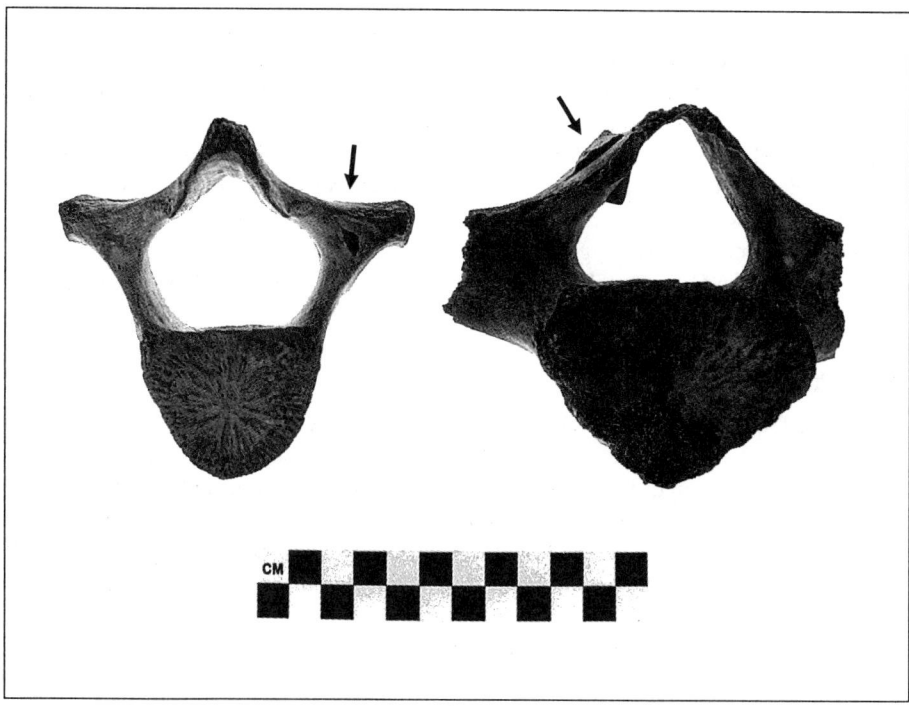

Figure 4. Posterior view of two mid-thoracic beluga vertebrae from Richards Island. Left: a vertebra from the Gupuk Area 2 midden retains the tip of a ground slate harpoon end blade in the posterior surface of the right transverse process. Right: a vertebra from the Pond Site retains the distal third of a ground slate harpoon end blade projecting through the neural arch into the vertebral foramen.

higher frequencies of a specific age, such as juveniles or adults (Jarman and Wilkinson 1972).

The establishment of an accurate age at death for mammalian species is most frequently accomplished through observation of annual growth layers in teeth (see e.g., Hillson 1986). However, beluga whales and other toothed cetaceans present a special difficulty to archaeologists. Although beluga teeth contain well-defined growth rings (Brodie 1982; Brodie et al. 1990; Goren et al. 1987), individual teeth lack morphological differentiation, and therefore cannot be identified to side or position. To add to this problem, beluga teeth are gripped in mandibles and maxillae only loosely, and tend to fall out shortly after death. Most beluga skulls and mandibles recovered from Gupuk, for example, did not retain a single tooth. Therefore, beluga teeth from archaeological faunal assemblages cannot be used to establish a mortality profile, because the analyst can never determine the number of individual whales from which the teeth were derived. In other words, 20 beluga teeth might be derived from anywhere between one and 20 individual animals.

The difficulty in using beluga teeth is offset by the fact that the periosteal bone of the mandibles of beluga whales (Brodie 1969), the closely related narwhal (Hay 1980), and several other cetacean species (Laws 1960; Nishiwaki et al. 1961) also contains annual growth layers. The growth of periosteal bone is uneven across the mandible, with the thickest and, therefore, most clearly observable layers located on the mid-labial and ventral aspects of the mandible, near the middle of the tooth row. Certain potential difficulties with interpretation of mandible thin sections exist. Hay (1980:130) observed that older narwhals, particularly females, exhibited resorption of periosteal bone, making growth layer counts less precise. In addition, the weathering or rootlet etching which frequently occurs in archaeological contexts can obscure growth layers. However, because beluga mandibles can be sided and identified accurately, they offer the best opportunity to establish a mortality profile.

Beluga Mortality at Gupuk

House 1 at Gupuk yielded a total of 23 mandibles, 15 right and eight left, which were complete enough to allow sectioning. Only the 15 right mandibles were used to construct the mortality profile, in order to ensure that each mandible represented a separate whale. Each mandible was sectioned near its anterior end, midway along the tooth row. The section was mounted on a glass slide and polished to a thickness of approximately 75 microns. Staining was not necessary for observation of the growth layer groups, which consist of alternate layers of opaque and translucent bone. The growth layers were counted using a binocular microscope at 40x and 100x magnification.

The mortality profile at Gupuk is predicted to resemble catastrophic death, since entire pods of beluga whales are hypothesized to have been driven into the shallows and killed. This prediction must be modified, however, to incorporate the fact that, during drives, older and larger belugas tend to break from the pod earlier than the young (Brodie 1989:134) and thereby escape. Therefore, a population hunted using a large-scale drive should contain fewer individuals in older age categories than are present in an idealized catastrophic profile.

Table 1 presents five beluga whale mortality profiles: (a) the Gupuk sample described herein, (b) a living (catastrophic) profile calculated by Burns and Seaman (1985:42) for the Bering Sea beluga population, (c) a sample of beluga whales selectively harvested by Inuvialuit in the Mackenzie Delta between 1974 and 1976, aged using tooth growth layers (Fraker et al. 1978:46), (d) a sample of beluga whales selectively harvested by Inuvialuit in the Mackenzie Delta between 1955 and 1961, aged using tooth growth layers (Sergeant 1973:1088), and (e) an aggregate selective profile, consisting of samples (c) and (d) combined.

Clearly, the sample size of 15 individuals from Gupuk, while considerable for an archaeological site, is too small to create a "smooth" mortality profile (Fig. 5a). However, at a general level it is comparable to the catastrophic profile (Fig. 5b), in that both exhibit highest frequencies in younger age categories and progressively fewer in older age categories. In contrast, the sample derived from recent Inuvialuit selective hunting indicates a distinct preference for older individuals, with a peak between 10 and 17 years of age (Fig. 5c). Modern Inuvialuit choose older male individuals for a number of reasons, including the larger meat and *muktuk* yields (Bruemmer 1987:47) and the greater ease of sighting the larger, lighter-colored adults which create more obvious wakes (cf. Burns and Seaman 1985:16). This differential harvest of older males contributes significantly to the long-term conservation of the beluga population (see e.g., Hazard 1988:218).

Prehistoric Beluga Whale Hunting at Gupuk   119

Table 1. Age at death information for five beluga populations (see text for descriptions and sources).

Age	(A) Gupuk (this report)		(B) Catastrophic Mortality (Burns and Seaman 1985)		(C) Selective Hunt (Fraker et al. 1978)		(D) Selective Hunt (Sergeant 1973)		(E) Selective Hunts Combined	
	n	%	n	%	n	%	n	%	n	%
0	1	6.7	50	9.5	-	-	1	0.9	1	0.7
1	1	6.7	35	6.6	-	-	1	0.9	1	0.7
2	2	13.3	31	5.9	-	-	-	-	-	-
3	-	-	29	5.5	-	-	-	-	-	-
4	2	13.3	26	4.9	1	2.8	-	-	1	0.7
5	2	13.3	25	4.7	1	2.8	2	1.7	3	2.0
6	1	6.7	23	4.4	1	2.8	2	1.7	3	2.0
7	1	6.7	22	4.2	-	-	4	3.5	4	2.6
8	1	6.7	21	4.0	-	-	6	5.2	6	3.9
9	-	-	19	3.6	1	2.8	5	4.3	6	3.9
10	1	6.7	18	3.4	3	8.3	14	12.1	17	11.2
11	1	6.7	17	3.2	1	2.8	11	9.5	12	7.9
12	1	6.7	16	3.0	4	11.1	15	12.9	19	12.5
13	-	-	16	3.0	2	5.6	5	4.3	7	4.6
14	-	-	15	2.8	2	5.6	10	8.6	12	7.9
15	-	-	14	2.7	2	5.6	5	4.3	7	4.6
16	-	-	13	2.5	-	-	7	6.0	7	4.6
17	1	6.7	12	2.3	4	11.1	9	7.8	13	8.5
18	-	-	12	2.3	1	2.8	5	4.3	6	3.9
19	-	-	11	2.1	2	5.6	3	2.6	5	3.3
20	-	-	10	1.9	2	5.6	3	2.6	5	3.3
21	-	-	10	1.9	2	5.6	-	-	2	1.3
22	-	-	9	1.7	3	8.3	5	4.3	8	5.3
23	-	-	9	1.7	1	2.8	1	0.9	2	1.3
24	-	-	8	1.5	-	-	1	0.9	1	0.7
25	-	-	7	1.3	1	2.8	1	0.9	2	1.3
26	-	-	7	1.3	-	-	-	-	-	-
27	-	-	6	1.1	-	-	-	-	-	-
28	-	-	6	1.1	-	-	-	-	-	-
29	-	-	5	0.9	-	-	-	-	-	-
30	-	-	5	0.9	1	2.8	-	-	1	0.7
31	-	-	4	0.8	-	-	-	-	-	-
32	-	-	4	0.8	-	-	-	-	-	-
33	-	-	3	0.6	1	2.8	-	-	1	0.7
34	-	-	3	0.6	-	-	-	-	-	-
35	-	-	2	0.4	-	-	-	-	-	-
36	-	-	2	0.4	-	-	-	-	-	-
37	-	-	2	0.4	-	-	-	-	-	-
38	-	-	1	0.2	-	-	-	-	-	-
Totals:	15	100.2	528	100.1	36	100.4	116	100.2	152	100.1

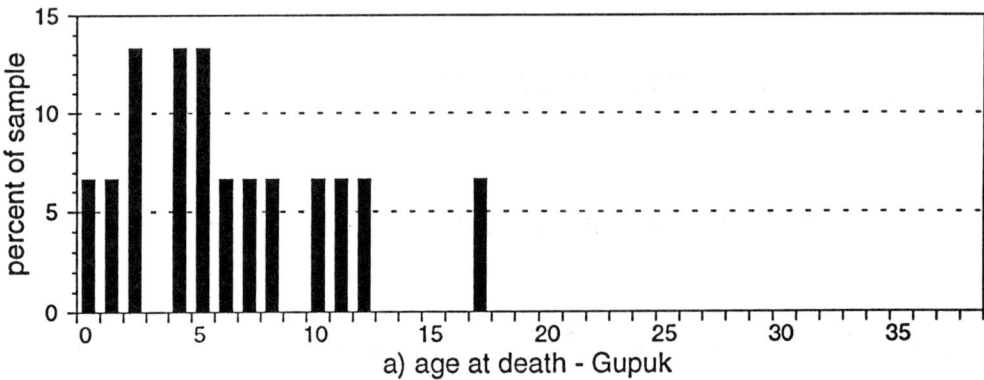
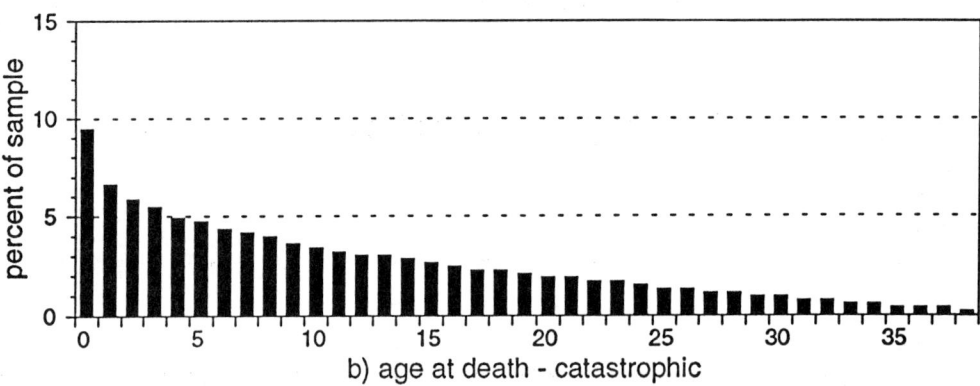
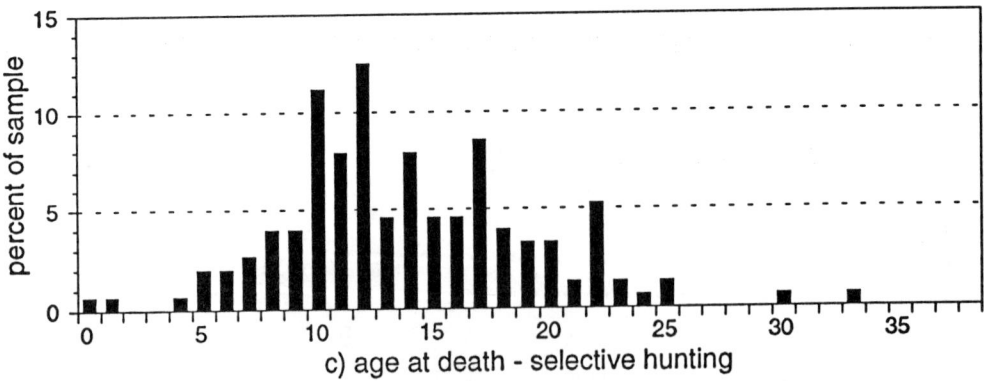

Figure 5. Selected beluga whale mortality profiles: (a) Gupuk, based on mandible thin-sections reported herein; (b) catastrophic mortality, calculated for the Bering Sea beluga population (Burns and Seaman 1985); (c) recent selectively hunted populations from the Mackenzie Delta (Fraker et al. 1978; Sergeant 1973).

DISCUSSION AND CONCLUSION

In order to clarify the mortality pattern represented by the Gupuk belugas, and to partially diminish the effects of a small sample size, the mortality profiles presented above can be collapsed into fewer age categories. In this case, the age data were arbitrarily collapsed into eight 5-yr increments (Fig. 6). The results, when viewed in these categories, indicate a significant degree of similarity between the Gupuk and catastrophic profiles. For both samples, the highest frequency is found in the first (0-4 year) age category, and each successive age category contains fewer individuals. The lower than expected frequencies in older age categories at Gupuk probably result from the tendency of older individuals to escape when being pursued, as noted above (Brodie 1989:134), although they could also result from the small sample size. As expected, the collapsed mortality profile resulting from the selectively hunted beluga samples contains a preponderance of prime-aged adult animals. This pattern, which includes few young animals and a distinct peak in the 10-14 year age set, is clearly differentiated from the Gupuk and catastrophic mortality profiles.

In conclusion, hunting methods employed at Gupuk were probably very similar to those recorded ethnographically at Kittigazuit, involving large-scale drives of entire pods of beluga whales into shallow water. Several lines of circumstantial evidence which

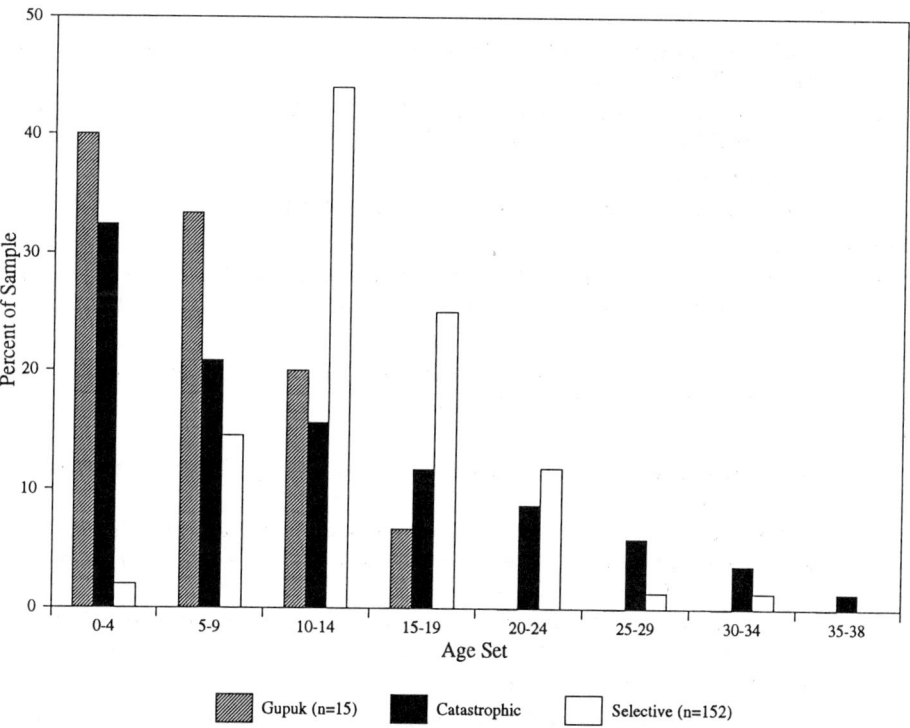

Figure 6. Comparison of mortality profiles, in five-year increments, for the Gupuk, catastrophic, and selectively hunted samples (see text for further description).

support this inference are reinforced by the beluga mortality profile which has been established on the basis of mandible thin sections.

This study is intended to provide a baseline of data for comparative studies of small whale hunting in the circumpolar North. In the case of Gupuk, the combination of a rich ethnohistoric record and a large sample of beluga bones allows a high degree of confidence regarding interpretations of hunting techniques. It is hoped that this will allow investigators of sites with fewer beluga bones, or without direct ethnohistoric information, to infer beluga hunting methods with greater certainty.

Acknowledgments. Major funding for the archaeological fieldwork at Gupuk was provided by the Government of the Northwest Territories, with additional funding provided through the Northern Oil and Gas Action Plan. Logistic support was provided by the Polar Continental Shelf Project (Department of Energy, Mines, and Resources). The authors wish to thank George Panagiotidis for his preparation of the mandible thin sections, Paul Brodie for information on beluga behavior and ageing techniques, and Darlene Balkwill of the Canadian Museum of Nature for her assistance with access to comparative osteological collections. We also thank Allen McCartney and two anonymous reviewers for their helpful comments on an earlier version of this paper.

REFERENCES

Arnold, C. D.
- 1988 Vanishing Villages of the Past: Rescue Archaeology in the Mackenzie Delta. *The Northern Review* 1:40-58.
- 1994 Archaeological Investigations on Richards Island. *Canadian Archaeological Association Occasional Paper* No. 2, edited by J.-L. Pilon, pp. 85-93.

Arnold, C. D. and E. J. Hart
- 1992 The Mackenzie Inuit Winter House. *Arctic* 45(2):199-200.

Birket-Smith, K.
- 1924 Ethnography of the Egedesminde District, With Aspects of the General Culture of West Greenland. *Meddelelser om Grønland* 66. Copenhagen.

Boas, F.
- 1888 The Central Eskimo. *Sixth Annual Report of the Bureau of American Ethnology of the Years 1884-1885*, pp. 399-669. Washington, D.C.

Brodie, P. F.
- 1969 Mandibular Layering in *Delphinapterus leucas* and Age Determination. *Nature* 221(5184):956-958.
- 1982 The Beluga (*Delphinapterus leucas*); Growth at Age Based on a Captive Specimen and a Discussion of Factors Affecting Natural Mortality Estimates. *Report of the International Whaling Commission* 32:445-447.
- 1989 The White Whale *Delphinapterus leucas* (Pallas, 1776). In: *Handbook of Marine Mammals,* Vol. 4, edited by S. Ridgway, pp. 119-144. Academic Press, London.

Brodie, P. F., J. R. Geraci, and D. J. St. Aubin
- 1990 Dynamics of Tooth Growth in Beluga Whales, *Delphinapterus leucas*, and Effectiveness of Tetracycline as a Marker for Age Determination. In: Advances in Research on the Beluga Whale, *Delphinapterus leucas*, edited by T. Smith, D. St. Aubin, and J. Geraci, *Canadian Bulletin of Fisheries and Aquatic Sciences* 224:141-148.

Bruemmer, F.
1987   Beluga Hunters. *Equinox* 6(5):44-53.

Burns, J. J. and G. A. Seaman
1985   *Investigations of Belukha Whales in Coastal Waters of Western and Northern Alaska: Biology and Ecology.* Alaska Department of Fish and Game, Fairbanks.

de Sainville, E.
1984   Journey to the Mouth of the Mackenzie River (1889-1894). *Fram: The Journal of Polar Studies* 1:541-550.

Fraker, M. A.
1980   Status and Harvest of the Mackenzie Stock of White Whales (*Delphinapterus leucas*). *Report of the International Whaling Commission* 30:451-458.

Fraker, M. A., C. D. Gordon, J. W. McDonald, J. K. Ford, and G. Cambers
1979   White Whale (*Delphinapterus leucas*) Distribution and Abundance and the Relationship to Physical and Chemical Characteristics of the Mackenzie Estuary. *Fisheries and Marine Service Technical Report* 863.

Fraker, M. A., D. E. Sergeant, and W. Hoek
1978   *Bowhead and White Whales in the Southern Beaufort Sea.* Beaufort Sea Project, Department of Fisheries and the Environment, Sidney, B.C.

Friesen, T. M. and C. D. Arnold
n.d.   Zooarchaeology of a Focal Resource: Dietary Importance of Beluga Whales to the Precontact Mackenzie Inuit. *Arctic* 48(1). In press.

Gordon, B. H. C.
1972   Activities of the Mackenzie Delta Archaeological Project. MS. 2044, Archaeological Survey of Canada, Canadian Museum of Civilization, Hull.

Goren, A. D., P. F. Brodie, S. Spotte, G. C. Ray, H. W. Kaufman, A. J. Gwinnett, J. J. Sciubba, and J. D. Buck
1987   Growth Layer Groups (GLGs) in the Teeth of an Adult Belukha Whale (*Delphinapterus leucas*) of Known Age: Evidence for Two Annual Layers. *Marine Mammal Science* 3:14-21.

Hay, K. A.
1980   Age Determination of the Narwhal, *Monodon monoceros* L. *Report of the International Whaling Commission, Special Issue* 3:119-132.

Hazard, K.
1988   Beluga Whale. In: *Selected Marine Mammals of Alaska: Species Accounts with Research and Management Recommendations*, edited by J. Lentfer, pp. 195-235. Marine Mammal Commission, Washington, D.C.

Hillson, S.
1986   *Teeth.* Cambridge University Press, Cambridge.

Jarman, M. R. and P. F. Wilkinson
1972   Criteria of Animal Domestication. In: *Papers in Economic Prehistory*, edited by E. Higgs, pp. 83-96. Cambridge University Press, Cambridge.

Klein, R. G.
1982   Age (Mortality) Profiles as a Means of Distinguishing Hunted Species From Scavenged Ones in Stone Age Archaeological Sites. *Paleobiology* 8(2):151-158.

Kleivan, I.
1984   West Greenland Before 1950. In: *Handbook of North American Indians*, Vol. 5, *Arctic*, edited by D. Damas, pp. 595-621. Smithsonian Institution, Washington, D.C.

Krech, S.
  1989   *A Victorian Earl in the Arctic: The Travels and Collections of the Fifth Earl of Lonsdale 1888-89*. University of Washington Press, Seattle.

Laws, R. M.
  1960   Laminated Structure of Bones from Some Marine Mammals. *Nature* 169: 972-973.

MacNeish, R. S.
  1956   Archaeological Reconnaissance of the Delta of the Mackenzie River and Yukon Coast. *Annual Report of the National Museum of Canada for the Fiscal Year 1954-1955*, pp. 46-81. Ottawa.

McGhee, R.
  1974   Beluga Hunters: An Archaeological Reconstruction of the History and Culture of the Mackenzie Delta Kittegaryumiut. *Memorial University of Newfoundland, Newfoundland Social and Economic Studies* 13.

Morrison, D. A.
  1990   Iglulualumiut Prehistory: The Lost Inuit of Franklin Bay. *Canadian Museum of Civilization, Mercury Series, Archaeological Survey of Canada Paper* 142.

Nelson, E. W.
  1899   The Eskimo About Bering Strait. *Eighteenth Annual Report of the Bureau of American Ethnology, 1896-1897*. Washington, D.C.

Nishiwaki, M., S. Ohsumi, and T. Kasuya
  1961   Age Characteristics in the Sperm Whale Mandible. *Norsk Hvalfangsttid* 50:499-507.

Nuligak
  1966   *I, Nuligak*. Edited by M. Metayer. Peter Martin Associates, Toronto.

Saladin d'Anglure, B.
  1984   Inuit of Quebec. In: *Handbook of North American Indians,* Vol. 5, *Arctic,* edited by D. Damas, pp. 476-507. Smithsonian Institution, Washington, D.C.

Savelle, J. M. and T. M. Friesen
  n.d.   An Odontocete (Cetacea) Meat Utility Index. MS.

Seaman, G. A., K. J. Frost, and L. F. Lowry
  1985   *Investigations of Belukha Whales in Coastal Waters of Western and Northern Alaska: Distribution, Abundance, and Movements*. Alaska Department of Fish and Game.

Sergeant, D. E.
  1973   Biology of White Whales (*Delphinapterus leucas*) in Western Hudson Bay. *Journal of the Fisheries Research Board of Canada* 30:1065-1090.

Spencer, R. F.
  1959   The North Alaskan Eskimo: A Study in Ecology and Society. *Bureau of American Ethnology Bulletin* 171.

Stefansson, V.
  1919   The Stefansson-Anderson Arctic Expedition of the American Museum: Preliminary Ethnological Report. *Anthropological Papers of the American Museum of Natural History* 14(1).

Stiner, M. C.
  1991   Introduction: Actualistic and Archaeological Studies of Prey Mortality. In: *Human Predators and Prey Mortality*, edited by M. Stiner, pp. 1-14. Westview Press, Boulder.

Taylor, J. G.
    1984   Historical Ethnography of the Labrador Coast. In: *Handbook of North American Indians*, Vol. 5, *Arctic*, edited by D. Damas, pp. 508-521. Smithsonian Institution, Washington, D.C.

Voorhies, M.
    1969   Taphonomy and Population Dynamics of an Early Pliocene Vertebrate Fauna, Knox County, Nebraska. *University of Wyoming Contributions to Geology, Special Paper* No. 1.

Whittaker, C. E.
    1937   *Arctic Eskimo*. Seeley, Service & Co., London.

Wilkinson, P. F.
    1976   "Random" Hunting and the Composition of Faunal Samples from Archaeological Excavations: A Modern Example from New Zealand. *Journal of Archaeological Science* 3:321-328.

Wrangell, F.
    1980   *Russian America: Statistical and Ethnographic Information*. Translated by M. Sadouski. Limestone Press, Kingston, ON. (1839)

An Ethnoarchaeological Investigation of Inuit Beluga Whale and Narwhal Harvesting

James M. Savelle
Department of Anthropology
McGill University
855 Sherbrooke St. West
Montreal, Quebec
Canada H3A 2T7

Abstract. *Ethnoarchaeological observations of beluga whale and narwhal harvesting at Creswell Bay, Somerset Island, Arctic Canada, are described. The whale hunting takes place within the context of a hunter-gatherer collecting system, and generates characteristic collecting system site-types: residential bases, field camps, stations, caches, and high-bulk locations. Differences in mortality profiles of retrieved animals of the two species is consistent with differences in hunting strategies. The nature and degree of processing and transport of whale products is found to be based on (a) animal part economic utility, (b) ease in removal of mattak, blubber, and meat from bone elements, and (c) variability in logistical context.*

INTRODUCTION

Research into prehistoric whaling in the Western Arctic has focused almost exclusively on the nature and extent of the use of large baleen whales, primarily the bowhead (*Balaena mysticetus*) and secondarily other species such as the humpback (*Megaptera novaeangliae*) and gray (*Eschrichitius robustus*; see, for example, the reviews by McCartney 1980, 1984 and Stoker and Krupnik 1993). On the other hand, investigation of the prehistoric use of smaller whales in this area, with the exception of McGhee (1974) and the recent studies by Friesen and Arnold (this volume), has received relatively little attention (although see Savelle 1994 for a discussion of prehistoric small whale use in the Eastern Arctic).

This relative slighting of smaller whales is unfortunate, as such animals, in particular the beluga whale (*Delphinapterus leucas*), may have significantly influenced subsistence-settlement systems and associated social characteristics of many prehistoric western arctic societies. Beluga whales are an important subsistence resource in many modern western arctic native settlements, with 500-600 animals taken annually (International Whaling Commission 1982:14). Nelson (1969:205-219) provides a detailed description of modern beluga whale hunting and use in northern Alaska, while Spencer (1959:33-34), McGhee

(1974), Ray (1975:113), Burch (1981:26), and various papers in Damas (1984), among others, provide details about aboriginal beluga whale use throughout the Western Arctic. Furthermore, beluga whale remains have been recovered from a number of sites associated with various prehistoric traditions, cultures, and/or phases throughout much of the Western Arctic. These include Choris (Giddings and Anderson 1986:228), Norton (Giddings 1964:186), Ipiutak (Giddings and Anderson 1986:154), Nukleet (Giddings 1964:96), Birnirk (Stanford 1976:69-71; the "small whale" reference is probably to beluga whale), Cape Nome (Bockstoce 1979:84), Western Thule (McGhee 1974; Stanford 1976:69-71), and Kachemak (Yesner 1992:173; the "small whale" reference is probably to beluga whale).

Allied to the relative paucity of research on prehistoric small whale hunting, we lack detailed information on its archaeological correlates, other than in a general sense from ethnographical research. This paper represents an initial attempt to address this deficiency by presenting the results of a recent ethnoarchaeological investigation of Inuit beluga whale and narwhal (*Monodon monoceros*) harvesting. Although the study was conducted in the Central Arctic, the results should nevertheless be equally applicable to many Western Arctic contexts.

GEOGRAPHICAL AND CULTURAL SETTING

The investigations were undertaken in 1989 and 1993 to determine the feasibility of a long-term study of the ecology of Inuit beluga and narwhal harvesting. The research was centered at the modern outpost camp of Kuvinaluk at the mouth of the Union River, Creswell Bay, Somerset Island, N.W.T. (Fig. 1), and followed previous ethnoarchaeological investigations there in 1980 (Savelle 1984). The Creswell Bay area has been continuously inhabited since 1925, when a group of approximately 45 Inuit moved there from northwestern Baffin Island (Kemp et al. 1977). From 1950 to the mid-1970s, only one family maintained a residence there, but since the mid-1970s the population has varied on a seasonal and annual basis, with usually two or three families in residence. In addition, the camp is frequently visited by Inuit from both Resolute, a settlement approximately 200 km to the north, and Spence Bay, a settlement approximately 350 km to the south. Whales have traditionally figured prominently in the Kuvinaluk subsistence economy and more recently in the monetary economy (Kemp et al. 1977; see also Fig. 2).

Each of the 1989 and 1993 studies were undertaken over a six-day period. The 1989 study involved my participation in one beluga whale hunt and one narwhal hunt, associated carcass processing, caching, and transport. In 1993, the study was restricted to the investigation of narwhal processing and caching. Accordingly, these data are limited, but nevertheless may potentially prove instructive in the investigation of prehistoric Inuit whaling practices.

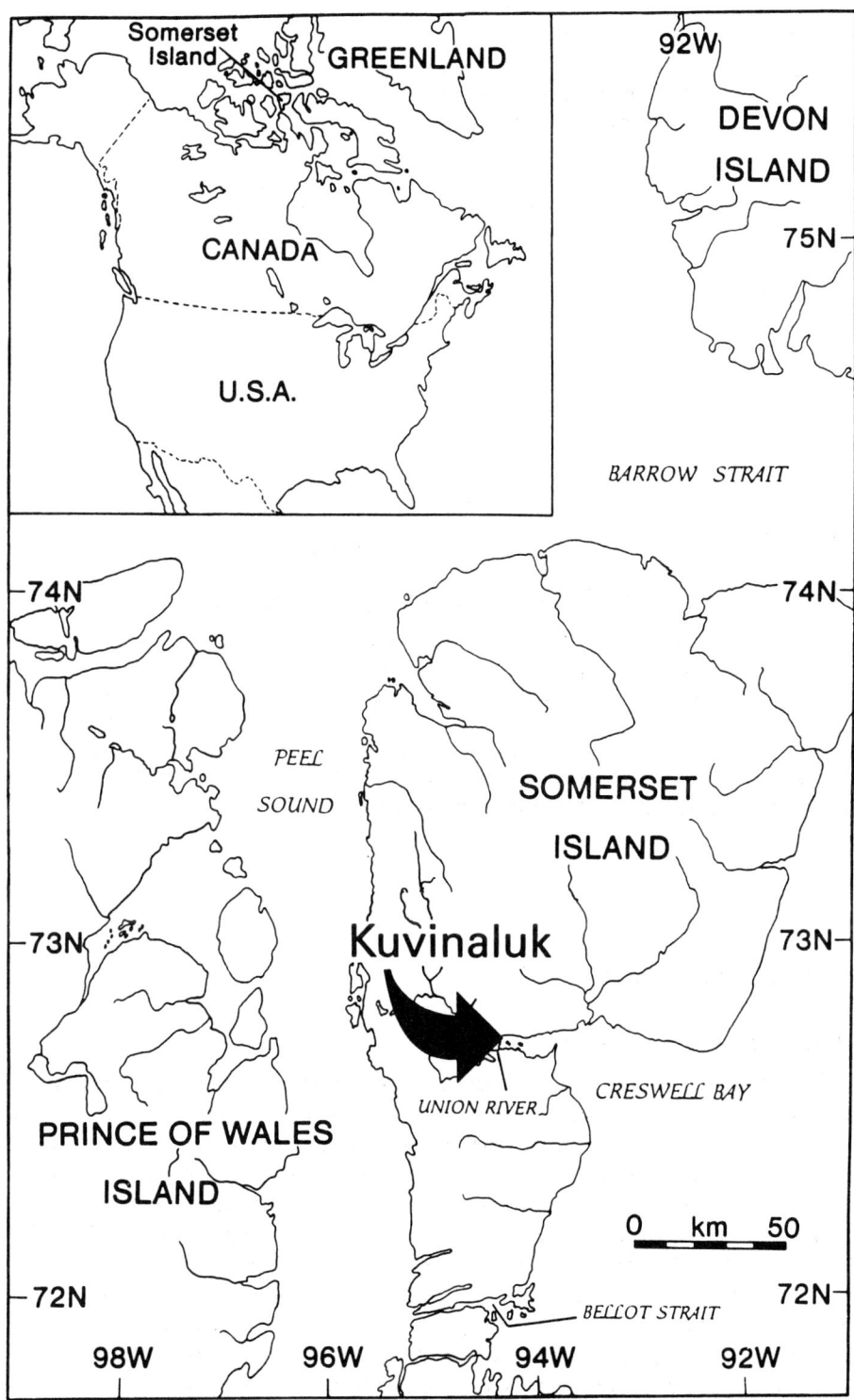

Figure 1. Location of Kuvinaluk, Somerset Island, Arctic Canada.

130 James M. Savelle

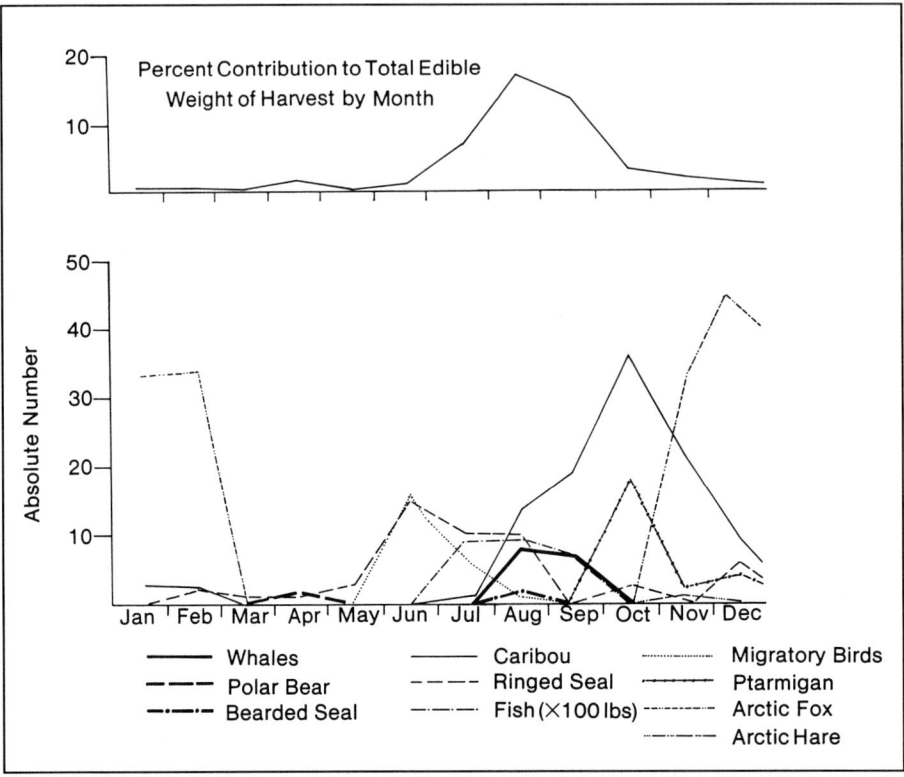

Figure 2. Monthly harvest totals for Kuvinaluk during 1976 (from Savelle 1984:Fig. 2; copyright by the Society for American Archaeology).

THE BELUGA WHALE HUNT

Beluga Whales at Creswell Bay

Beluga whales are small by whale standards; males are slightly larger than females, and at physical maturity can attain lengths of up to 5.0 m and weigh up to 2000 kg. Average adults, however, are approximately 3.0-4.5 m in length and 400-1400 kg in weight (Reeves and Mitchell 1987). Beluga whales migrate in groups to the High Arctic during the summer, following the retreating ice edge. They typically enter the Prince Regent Inlet-Creswell Bay region in late July and early August (Sergeant and Brodie 1975; Reeves and Mitchell 1987). After entering the latter bay, they congregate primarily in the warmer, shallow waters of the Creswell River estuary (Fig. 3). There, the suckling of calves and yearlings and bottom rubbing associated with seasonal epidermal moulting are presumably the predominant behaviors (Smith et al. 1992, 1994). Concentrations typically number up to 1000 animals (Sergeant and Brodie 1975), although an estimated 3900 animals were observed in Creswell Bay on 14 August 1975 (Finley 1982).

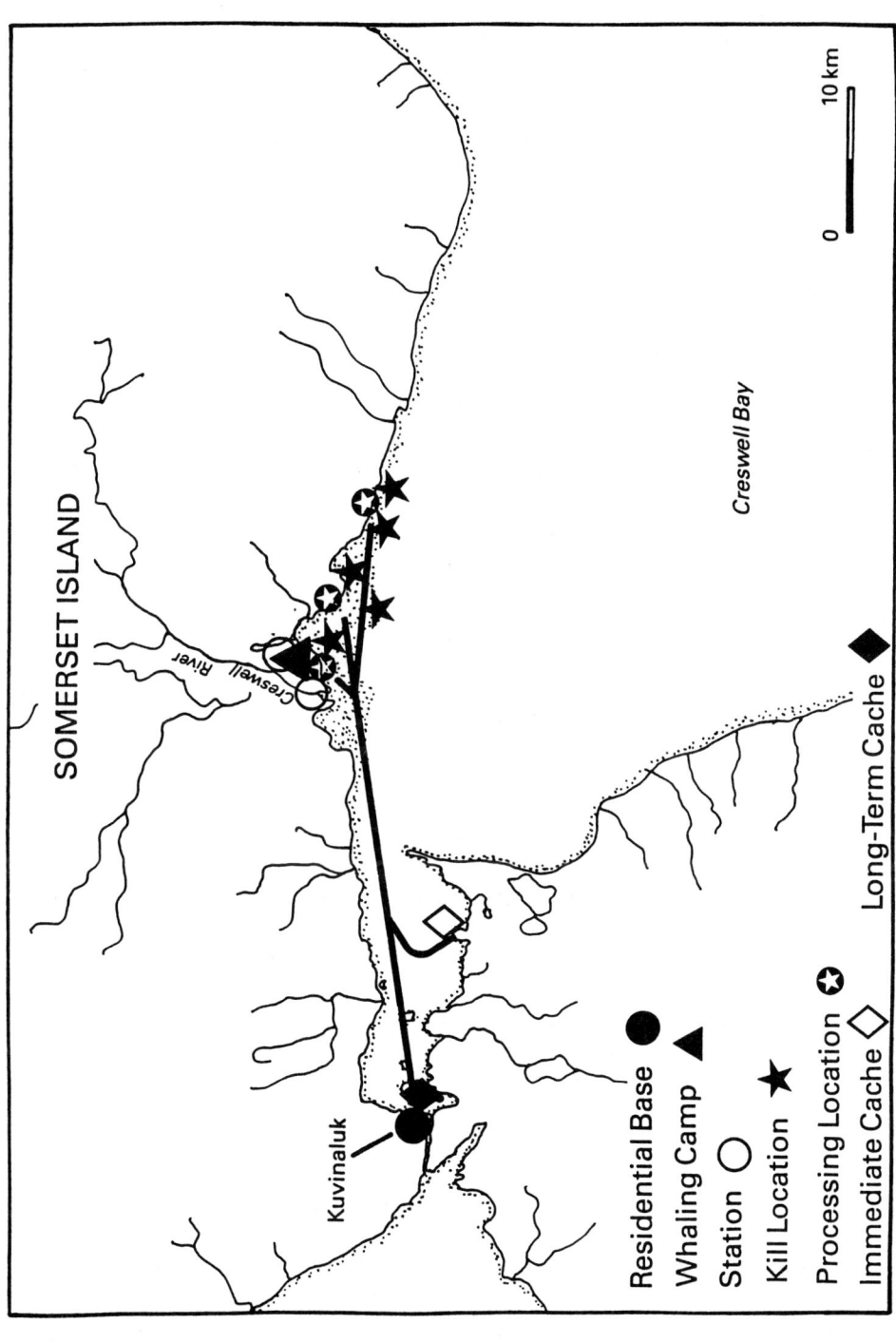

Figure 3. Locations of sites and activities associated with the beluga whale hunt. Solid lines indicate routes of movement of whale products from processing locations to cache sites.

Hunting Procedures

Given the characteristics and behavior of beluga whales noted above, the ideal location to intercept the whales is at or near the Creswell River estuary, and it is here that the 1989 whaling camp was set up (Fig. 3). The camp (field camp *sensu* Binford 1980), situated approximately 35 km east of Kuvinaluk, was occupied by seven adult male hunters — two from Kuvinaluk and five from Spence Bay — living in three tents. A low ridge and knoll located approximately 500 m southwest of the camp served as a primary lookout station, while a low ridge located approximately 200 m north of the camp served as a secondary lookout station.

Since a chartered Twin Otter aircraft was waiting at Kuvinaluk to transport most of the products of the hunt to Spence Bay, time was a critical factor. The primary lookout station was generally occupied by at least one person on a continuous basis (a 24-hour watch is possible, of course, because the region experiences 24 hours daylight throughout the summer), and the secondary station was manned on an intermittent basis.

The first whales sighted, a group estimated at approximately 50, traveling in several separate smaller pods, were observed in the shallows that are located approximately 5 km southeast of the camp. Based on coloration and relative size, the group included at least adults and juveniles; calves may have been present, but none were definitely identified. At the first sighting, the hunters manned two outboard motorboats, three in one and four (plus myself) in the other. Both boats flanked the whales by traveling to the south, then to the east, and then turned and approached from the seaward side. While some of the whales escaped to seaward, many immediately fled shoreward, toward shallow water, and were herded toward shore by the two boats working in unison. This "flee" response toward shore is apparently typical of beluga whales (Finley et al. 1990).

At this point, the boat operators isolated individual or smaller groups of whales, and continued to drive them into increasingly shallower water. The whales were not fired upon until they were in water of less than about 2 m depth or when they attempted to make a break toward deeper water. In either instance, shots were fired at the whales' heads as they broke the surface. While larger animals were presumably preferred, the hunters shot at any whale that was close enough for a reasonable chance of being hit. Firing continued at an animal until it was determined to have been killed. The boat crews then immediately gave chase to another whale or group of whales.

After all remaining whales had left the area and headed into the deeper waters of the bay, efforts were made to retrieve the whales that had been killed. This was done using gaffs, as all whales sank to the bottom immediately after being killed. Once the carcass had been gaffed, it was towed behind the boat to a previously designated coastal location for butchering. The kill and butchering locations for the five whales retrieved during the hunt are shown in Figure 3.

Processing and Use

Butchering was done immediately after the hunt and as close to shore as the whale could be towed. This was within a few meters for the larger whales (Fig. 4) and right at the water's edge in the case of smaller whales. As the tide was receding at this time, making it more difficult to navigate inshore back to the field camp, the processing was done as quickly as possible. Detailed observations were recorded for three of the five beluga whales processed. In each instance, a slit was first made through the skin (*mattak*) and

Figure 4. Processing a large (4.67 m) adult male beluga whale close to shore. Almost all body *mattak* and associated blubber and the flippers have been removed at this point.

blubber along the ventral surface from the head to a point immediately anterior of the flukes. The carcass was then rolled on to one side, and a series of slits made perpendicular to the initial slit. The *mattak* on that side, with attached blubber, was then removed in large rectangular sections. The flipper on that side was also removed at this time. The carcass was then rolled over onto the other side, and the procedure repeated. Finally, each fluke was removed. Normally, the remainder of each carcass was left at the butchering site (Fig. 4). However, four of the heads were collected and later processed for a biological study by Thomas G. Smith (Department of Fisheries and Oceans, Canada).

Based on ratios on whale length:body weight and whale length:*mattak* weight given in Doidge (1990) and Heide-Jorgensen (1994), and adding flippers and flukes, the total weight of *mattak* and associated blubber obtained as a result of the hunt is estimated to have been approximately 450-550 kg. This is approximately the weight that the Twin Otter pilot estimated that he could transport to Spence Bay after factoring in the hunters, gear, and typical off-strip takeoff conditions.

Following the processing, the field camp was dismantled and the boats, with the *mattak*, returned to the main residential base camp at Kuvinaluk. On the way to the base camp, however, most of the *mattak* was deposited in a temporary cache located approximately 10 km east of Kuvinaluk, on the assumption that this locality would provide a greater off-strip takeoff distance than that available at Kuvinaluk. The remainder (less than 50 kg) was eventually brought to Kuvinaluk, most of which was stored in a permanent cache located approximately 1.5 km from the base camp (Fig. 3). Some was also held for immediate consumption at the base camp. *Mattak* at the temporary cache was eventually placed aboard the Twin Otter and, along with the five Spence Bay hunters, flown to Spence Bay.

THE NARWHAL HUNT

Narwhals at Creswell Bay

Narwhals are similar in size to beluga whales, with adult males measuring 4.7 m in length and weighing 1600 kg and adult females reaching 4.15 m and 1000 kg at physical maturity (Reeves and Mitchell 1987). Each animal has two maxillary teeth, both of which are usually unerupted in females. In males, however, the left tooth erupts into a prominent, spiralled tusk that may attain lengths of up to 3 m.

Narwhals, like beluga whales, follow the retreating ice margin north during the summer, and enter Creswell Bay in late July and August. Unlike beluga whales, however, they do not congregate in the shallow water at the mouth of the Creswell River. Instead, they prefer deeper waters and fjords (Mansfield et al. 1975; Reeves and Mitchell 1987; Kingsley et al. 1994), and, accordingly, tend to follow the deeper channels, penetrating as far into the bay as the Kuvinaluk settlement (Fig. 5).

Hunting Procedures

Given the proclivity of narwhals to enter the inner part of Creswell Bay, narwhal hunting is usually conducted from Kuvinaluk itself, with a series of knolls located between 100-500 m immediately north of the settlement being used as a lookout station. Once the narwhals, approximately 20 in number, were spotted heading toward the settlement, two boats were launched, one with two male hunters and the other with one male hunter and myself. Based on animal size, presence or absence of tusks, and length of tusks when present, both immature and mature males and, at least, mature females appear to have been present in the whale group.

The overall procedure for hunting narwhal was similar to that for beluga whales, that is, they were driven into shallow water through the coordinated efforts of the boat operators before they were killed. However, it is far more difficult to harvest narwhal in this fashion, since they typically display much different avoidance behavior.

First, unlike beluga whales which are very easy to track at or near the surface as they flee, the immediate response of narwhals to a perceived threat is to lie motionless beneath the surface of the water (Finley et al. 1990). As a result, the boats are required to establish informal search "grids" until the whales are relocated. If the whales are sighted between the boat and shore, they can potentially be driven toward the shallows. If, on the other hand, the whales are relocated seaward of the boat, the chances of driving them back inshore are slight.

Second, even if the whales are sighted between the boat and shore, they are still difficult to drive into the shallows. This is because they flee toward deeper water when they are directly pursued, unlike belugas. During this part of the hunt, the fleeing whales turned ventral surface up and swam inverted as fast as possible, immediately above the sea bed.

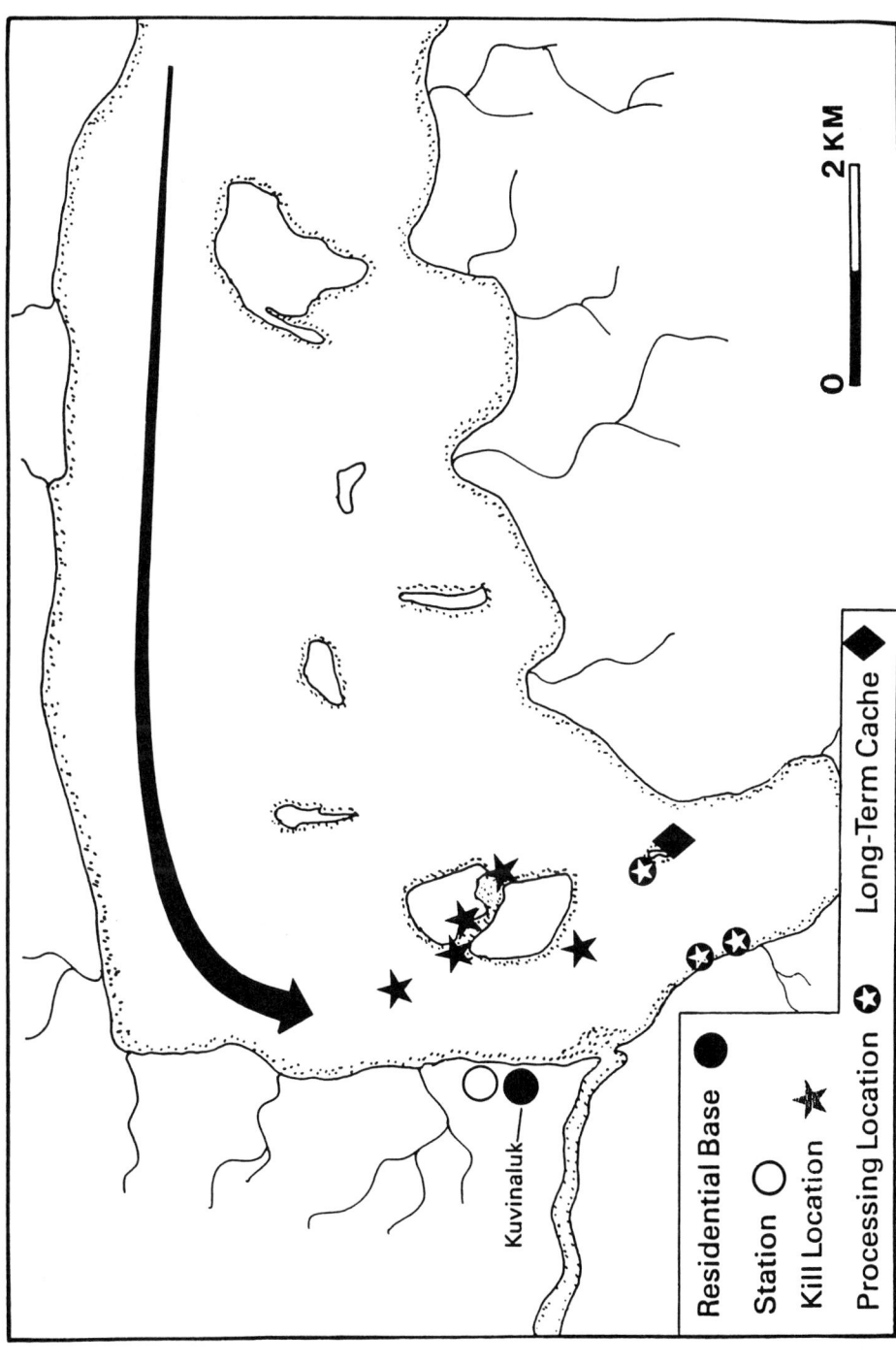

Figure 5. Locations of sites and activities associated with the narwhal hunt. The solid arrow represents the route of narwhals as they enter the inner part of the bay.

Figure 6. Adult male narwhal prior to processing, showing prominent tusk.

While narwhal *mattak* and meat are desired products, the most valuable is the tusk, from the point of view of the present combined hunting and cash economy. Therefore, adult males with longer and preferably undamaged tusks are preferred (Fig. 6). The procedure, then, is to first select a male with a large tusk that is in good condition. Once the animal has been selected, the group the animal is in is isolated and driven toward shallow water, and the selected animal is then repeatedly shot in the body until it is very weak and floating on the surface. At this point, it is slowly maneuvered into very shallow water (<2 m) where it is killed outright with a fatal head shot. Most animals were driven into a shallow channel between two islands immediately east of Kuvinaluk (Fig. 5). While this is the ideal method, occasionally the whales are killed in deeper water. This happened once during the hunt, and the whale was immediately harpooned and held near the surface by a plastic 5 gal jerry can float attached to the harpoon line.

As in the beluga whale hunt, once a whale had been killed, it was left at the kill location and the boat operator immediately went on to pursue other whales. The hunt ended when all remaining whales had escaped into deeper water. A total of four whales were subsequently retrieved with gaffs and a fifth from the harpoon and float.

Processing and Use

The following processing description incorporates data from observations in both 1989 (hunt and processing) and 1993 (processing only). Narwhal butchering procedures were initially similar to those for the beluga whales. That is, the whales were first towed to a designated location on the coastline (Fig. 5). Most were towed as near shore as possible during high tide, left where they grounded, and processed during several following low tides.

The heads were first removed and chopped through with a hatchet to allow for the extraction of the tusk. The *mattak* and associated blubber were then slit along the ventral surface from the cervical vertebrae to the flukes, in a fashion similar to that used for beluga whales. The animal was then rolled on to one side, and the flipper on that side was removed. Slits were then made perpendicular to the initial incision, and rectangular slabs of *mattak* and attached blubber were peeled off in strips (Fig. 7). The carcass was then rolled over and the procedure repeated. Each fluke was then removed.

Unlike the processing pattern for beluga whales, meat was removed from the narwhal carcasses. The meat, to be used primarily as dog food, was removed in pieces from a series of long strips running along the left and right dorsal and left and right ventral sides of the vertebral transverse processes (Fig. 8). These muscles are described in detail by Pabst (1990).

In some instances (two of the 10 butchered narwhals observed in 1993), the peduncle (posterior-most caudal section to which the flukes are attached) was also detached from the main axial column (Fig. 9). At least one of these, however, was discarded close to the carcass. This portion contained a small amount of meat and *mattak*, and was typically removed only when, according to an informant, the base camp was "desperate" for *mattak*.

Based on data on narwhal biology in Reeves and Tracy (1980) and narwhal weight:length ratios presented in Hay (1984), the five animals yielded a total of approximately 500 kg of *mattak*, 1200-1500 kg of associated blubber, and 1000 kg of meat, in addition to five tusks.

Figure 7. Processing adult male narwhals at low tide, showing removal of body *mattak* in rectangular slabs.

138 James M. Savelle

Traditional processing practices for both narwhals and belugas at Kuvinaluk, no longer followed, included removal of the oil and blubber from the melon on the head for use in lamps and the removal of sinew from along the dorsal side of the transverse vertebral processes.

All products which were removed, with the exception of the tusks, were stored in the series of long-term caches located approximately 1.5 km from Kuvinaluk (Figs. 5 & 10). The tusks and a small amount of *mattak* were taken directly to Kuvinaluk. It should be noted that in 1989, the skulls were also taken to Kuvinaluk. However, these were collected specifically for biological study by Thomas G. Smith, as noted above.

DISCUSSION

This study underscores a number of behaviors relating to logistical organization, processing and transport, and the generation of prey animal mortality profiles that tend to be characteristic of hunter-gatherers in general. At the same time, it also identifies several factors or characteristics that would appear to be unique to large marine mammal harvesting. Each of the above will be discussed in turn.

Logistical Organization

The logistical organization of beluga and narwhal hunting and processing can be considered a classic example of a collecting system, as defined by Binford (1980). Intercept strategies were employed by specialized task groups (hunting crews), and the five collecting system site types — residential bases, caches, "high-bulk" processing locations, stations, and field camps — were all generated, although the field camp was only

Figure 8. Narwhal carcass after processing, with almost all *mattak* and easily stripped meat removed.

Figure 9. Narwhal carcass after processing, with almost all *mattak* (including that associated with the peduncle) and easily stripped meat removed.

used in the case of the beluga whale hunt. In addition, the beluga whale hunt incorporated various logistically-related sites (field camp, caches, processing locations) and activities within a well-defined and geographically separated logistical zone.

Whale Processing and Transport

Animal processing and transporting activities very closely followed "ideal" economic patterns (see e.g., Binford 1978; Lyman 1992; Metcalfe and Jones 1988, among others), when considering both anatomical part utility (that is, the relative utility of individual anatomical parts based on associated edible tissues) and situational variables (e.g., distance from residential base, time constraints, transport capacity).

Unfortunately, there are no utility indices (i.e., a ranking of individual anatomical parts based on known relative weights of edible tissue for each part) that have been derived specifically for beluga whales or narwhals. However, utility indices for the harbor porpoise (*Phocoena phocoena*), a much smaller cetacean but one that is nevertheless similar in morphology to both larger species, have recently been constructed (Savelle and Friesen n.d.). The rank order of individual parts according to flesh weight (including muscles, tendons, and other fibers) and the sculp weight (hide and associated blubber) is summarized in Table 1.

The removal of the body sculp, flippers, and flukes in both beluga whales and narwhals is consistent with their respective ranks and with the fact that *mattak* is the most highly prized edible portion of these species. The one anatomical part from which significant amounts of *mattak* which was not removed was the head. The *mattak*

Table 1. Rank of harbor porpoise (*Phocoena phocoena*) body parts based on flesh weights (organs excluded).

Body Part	Rank	Primary Associated Material
Sculp	1	*Mattak*
Lumbar Vert.	2	Meat
Caudal Vert.	3	Meat
Thoracic Vert.	4	Meat
Ribs	5	Meat
Head	6	*Mattak*/Meat
Flukes (2)	7	*Mattak*
Flipper	8	*Mattak*
Scapula	9	Meat
Cervical Vert.	10	Meat
Sternum	11	Meat
Pelvis (remnant)	12	Meat

Figure 10. Narwhal *mattak* and associated blubber in cache.

associated with the head was ignored by the modern Inuit in all cases observed, because it is extremely difficult to remove, unlike the *mattak* from other regions where it peels off very easily (see Fig. 7). Finally, meat removal from narwhal carcasses is also consistent with the ranks of the various parts.

While the use of the various anatomical parts is in accordance with their respective utility ranks, the bone elements remaining at the processing site do not reflect this. That is, with the exception of narwhal tusks, the only bone elements actually removed from the processing sites were those associated with the flippers (all bones distal from, and in most cases including, the humerus) and, rarely (in the case of narwhal *mattak*), the vertebrae associated with the peduncle (approximately Ca 16/17-26/27). In particular, none of the vertebrae from which the large meat strips were removed in the narwhal were transported away from the processing sites. This is consistent with the study by O'Connell et al. (1988), in which the ease that meat could be stripped from an individual bone was shown to be a primary determinant of that bone being removed and transported. Accordingly, the whale bone elements remaining at the processing site are not a direct reflection of the anatomical parts that were utilized, when considered in the context of utility indices.

Finally, and as discussed in considerable detail by Binford (1978), "situational" variables can be seen to determine processing and transporting of parts to a considerable degree. For example, one variable generally assumed to influence these factors is distance between the residential base camp and the processing/caching locality. The greater the distance, the greater the selection for high utility parts. With the beluga whales being processed 35-45 km from the base camp, only *mattak* and associated blubber (approximately 30% of the total edible weight of each animal) was removed and transported. In the case of narwhals, on the other hand, they were processed and cached only 1.5-3.0 km from the residential base camp, and approximately 70-75% of the edible parts of each animal were removed.

Another situational variable was the time factor. During the beluga whale hunt, time was a critical factor, due to the Twin Otter waiting at Kuvinaluk to take the whale products and hunters back to Spence Bay (minimum rates are charged per day, whether or not the aircraft is actually used). Consequently, all processing was conducted immediately after the animals had been retrieved and during less than ideal tidal conditions. There was no possibility of caching any parts in anticipation of returning after a period of hours or days. With the narwhal hunt, on the other hand, because of the proximity to the residential base camp, processing and caching could be conducted over a period of several days according to tidal conditions, with repeated trips between the processing and caching sites and the residential base camp.

Mortality Profiles and Hunting Strategies

Stiner (1990, 1991, and papers therein) has recently summarized studies investigating the relationship between mortality profiles and hunting strategies. While such studies typically rely on age profiles to interpret hunting strategies, size, especially in the case of cetaceans, can also be used (e.g., Savelle and McCartney 1991, 1994).

Length measurements were taken of all beluga whales and narwhals retrieved in 1989, and are summarized in Table 2. Although the samples are admittedly small, they nevertheless show a direct relationship to the hunting strategies employed. Beluga whale hunting, in which animal selection was essentially random, produced a wide range of

Table 2. Length and sex of measured retrieved beluga whales and narwhals, 1989.

Species	Sex	Length (m)
Beluga Whale	Male	4.67
Beluga Whale	Male	4.52
Beluga Whale	Male	3.86
Beluga Whale	Male	3.27
Beluga Whale	Male	3.02
		Average = 3.87
Narwhal	Male	4.78
Narwhal	Male	4.62
Narwhal	Male	4.57
Narwhal	Male	4.44
Narwhal	Male	4.16
		Average = 4.51

sizes, including both juvenile and adults. Narwhal hunting, in which selection was for older males with larger tusks, produced a restricted distribution toward the upper size limit.

Unique Characteristics of Marine Mammal Hunting

This study has identified a number of unique characteristics of marine mammal hunting that may be considered significant. Most of these will be intuitive and/or have been discussed in the literature previously, but it is appropriate to reiterate them here.

First, coastal processing locations are not necessarily immediately adjacent to the kill location. That is, due to the ease in transporting dead animals in an aquatic environment, such locations may in fact be several kilometers distant from the nearest accessible point of land (see Figs. 3 & 5).

Second, processing locations will often contain far fewer whale remains than were originally deposited, and many locations may not be identifiable at all. Carcasses that are left at these locations within the intertidal zone or below low tide, especially in the case of smaller whales, will likely be removed through water action or ultimately ice action. For example, all three of the beluga whale processing locations used in 1989 were well within the intertidal zone, and none contained any whale remains two years later, when they were examined during a helicopter survey. On the other hand, many of the narwhal remains, which had been pulled up to the highest possible position (high intertidal/supratidal) were clearly in evidence several years later (see Fig. 11).

Third, because of whale anatomy, the vast majority of edible products can be recovered with little bone removal. In the present study, 30% of the edible beluga whale and 70-75% of the edible narwhal products were removed. However, for both species the only bones associated with these removed edible products were those associated with the flippers (Fig. 8) and, secondarily, the posterior caudal vertebrae (Fig. 9). Consequently, the amount and type of bone present at processing, caching, or residential sites may bear very little relation to the actual amount of the whale utilized for subsistence purposes.

A final characteristic relates to carcass retrieval rates. Almost all forms of hunting result in some wounded animals that eventually die elsewhere. However, in marine situations, the retrieval rate of animals *killed outright* can be expected to be lower, since many sink in waters too deep to be retrieved with a hook or other means (the loss rate will also be determined by the technology employed). By the time the carcasses eventually surface, tides and currents will, in many cases, have carried them considerable distances from the kill sites, and they are unlikely to be retrieved. In addition, most surfaced carcasses will have decomposed to the extent that the *mattak* and meat is unfit for consumption.

Figure 11. Narwhal remains at processing site after several years exposure to taphonomic processes.

IMPLICATIONS FOR TRADITIONAL AND PRECONTACT WHALE HUNTING

While it should be stressed again that this study was conducted within the context of a semitraditional economy and the use of modern technologies, the resulting data are nevertheless applicable in several respects to traditional and precontact whaling in the Western Arctic and Subarctic.

First, the hunting of even the smallest whales can be expected to be conducted within the context of a fully logistically-organized collecting system, with the characteristic site types — residential bases, field camps, stations, high-bulk locations, and caches — being generated. Such systems are more typically associated with bowhead hunting, such as that practiced by many prehistoric Western Thule and historic Inupiat societies. However, they may also have been characteristic, at least seasonally, of many other groups without regular access to bowheads but with regular access to migrating and/or summer concentrations of belugas. For example, this is almost certainly the case for protohistoric and historic Mackenzie Inuit (see e.g., McGhee 1974; Friesen and Arnold, this volume) and possibly late prehistoric and historic Inuit groups in Norton Sound (Ray 1975, 1984), Kuskokwim Bay (VanStone 1984), and other Alaskan coast localities, as well as prehistoric and historic Athapaskan groups in Cook Inlet (Townsend 1981).

Second, taphonomic processes associated with marine environments may be much more detrimental to the preservation of high-bulk processing sites and, thus, unlike those of terrestrially-based collecting systems. Otherwise, other logistical site types (e.g., whaling camps, caches) should often be recognizable. Their identification, however, would in many instances require site surveys within logistical zones associated with the more easily recognizable residential bases.

Third, the processing of small whales can be expected to be consistent with the economic utility of individual anatomical parts. However, and again unlike the situation for terrestrial mammals, the transport of bone elements, although consistent with economic utility, is influenced to a much greater extent by the ease in which the sculp and flesh can be removed. Consequently, very few small whale bones are likely to be recovered from typical residential sites, since they have very little architectural utility, unlike bones from larger whales such as bowheads. Exceptions to this pattern can be expected in situations where the kill and processing sites are immediately adjacent to residential bases. This was apparently the case at Kittigazuit and Radio Creek (McGhee 1974) and Gupuk (Friesen and Arnold, this volume) in the Mackenzie Delta, where high concentrations of beluga bones were recorded at prehistoric and early historic Mackenzie Inuit sites.

Finally, it is unlikely that beluga whale mortality profiles will indicate selection for specific animal sizes, unlike the situation for bowheads (see e.g., McCartney, this volume), because even the largest belugas are relatively easy to kill and process. Both "random" hunting, as in the Creswell Bay situation, or mass drive techniques, wherein whole groups are driven into shallow water and become stranded at low tide, will produce "live population" profiles.

CONCLUSIONS

It should be emphasized that this study represents a preliminary, pilot project only. Nevertheless, the data have been instructive in several respects. Major observations are as follows:

(a) modern Inuit beluga whale and narwhal harvesting at Creswell Bay is conducted within the context of a logistically-organized collecting system, with the characteristic site types of residential bases, field camps, stations, high-bulk locations, and caches being generated;

(b) mortality profiles are consistent with the harvesting strategies for each species: essentially random in the case of beluga whales and selective in the case of narwhals;

(c) processing is consistent with the economic utility of individual body parts for each species;

(d) transportation of bone elements of marine mammals, although consistent with economic utility, is influenced to a much greater extent by the ease in skin and flesh removal than for most other mammals; and,

(e) relative to terrestrial environments, taphonomic processes associated with marine environments may be much more detrimental to the preservation of high-bulk processing sites, such as whale butchering locations.

Acknowledgments. Sincere appreciation is extended to Nathaniel Kalluk and Andrew Atagotaluk for providing the opportunity to conduct the research described in this paper, and to other members of the Kuvinaluk camp for their very generous hospitality.

The field research was funded by the Social Sciences and Humanities Research Council of Canada and the Social Sciences Research Committee at McGill University, and was supported logistically by the Polar Continental Shelf Project (Energy, Mines and Resources, Canada). Thomas G. Smith (Fisheries and Oceans, Canada) provided valuable input into the research from both an academic and logistical standpoint.

Finally, the many valuable comments and suggestions by Allen McCartney and an anonymous referee on earlier versions of this paper are sincerely appreciated.

REFERENCES

Binford, L. R.
    1978   *Nunamiut Ethnoarchaeology.* Academic Press, New York.
    1980   Willow Smoke and Dog's Tails: Hunter-Gatherer Settlement Systems and Archaeological Site Formation. *American Antiquity* 45(1):4-20.

Bockstoce, J.
    1979   The Archaeology of Cape Nome, Alaska. *University of Pennsylvania, University Museum Monograph* 38.

Burch, E. S., Jr.
    1981   *The Traditional Eskimo Hunters of Point Hope, Alaska: 1800-1875.* North Slope Borough, Barrow.

Damas, D. (editor)
    1984   *Handbook of North American Indians*, Vol. 5, *Arctic.* Smithsonian Institution, Washington, D.C.

Doidge, D. W.
  1990   Age-Length and Length-Weight Comparisons in the Beluga, *Delphinapterus leucas*. *Canadian Bulletin of Fisheries and Aquatic Sciences* 224:97-117.

Finley, K. J.
  1982   The Estuarine Habitat of the Beluga or White Whale *Delphinapterus leucas*. *Cetus* 4(2):4-5.

Finley, K. J., G. W. Miller, R. A. Davis, and C. R. Greene
  1990   Reactions of Belugas, *Delphinapterus leucas*, and Narwhals, *Monodon monoceros*, to Ice-Breaking Ships in the Canadian High Arctic. *Canadian Bulletin of Fisheries and Aquatic Sciences* 224:97-117.

Giddings, J. L.
  1964   *The Archaeology of Cape Denbigh*. Brown University Press, Providence.

Giddings, J. L. and D. D. Anderson
  1986   Beach Ridge Archaeology of Cape Krusenstern. *National Park Service, Publications in Archaeology* 20.

Hay, K. A.
  1984   *The Life History of the Narwhal* (Monodon monoceros L.) *in the Eastern Canadian Arctic*. Ph.D. thesis, Institute of Oceanography, McGill University.

Heide-Jorgensen, M. P.
  1994   Distribution, Exploitation, and Population Status of White Whales (*Delphinapterus leucas*) and Narwhals (*Monodon monoceros*) in West Greenland. *Meddelelser om Grønland, Bioscience* 39:135-149.

International Whaling Commission
  1982   Aboriginal/Subsistence Whaling (with special reference to the Alaska and Greenland Fisheries). *Reports of the International Whaling Commission, Special Issue* No. 4. Cambridge.

Kemp, W. B., G. Wenzel, N. Jensen and E. Val
  1977   *The Communities of Resolute and Kuvinaluk*. Polar Gas Socioeconomic Program.

Kingsley, M. C. S., H. J. Cleator, and M. A. Ramsay
  1994   Summer Distribution and Movements of Narwhals (*Monodon monoceros*) in Eclipse Sound and Adjacent Waters, North Baffin Island, N.W.T. *Meddelelser on Grønland, Bioscience* 39:163-174.

Lyman, R. L.
  1992   Anatomical Considerations of Utility Curves in Zooarchaeology. *Journal of Archaeological Science* 19:7-22.

Mansfield, A. W., T. G. Smith, and B. Beck
  1975   The Narwhal, *Monodon monoceros*, in Eastern Canadian Waters. *Journal of the Fisheries Research Board of Canada* 32:1041-1046.

McCartney, A. P.
  1980   The Nature of Thule Eskimo Whale Use. *Arctic* 33(3):517-541.
  1984   History of Native Whaling in the Arctic and Subarctic. In: *Arctic Whaling, Proceedings of the International Symposium,* edited by H. K. s'Jacob, K. Snoeijing and R. Vaughan, pp. 79-111. Arctic Centre, University of Groningen.

McGhee, R.
  1974   Beluga Hunters: An Archaeological Reconstruction of the History and Culture of the Mackenzie Delta Kittegaryumiut. *Memorial University of Newfoundland, Newfoundland Social and Economic Studies* 13.

Metcalfe, D. and K. T. Jones
   1988   A Reconsideration of Animal Body-Part Utility Indices. *American Antiquity* 53(3):486-504.

Nelson, R. K.
   1969   *Hunters of the Northern Ice*. University of Chicago Press, Chicago.

O'Connell, J. F., K. Hawkes, and N. B. Jones
   1988   Hadza Hunting, Butchering, and Bone Transport and Their Archaeological Implications. *Journal of Anthropological Research* 44(2):113-161.

Pabst, D. A.
   1990   Axial Muscles and Connective Tissues of the Bottlenose Dolphin. In: *The Bottlenose Dolphin*, edited by S. Leatherwood and R. R. Reeves, pp. 51-67. Academic Press, New York.

Ray, D. J.
   1975   *The Eskimo of Bering Strait, 1650-1898*. University of Washington Press, Seattle.
   1984   Bering Strait Eskimo. In: *Handbook of North American Indians*, Vol. 5, *Arctic*, edited by D. Damas, pp. 285-302. Smithsonian Institution, Washington, D.C.

Reeves, R. R. and E. Mitchell
   1987   *Cetaceans of Canada*. Department of Fisheries and Oceans, Ottawa.

Reeves, R. R. and S. Tracey
   1980   Monodon monoceros. *American Society of Mammalogy, Mammalian Species* No. 127.

Savelle, J. M.
   1984   Cultural and Natural Formation Processes of a Historic Inuit Snow Dwelling Site, Somerset Island, Arctic Canada. *American Antiquity* 49(3):508-524.
   1994   Prehistoric Exploitation of White Whales (*Delphinapterus leucas*) and Narwhals (*Monodon monoceros*) in the Eastern Canadian Arctic. *Meddelelser om Grønland, Bioscience* 39:101-117.

Savelle, J. M. and T. M. Friesen
   n.d.   An Odontocete (Cetacea) Meat Utility Index. MS.

Savelle, J. M. and A. P. McCartney
   1991   Thule Eskimo Bowhead Whale Procurement and Selection. In: *Human Predators and Prey Mortality*, edited by M. Stiner, pp. 201-216. Westview Press, Boulder.
   1994   Thule Inuit Bowhead Whaling: A Biometrical Analysis. In: Threads of Arctic Prehistory: Papers in Honour of William E. Taylor, Jr., edited by D. Morrison and J.-L. Pilon, pp. 281-310. *Canadian Museum of Civilization, Mercury Series, Archaeological Survey of Canada Paper* No. 149.

Sergeant, D. E. and P. F. Brodie
   1975   Identity, Abundance, and Present Status of Populations of White Whales, *Delphinapterus leucas*, in North America. *Journal of the Fisheries Research Board of Canada* 32:1047-1054.

Smith, T. G., D. J. St. Aubin, and M. O. Hammill
   1992   Rubbing Behavior of Belugas, *Delphinapterus leucas*, in a High Arctic Estuary. *Canadian Journal of Zoology* 70:2405-2409.

Smith, T. G., M. O. Hammill, and A. R. Martin
   1994   Herd Composition and Behavior of White Whales *Delphinapterus leucas*) in Two Canadian Arctic Estuaries. *Meddelelser om Grønland, Bioscience* 39:175-184.

Spencer, R. F.
   1959   The North Alaskan Eskimo: A Study in Ecology and Society. *Bureau of American Ethnology Bulletin* 171.

Stanford, D. J.
   1976   The Walakpa Site: Its Place in the Birnirk and Thule Cultures. *Smithsonian Contributions to Anthropology* 20.

Stiner, M. C.
   1990   The Use of Mortality Studies in Archaeological Studies of Hominid Predatory Patterns. *Journal of Anthropological Archaeology* 9(4):305-351.

Stiner, M. C. (editor)
   1991   *Human Predators and Prey Mortality*. Westview Press, Boulder.

Stoker, S. W. and I. I. Krupnik
   1993   Subsistence Whaling. In: The Bowhead Whale, edited by J. J. Burns, J. J. Montague, and C. J. Cowles, pp. 579-629. *Society for Marine Mammalogy, Special Publication* No. 2.

Townsend, J. B.
   1981   Tanaina. In: *Handbook of North American Indians*, Vol. 6, *Subarctic*, edited by J. Helm, pp. 623-640. Smithsonian Institution, Washington, D.C.

VanStone, J. W.
   1984   Mainland Southwest Alaska Eskimo. In: *Handbook of North American Indians*, Vol. 5, *Arctic*, edited by D. Damas, pp. 224-242. Smithsonian Institution, Washington, D.C.

Yesner, D. R.
   1992   Evolution of Subsistence in the Kachemak Tradition: Evaluating the North Pacific Maritime Stability Model. *Arctic Anthropology 29(2):167-181*.

Whales, Mammoths, and Other Big Beasts: Assessing their Roles in Prehistoric Economies

David R. Yesner
Department of Anthropology
University of Alaska Anchorage
3211 Providence Drive
Anchorage, AK 99508

Abstract. *This paper examines commonalities in patterns of human exploitation of the largest mammals, specifically whales and very large terrestrial pachyderms. The exploitation of these animals offers high nutrient yields but also poses significant risks. Challenges are offered to the straight-line "scavenging to hunting" hypothesis, based strictly on energetic maximization, and support is garnered for a risk-reduction model in which both scavenging and various levels of hunting are likely to occur depending on regional environmental, technological, and demographic factors. In order to test this model, our ability to identify scavenging versus hunting in the zooarchaeological record of marine as well as terrestrial mammals must be enhanced, and correlates with other archaeological data must be provided.*

INTRODUCTION

Exploitation of the largest mammals, such as whales or elephants, has historically provided for human populations the largest return for energy invested in hunting strategies. For this reason, the largest mammals are often considered "preferred" prey, from both an energetic and psychological viewpoint. In addition to providing an important food supply, the capture of these animals allows for extensive food sharing and cooperative social endeavors in their stalking, immobilization, and retrieval. A cultural-evolutionary paradigm that emphasizes resource maximization as a corollary of Darwinian selection would suggest that, over time, human populations should move from scavenging to hunting of the largest animals, and should increase the level of their exploitation of these species as technology allows. However, the capture of the largest animals entails high risk as well as high reward, and frequently requires complex technology and extensive social networks. Cultural-evolutionary paradigms based on risk reduction as well as maximization (see e.g., Hayden 1981) would, therefore, suggest that both the degree of emphasis placed on scavenging as opposed to hunting and the overall utilization of these animals might depend on a host of environmental, technological, and demographic factors. Critical to assessing these hypotheses is our ability to recognize scavenging and hunting from the archaeological record, and to determine under what sets of conditions they are likely to occur.

HUNTING AND SCAVENGING THE LARGEST MAMMALS

Although one commonly hears the terms "megafauna" and "megaherbivores" applied to large mammals, there is disagreement as to what attributes should be used to characterize such a unit of mammals and what taxa should be comprised by such a unit. In part, the differences stem from whether the classification is directed towards the isolation of physiological and behavioral features distinguishing the animals or towards questions of human-animal relationships (since, after all, the perception of what constitutes a "large" mammal is from the human viewpoint). For example, in his recent book on "megaherbivores," the zoologist Owen-Smith (1988:1) defines these as "mammals that typically attain an adult body mass in excess of one megagram, i.e., ... 1000 kg, one metric ton," whereas the paleontologist Paul Martin (1967:77) defines them as "'big-game' mammalian and avian herbivores of over 50 kg [approximately 100 pounds] adult body weight." Owen-Smith (1988:1) finds the latter distinction "arbitrary" and without "functional basis," but it is clear that, from the human viewpoint, all mammals in excess of even 50 kg body mass do have a certain number of commonalities. In cultural terms, they yield a high potential for large nutrient, particularly caloric, contributions to the human diet. In the terms of optimal foraging theory, this represents a high "value" from each animal unit, defined as the ratio between energy gained and energy expended in the process of harvesting these resources, with the latter including searching, stalking, immobilization, retrieval, and processing costs (cf. Yesner 1981; Smith 1991). This is particularly true for large game animals which also aggregate and/or migrate in herds. In fact, it is the high "value" of such species that has led to a tendency for humans to concentrate on their exploitation, until such time as their abundance falls (for either natural or cultural reasons, with the latter including human technological efficiency) and search costs become too high. This phenomenon is well illustrated in the megafaunal extinctions that occurred in the terminal Pleistocene.

However, at the level of truly large mammals (those in excess of 1000 kg body weight, as defined by Owen-Smith), a number of other factors come into play. These commonalities are of interest to anthropologists and archaeologists, because they impact the technologies and strategies employed to obtain animals of this size, as well as the circumstances under which different technologies and strategies are employed. Animals included in this classification are limited to the terrestrial pachyderms, including the elephants, rhinoceroses, and hippopotami, and the marine cetaceans and sirenians. These animals yield an obvious quantum increase in nutrient yields, particularly when they are found in herds or pods. This may be particularly important in cold, seasonal, high latitude environments, where they also yield a potential for storage of large food masses, as well as bones for tools and dwelling construction where wood is not readily available. In addition, these very large animals have few natural predators and, therefore, suffer little competition in their exploitation by humans, at least in the "fresh" state. However, because of the increased momentum attached to their escape movements and relative impenetrability of their hides, they may be much more difficult to hunt. In addition, their large body mass means that they are frequently more dangerous to hunt; "hunting large mammals is an occupation during which there is a higher than normal potential for bodily injury" (Frison 1991a:22). Difficulty in obtaining the animals and limitations on their temporal availability imposed by herd migrations means that high labor costs are involved in their exploitation, and in the butchering process associated with necessary community-wide food sharing. Furthermore, once immobilized or killed, these animals are difficult to move, which places additional costs on the retrieval and butchering process. An upper limit on labor availability, however, is set by the total food supply and other constraints faced by hunter-gatherer populations, and thus reductions in labor costs are sought.

Reducing labor costs in exploiting very large mammals frequently involves some form of "scavenging," by which I mean the taking of animals in which implements are not used in every stage of the process. In some cases, implements may be used to immobilize prey but not retrieve them, while in other cases implements are not used at all. Thus, scavenging would comprise a variety of situations: obtaining prey from other predators which have previously immobilized and/or consumed them, obtaining prey from natural traps in which they have either been killed or confined, individually or en masse, and obtaining animals wounded or killed by other groups of hunters. The detailed examination of human hunting and scavenging patterns is of great interest, because their study may ultimately lead to a better understanding of an evolutionary process by which, during the long span of human history, scavenging was largely replaced by active hunting (Binford 1981, 1985; Shipman 1983, 1986; Potts 1984; Blumenschine 1986a,b; Klein 1987; Lupo 1994) and was, to some extent, retained in the exploitation of the largest animal species.

As a part of this analysis, we are desperately in need of a more comprehensive taxonomy of exploitation patterns, which would distinguish between a variety of different types of big game scavenging and hunting, each with characteristic archaeological signatures in terms of associated settlements, artifacts, and faunal remains. Thus, the two types of scavenging outlined above could be distinguished from "coarse-grained" hunting techniques, in which animals are taken "as encountered" (Yesner 1981; Winterhalder 1981), and "fine-grained" hunting techniques, in which some animals are passed over in the process of selecting individuals of particular ages, sexes, body size, or other characteristics. At least three different types of "coarse-grained" hunting techniques could be further distinguished: (a) "opportunistic" hunting, in which individual animals are approached, giving others an opportunity to escape, and in which some individuals may be wounded and may die at some distance from the initial approach (Frison 1987, 1991a); (b) "herd confrontation" hunting, in which larger family or matrifocal units are approached, contained, and killed (Saunders 1980); and (c) animal drives, including simple stampedes, natural traps (dunes, talus slopes, box canyons), and impoundments (corrals, artificial ramps, etc.). To date, biologists have paid relatively little attention to commonalities in large terrestrial and marine mammals. In part, this has been due to specialization within zoology (Owen-Smith 1988:1). Paleontological discussion of marine mammals is also rare, in part due to a paucity of fossils. For example, only sirenians are covered in published Pleistocene "bestiaries" (Martin and Guilday 1967; Anderson 1984), while sirenians and pinnipeds, but not cetaceans, are covered in Kurten and Anderson's (1980) more comprehensive treatment. Archaeologists working in coastal environments have been equally slow to recognize the potential in assessing archaeological signatures associated with different types of large sea mammal (i.e., whale) scavenging and hunting. In part, this may also be a matter of training; relatively few zooarchaeologists, for example, work extensively on both marine and terrestrial animal assemblages (Lyman 1992; Yesner et al. 1993).

Although it is difficult to apply directly the animal drive or herd confrontation models from terrestrial to marine megafauna, there are a number of insights to be gained in a similar pattern of analysis in which exploitation techniques are linked to the behavioral patterns of large mammals (McCartney and Mitchell 1988; Savelle and McCartney 1991; Krupnik 1993a,b). For example, in northern Alaska, hunting of migratory bowhead whales during the period from April through June involves ice lead hunting, which requires relatively few boats and primary butchering on ice floes. In contrast, in the Bering Sea region, open water whaling characteristically predominates, more boats are needed for landing whales, and animals are butchered only after hauling them to shore. In the latter situation, where visibility is greater, the possibility of culling individuals is enhanced, and the taking of smaller whales might be expected, because of the ease of long-distance transport to base camps. Also, in more open water situations, whale calves often feed near shore at the mouths of rivers and lagoons, making them easier prey (Krupnik 1993a,b). Finally, since spring whaling is possible earlier in the central Bering Strait than further north, more immature individuals are likely to be taken. Thus, the differences in these environmental situations lead both to differences in patterns of herd encounter, and to differences in animal selection, retrieval, and butchery (in this case, involving whale calves). The above example also points out another fact: the nature of the technology required to immobilize, kill, and retrieve large sea mammals such as baleen whales, particularly the use of boats, is so different from that used to obtain large terrestrial mammals as to limit the comparability of harvesting strategies. For example, in open water hunting of whales where field butchery is impossible, any selectivity that might occur will be at the level of the size of animal to be retrieved, rather than at the level of carcass sections or meat portions to be retrieved.

In addition, differences in the anatomy of terrestrial pachyderms and cetaceans result in critical differences in the patterning and interpretation of bone assemblages. For example, the nature of the appendicular elements of marine and terrestrial animals differs greatly. The appendicular elements of terrestrial animals contain marrow which possesses a relatively high fat content, a resource which may be of critical importance when other resources are unavailable. This is particularly the case among seasonal hunter-gatherers for whom the retention of appendicular limbs may serve as a kind of "storage" device for obtaining marrow at resource-poor times of year. This would be very important for terrestrial hunter-gatherers exploiting a lower diversity of species, for whom bone soup, bone grease, and bone marrow may have been critical resources in lean times. In contrast, the appendicular elements in marine mammals do not contain marrow, at least in sufficient quantities to be useful to human hunter-gatherers. For whales, the use of stored blubber or rendering of storable oil may have served a similar function to dried meat and crackable bones from terrestrial species.

Whether or not groups return appendicular skeletal elements to base camps must be considered in the light of both retrieval costs and benefits. For coastal Eskimo groups, living in a highly seasonal environment, the bones come to shore with the whale, and are used as structural elements and sources of tools, even if they have relatively limited nutritional value. In contrast, the Hadza of Tanzania, as low-latitude hunter-gatherers with limited environmental seasonality, hunt elephants and butcher them in the field, removing the largest, heaviest appendicular bones and leaving them in the field (O'Connell et al. 1988, 1990; Bunn et al. 1988). In fact, many of the appendicular elements are cracked and the marrow is consumed in the field. Similarly, Fisher (1987) indicates that, in the case of Pygmy groups that actively hunt elephants, skulls, scapulae, and pelvic bones are almost always left behind, upper limb bones are almost always retrieved, and lower limb bones show intermediate rates of retrieval. Thus, the failure to return appendicular bones to camp must be viewed in the context of not only the relative benefits of the marrow that they contain (i.e., storage on the hoof), but also the cost of transporting the bones. Other factors governing this process include seasonal variations in nutritional stress, especially relative demand for high-fat resources including bone marrow.

Another factor in the Hadza situation appears to involve time constraints in stripping meat from irregularly-shaped bones such as vertebrae. Here, it is useful to note that meat is probably easier to strip from the vertebrae of sea mammals because of differences in muscular attachments related to aquatic locomotion, and as a result vertebrae from large sea mammals such as sea lions, walrus, or small whales are infrequently transported to base camps (see Savelle, this volume). Large, adult whales may be an exception here, because the vertebrae or vertebral epiphyses are sometimes transported to the base camp, to be used as structural elements or as artifacts.

An additional variable that must be considered is the importance of environment in constraining the necessity for retrieving large mammal bones for constructional materials. Whale bones are regularly used for house construction in coastal environments where wood is scarce, whether they are arctic tundras or tropical deserts (e.g., along the Peruvian coast; cf. Bird and Hyslop 1985). Thus, the failure of the Hadza or Pygmies to retrieve large bones, in an environment where wood is abundant, can be contrasted with the mammoth hunters of the Upper Palaeolithic tundra/steppe, who required the bones both for fuel and house construction.

A more general contrast can also be drawn between the anatomy of large terrestrial and aquatic animals that is useful in understanding differences in butchering patterns. Large terrestrial animals such as elephants possess relatively large appendicular skeletal elements, including the pectoral and pelvic girdles. Whales, however, generally have small appendicular skeletal elements *relative to body mass*; this is true of all cetaceans, including both baleen and toothed whales of various sizes. As a result, these bones have little meat adhering to them out of the total package, and may be left behind as a part of the primary butchering process. Scavengers of older carcasses for whale bones useful in either house construction or tool-making are also likely to select against these appendicular elements in favor of elements such as mandibles or ribs. Therefore, a lack of short, appendicular whale elements in a site does not tell us much about the nature of whale utilization. Even where whale bones are not used for dwelling construction, this may still be the case. For example, in archaeological sites on Yukon Island, in Kachemak Bay, south-central Alaska, where I have recently been excavating (Yesner 1992), houses are constructed of wooden timbers without the use of whale bone elements, but the predominant elements from both toothed whales (porpoise, orca, beluga) and small baleen whales (sei, minke, and fin whales) found in the middens are vertebrae and rib fragments, while appendicular elements are correspondingly rare. It is important to remember, of course, that because they are difficult to move, the final resting position of either elephants or whales is at least a partial determinant of details of the butchery process.

Finally, it is hard to generalize about the role of non-human scavengers in altering whale bone assemblages in a fashion similar to that which has been argued for elephants or other large terrestrial game. It is largely environmentally- dependent. In most regions, at least some minor disturbance can be expected from a variety of agents such as wolves, wolverines, and foxes. In the High Arctic, ranging north of the Brooks Range in northern Alaska and eastward across Canada north of the tree line, polar bears are probably the major scavenger of beached whale carcasses, on a seasonal basis. However, in more subarctic environments and in oceanic situations like the Aleutian Islands where the terrestrial fauna is depauperate, few such scavengers exist. Therefore, it is in precisely those latter types of environments where one would predict a greater likelihood of scavenging natural kills and of hunting techniques designed to wound rather than kill immediately (resulting in scavengeable carcasses). This is not so much because of energetic considerations as a lack of competition from other scavengers.

Nevertheless, in spite of the above constraints, there are a number of commonalities in the treatment of large terrestrial mammals, such as mammoths, and large marine mammals, such as whales. Krupnik (1993a,b) has recently noted that these include the use of bones for dwelling construction and fuel and the necessity for caching or storage of meat supplies. In addition, he notes that both whaling and mammoth hunting techniques seem to have involved the minimization of retrieval cost by driving animals "as close to the village as possible ... Otherwise, dragging the meat, blubber, and bones back to the village from the hunting site would have consumed a disproportionate amount of labor" (Krupnik 1993b:244).

EVOLUTION OF LARGE MAMMAL HUNTING

After dealing with the above constraints in understanding how large game bone assemblages are generated, it is important to turn to the larger question of the evolution of these species' exploitation. In order to understand more generally the evolution of big game exploitation, including that of whale, two major questions need to be answered: (a) Under what situations does scavenging make sense as an adaptive strategy for human populations exploiting big game such as elephants or whales?, and (b) How do we clearly distinguish the signatures of different types of scavenging and active hunting of these species in the archaeological record?

The answers to both of these questions are fraught with a great deal of complexity. Clearly, the role of technology in both the hunting and butchering process is a critical one. It is widely assumed that in the course of human evolution, scavenging precedes hunting, particularly for large game, and that this occurred largely because the technology (and social organization) required for scavenging was simpler than that required for hunting. Scavenging is thought to be possible using only multipurpose butchering tools, many of which might be expedient tools and perhaps weapons to keep other predators at bay. The latter may not have been entirely necessary, however, as it is sometimes argued that at least the initial unique human niche may have consisted of a superior ability to utilize parts of the carcass not as easily utilizable by other carnivores or scavengers, such as the bone marrow (Blumenschine 1986a,b). Here elephants may have had a particularly important role, since it is often argued that it was with animals of this size that humans were at a distinct advantage in being able to crack apart their very large long bones and extract marrow from the trabecular bone. One argument justifying human contribution to the Pleistocene-aged Old Crow assemblage from the northern Yukon, and other putative early bone assemblages in Beringia, is that only humans are likely to have been able to produce spiral fractures of the type observed on mammoth bone (Morlan and Cinq-Mars 1982).

At any rate, it should by now be clear that no simple scavenger-to-hunter transition occurred in human history. Scavenging seems to have occurred at different times, under different sets of circumstances. Carcass density and competition are obvious factors: where carcass encounter rates are sufficiently high and other potentially competitive scavengers are few and far between, both pursuit costs and mortality risk would be reduced and the practice would be of greater selective "value." Blumenschine (1986a) has pointed this out for early man in Africa. During later human occupation of high latitude environments, however, a similar process may have reemerged due to the increased availability of frozen carcasses and the use of fire to prevent other predators from accosting retrieved carcasses. Again, the southwestern U.S. during the late Pleistocene may be an unusual situation in terms of availability of big game carcasses, with the drying up of numerous lakes and water holes as a radical change occurred between the pluvial climate of the glacial maximum and the arid climate of the early Holocene (Haynes 1991, 1993). Competition with other scavengers may have been low, in part, because the populations of many Ice Age carnivores appear to have crashed before those of their prey.

EVOLUTION OF WHALING IN THE CONTEXT OF LARGE MAMMAL HUNTING

A similar analysis could be applied to the evolution of whaling in arctic regions. As noted for elephants, scavenging of either naturally beached whales or struck whales that have washed ashore should be expected where (a) whale carcass density is higher, (b) winds and/or currents are favorable to drive carcasses to shore, and/or (c) competitive scavengers are few or absent. As noted above, one area in which these conditions were met is the Aleutian Islands, and in this region widespread scavenging of both naturally killed and struck animals did, in fact, take place. Ethnographic accounts (e.g., Veniaminov 1984) suggest that hunters, using slate-tipped darts equipped with slow-acting aconite poison, attempted to direct the carcasses to nearby shores or even villages. But, this was not by any means an assured process, and it required much spiritual manipulation. In any case, such active hunting was limited to island passes in the eastern Aleutians, where migratory whale densities were higher, while further west in the Aleutian chain carcass scavenging became more important as a source of food, fuel, tools, and constructional material. Unfortunately, whale bone measurements that might help to distinguish between a hunted and naturally-killed assemblage have never been collected from Aleutian sites.

Another region which showed widespread scavenging of whale carcasses is the Beagle Channel region of Tierra del Fuego. There, the ethnographic record and artifact inventory from the "Recent Phase" of Beagle Channel prehistory suggest that scavenging rather than active hunting of whales occurred, but that this comprised a minor portion of the overall marine economy during some 6000 years of regional occupation. However, recently acquired data from Valentine Bay in the eastern Beagle Channel region (Lanata 1990) suggest that there may have been an increase in scavenging of whale carcasses during historic times. It is interesting to speculate that increased hunting of whales by the early 19th century Europeans in the region may have increased the carcass density, thus creating a shift in subsistence and settlement patterns to accommodate this "new" resource (Yesner 1991, 1994).

Conversely, active whale hunting should be expected in regions where carcass density is lower, potentially competitive scavenger numbers are higher, and whales migrate closer to shore, thus reducing both search and pursuit costs. These condition apply equally to the High Arctic and southern Northwest Coast (west coast of Vancouver Island and northwest Olympic Peninsula). In the latter case, the greater focus of native groups on whaling is almost certainly related to the closer approach that whales make to shore, which corresponds to one of the narrowest regions of continental shelf north of Baja California. A similar argument could be raised for the reason why only eastern Aleuts were involved with active whale hunting. It was only there that whales came close to shore as they migrated through interisland passes. In both cases, following the tenets of optimal foraging theory, it appears that the primary reason for active hunting, rather than scavenging, of whales was related to the lower cost of these resources, rather than their greater abundance.

Pursuing this theme, Huelsbeck (1988a) has recently shown that prehistoric peoples of the Olympic Peninsula considered whaling as a highly "ranked" activity of economic value and not just as a "ritual/prestige activity." Using the comparative biomass approach that I developed for examining prehistoric Aleutian faunal data (Yesner 1981), Huelsbeck (1988b) demonstrates that whales are taken roughly in proportion to their abundance in the ecosystem and that whales are certainly included within the "optimal diet." However, whales are not as highly ranked as are other resources such as halibut or fur seals. One gets a sense that on the Olympic Peninsula active whaling is somehow related to the high population density there. However, in other areas of the Pacific coast, including the Aleutian Islands, population densities are high but active whaling was not undertaken. Perhaps the answer is more related to Huelsbeck's suggestion that other resources, such as salmon or halibut, were more easily overexploited.

Another factor that needs to be considered here is environmental change. Are there any data, for example, to indicate that a change took place relatively recently which made scavengeable carcasses less available or shifted migrating whale populations closer to shore? Donald Clark (1986), for example, has suggested that a large increase in the take of fur seals on Kodiak Island during the 19th century was a result of a shift in the fur seal migration route closer to the Kodiak shore. It would be intriguing to know if this analysis might also apply to the enigmatic Old Whaling culture of the Bering Sea region in western Alaska (see Mason and Gerlach, this volume). Although the bulk of Alaskan prehistoric data indicate a gradual increase in whaling during late prehistoric times (from around A.D. 1 to 1400), the Old Whaling assemblage, that includes some whale bones dating c. 1500 years earlier, has presented an interpretive dilemma for Alaskan archaeologists since their discovery by Giddings in the 1950s. Perhaps the Old Whaling assemblage represents either a period during which whales came closer to shore in this region or whale carcass density was unusually high. The former is suggested by paleoenvironmental data indicating a windier, stormier period (Mason and Gerlach, this volume). In any case, both scenarios would have led either to an increase in scavenging opportunities or a decrease in the cost of active whaling. However, since it is unlikely that the technology, demographic base, or social organization requisite for systematic large whale hunting existed at that time, an increase in scavenging is the best explanation for the Old Whaling culture.

ASSESSING HUNTING VERSUS SCAVENGING IN THE ARCHAEOLOGICAL RECORD

Finally, what types of archaeological signatures can be used to differentiate, in a rigorous fashion, the hunting and scavenging of large mammals, whether elephants or whales? There is, in fact, a great deal of overlap between various proposed models, nearly all of which are based on animal age distributions observable in the bone assemblages. For example, Saunders (1980) has concluded that a "herd confrontation" model has the closest fit to available zooarchaeological data on mammoths from Paleo-Indian sites. As evidence, he cites the large numbers of juvenile or subadult individuals in the mammoth age distribution, but an "opportunistic" model could theoretically produce a similar age distribution if, for example, juveniles became separated from the rest of the herd (by natural means or human confrontation) and were opportunistically picked off. Perhaps a "stampede" model would be likely to produce a larger number of adults, but this depends on whether only matriarchal units or units of bulls as well are being driven. In contrast, both the "scavenger" and "age-selective cull" models seem likely to mimic the "herd confrontation" model in age distributions. Clearly, what we need is a method of multiple working hypotheses, each of which clearly articulates archaeological expectations of various hunting or scavenging techniques in a mutually exclusive fashion.

Recently, Haynes (1992) has suggested that we must not only recognize that competing explanations might produce similar zooarchaeological age distributions, but that we must consider other types of data as indicative of hunting versus scavenging, particularly patterns of carcass utilization. For example, Paleo-Indian mammoth sites tend to show relatively "light" carcass utilization, including a great number of articulated or partly-articulated skeletons, and relatively few cut or bashed bones. Haynes argues that, when coupled with the age distribution data, these patterns reflect an abundance of scavengeable carcasses, precluding a necessity for active hunting (again, because once carcass abundance exceeds a certain level, it reduces the pursuit costs of even the best-equipped hunters!). It is unfortunate that sex data are not available for the mammoth remains, because they would help to settle the problem by differentiating between these hypotheses. For example, by sexing sea lion remains from Aleutian and Fuegian sites, it has been possible to show that, in both cases, pelagic sealing was being undertaken, because the nearly equal sex ratio in the archaeological remains precludes the harvesting of rookery harem populations that are numerically dominated by females (Yesner 1988; Schiavini 1986, 1993).

In much of their work, McCartney and Savelle (1985; Savelle and McCartney 1988) have demonstrated that the problem of differentiating between hunting and scavenging is equally as applicable to whales as it is to elephants. They also use age distribution patterns to decipher the nature of Thule Eskimo whaling. They find a large number of juveniles in their faunal inventories from Somerset Island in the Canadian High Arctic, and, in fact, find neither fully adult or calf whales represented. They conclude that neither a scavenging nor "opportunistic" hunting model will fit their Thule whale bone data, and suggest that an "age-selective culling" model offers the closest fit to the archaeological data. In accordance with Krupnik's (1993a,b) observations, McCartney and Mitchell (1988:4), citing W. S. Duval, note that:

> ... immature, growing whales, whose energy requirements are higher than adult animals, feed in numbers in shallow water estuaries where plankton yields are high, whereas adult whales feed in deeper offshore waters.

Could it not be that rather than undertaking age-selective culling of whale herds, Thule hunters were merely taking "as encountered" whales closer to shore, thus reducing their search and pursuit costs? If immature whales taste better and are (as argued above) easier to retrieve in open water hunting, such a strategy makes sense. Furthermore, the lack of calves in the faunal sample may be due to the fact that they are in deeper water with the adult females, and thus are not being encountered.

What types of criteria, then, could be used to differentiate between the active hunting and scavenging of whale carcasses? McCartney (1980) has considered a number of factors, and I include those in the following comparison, along with some additional factors. A comparative framework, then, for examining whale hunting versus scavenging might involve the criteria listed in Table 1.

Eventually, more specific criteria need to be developed that distinguish not only "whalers" versus "scavengers," but also "sedentary scavengers" versus "mobile scavengers" (e.g., Aleuts versus Fuegians), based on variations in permanence of site occupation among coastal hunter-gatherers (Yesner 1993). Even from the above, however, it is clear that two features stand out as more diagnostic of whale hunting as opposed to scavenging: whale butchery sites directly adjacent to and between settlements (because scavenged whale remains are not likely to be transported to settlements), and whaling gear present in the artifact inventory. If articulated whale carcasses can be located, their presence in direct association with villages suggests active whaling occurred, *unless* villages are actually moved to take advantage of scavenging beached carcasses. I know of at least one instance in which a 19th century camp of the Haush people on the Peninsula Mitre on southeastern Tierra del Fuego was apparently set up to do just that (Vidal, pers. comm. 1991).

Whaling gear in the archaeological record is potentially an even more powerful criterion, but in many ways it involves some of the most difficult problems to unravel. First, as noted by McCartney (1980:521), whaling gear is generally limited to large lance heads, toggling harpoons, and harpoon foreshafts, none of which are necessarily used exclusively for whaling, and few of which are identified on archaeological sites because of curation. While Frison (1991b) has shown that microwear can be used to determine unambiguously that Clovis points were, in fact, used as projectiles in mammoth hunting, it is difficult to apply similar criteria to whaling. For example, the Old Whaling culture has produced large chipped points or blades which may have served as whaling equipment, but microwear would not allow us to distinguish the use of such points between seal or whale hunting. Perhaps the analysis of blood residues, as has been done for

Table 1. Comparative criteria for distinguishing hunters from scavengers.

Criterion	Hunters	Scavengers
(a) Whales abundant in region	+	-
(b) Whales largest mammals available, yield large masses of meat/blubber	+	+
(c) Shore-edge site location, to view whales	+	+
(d) Settlements at "logistical" intervals	+	+
(e) Whale carcasses located adjacent to settlements, as well as between them	+	-
(f) Large winter settlements, sufficient for manning large skin boats and for beaching and butchering animals	+	+
(g) Whaling gear in archaeological record	+	-
(h) Storage pits/meat caches	+	?
(i) Utilization of whale bone for tool making and house construction	+	+
(j) Art depicting whaling	+	-
(k) Artifacts related to whaling rituals	+	-

mammoth blood on Paleo-Indian points (Dixon 1993), will eventually resolve some of these issues.

Artistic depictions of whaling are perhaps the *best* criterion for distinguishing hunting from scavenging, but are exceedingly rare. Similarly, unless artifacts related to whaling rituals specifically depict whales or whaling, an equally rare occurrence, this criterion may be of little additional help.

One possibility that remains in assessing the importance of active whaling versus scavenging is examining the overall pattern of the associated faunal assemblage. Savelle and McCartney (1988) have suggested that systematic (as opposed to opportunistic) whaling may be accompanied by a decrease in the utilization of other species. In order to test that notion, I reanalyzed the faunal data presented by Stanford (1976) from the Walakpa site in northern Alaska, which includes the transition from the Birnirk to the Thule culture when systematic whaling became fully developed. As Stanford noted, although the number of specimens is not large, whale remains (based on NISPs or numbers of identifiable specimens) more than double between Birnirk and Thule levels at the site. However, it is also clear that, although no species were dropped from the diet, there was a distinct decline in the use of all other marine species, particularly seals (from 60% to 37% of the faunal assemblage). Walrus, bearded seal, and sea birds also declined. This would indicate that active whaling was carried out by Thule people in northern Alaska, unlike the earlier Old Whaling culture.

Finally, refinement of these techniques should be possible with more thorough assessment of existing faunal data. Whether this criterion can be used in a more widespread fashion is unknown, but perhaps it can be added to the inventory of techniques that we use to distinguish between scavenging and systematic hunting, particularly for whales.

DISCUSSION

In sum, while a straight-line cultural-evolutionary paradigm of scavenging to hunting has often been proposed for human subsistence systems, the overall pattern appears to break down somewhat for the largest animals, whose exploitation is the most costly whether measured by requisite physical effort, technology, or social organization. For these animals, whether elephants or whales, scavenging may continue to occur even when sufficient technology is available in cases where animal numbers are low and yet requirements are high for large packages of storable meat, for dietary fats, and for the use of large bones for fuel or for house construction in treeless environments. Nevertheless, over time, increasingly efficient methods of exploiting these animals were continually discovered. The history of the Arctic, in particular, is one in which increasingly efficient methods of active whaling have been discovered and utilized. As Krupnik (1993b:193-194) has recently noted:

> The history of indigenous whaling ... shows a ... pattern of troughs and crests [in response to changing climatic conditions] ... But as ecological conditions inexorably came full circle, the hunting of large sea mammals once again recovered the ecological high ground ... as the result of intervening cultural progress. Archaeological data show distinct progress in hunting weapons, and increases in sheer amounts of whale bones ... Thus ... the role and efficiency of indigenous whaling clearly displays an absolute growth trend over time.

For mammoths, this increase in the efficiency curve may have been an important factor in their extinction (Martin 1967, 1984). For large whales, prehistoric extinction did not occur. The latter may be attributed to at least three factors: the relatively later date at which intensive whaling began, the lack of development of drive techniques for harvesting all locally available animals, and the difficulty in general of overharvesting marine mammals with premodern technology where visibility is limited and distances are vast.

Thus, while vastly different forms of technology and social organization were developed in the hunting of whales and elephants, there are some instructive insights to be gained in examining commonalities in the approaches to the harvesting of these very large creatures. In particular, they help to define the conditions under which scavenging and systematic hunting have occurred. Hopefully, this will shed some light on what has been a largely intractable problem in the evolution of human subsistence.

Acknowledgments. I sincerely thank James Savelle and Allen McCartney for their encouragement to write the original (somewhat different) version of this paper as the commentary to a symposium on archaeological approaches to megafauna which they organized for the 53rd Annual Meeting of the Society for American Archaeology in Phoenix, Arizona. I also thank Allen McCartney for his willingness to include my remarks with the current set of papers as well as for his suggestions for paper revision. I additionally thank Bill Workman, Kathryn Holland, Richard VanderHoek, George Frison, and Jim Savelle for comments or insights on these issues.

REFERENCES

Anderson, Elaine
  1984 Who's Who in the Pleistocene: A Mammalian Bestiary. In: *Quaternary Extinctions: A Prehistoric Revolution*, edited by P. S. Martin and R. G. Klein, pp. 40-89. University of Arizona Press, Tucson.

Binford, Lewis R.
  1981 *Bones: Ancient Men and Modern Myths.* Academic Press, New York.
  1985 Human Ancestors: Changing Views of Their Behavior. *Journal of Anthropological Archaeology* 4:292-327.

Bird, Junius B. and John Hyslop
  1985 The Preceramic Excavations at Huaca Prieta, Chicama Valley, Peru. *Anthropological Papers of the American Museum of Natural History* 62(1).

Blumenschine, Robert
  1986a Early Hominid Scavenging Opportunities. *British Archaeological Reports, International Series* 283.
  1986b Carcass Consumption Sequences and the Archaeological Distinction of Scavenging and Hunting. *Journal of Human Evolution* 15:639-659.

Bunn, Henry, L. Bartram, and E. Kroll
  1988 Variability in Bone Assemblage Formation from Hadza Hunting, Scavenging, and Carcass Processing. *Journal of Anthropological Archaeology* 7:412-457.

Clark, Donald W.
  1986 Archaeological and Historical Evidence for an 18th Century "Blip" in the Distribution of the Northern Fur Seal at Kodiak Island, Alaska. *Arctic* 39(1):39-42.

Dixon, E. James
  1993 *Quest for the Origins of the First Americans.* University of New Mexico Press, Albuquerque.

Fisher, Daniel C.
  1987 Mastodont Procurement by Paleoindians of the Great Lakes Region: Hunting or Scavenging? In: *The Evolution of Human Hunting*, edited by M. H. Nitecki and D. V. Nitecki, pp. 309-422. Plenum Press, New York.

Frison, George C.
  1987 Prehistoric, Plains-Mountain, Large-mammal, Communal Hunting Strategies. In: *The Evolution of Human Hunting,* edited by M. H. Nitecki and D. V. Nitecki, pp. 177-224. Plenum Press, New York.
  1991a Hunting Strategies, Prey Behavior, and Mortality Data. In: *Human Predators and Prey Mortality,* edited by Mary C. Stiner, pp. 15-30. Westview Press, Boulder.
  1991b *Prehistoric Hunters of the High Plains* (2nd edition). Academic Press, San Diego.

Hayden, Brian
    1981   Research and Development in the Stone Age. *Current Anthropology* 22(5):519-548.

Haynes, C. Vance, Jr.
    1991   Geoarchaeological and Paleohydrological Evidence for a Clovis-age Drought in North America and its Bearing on Extinction. *Quaternary Research* 35:438-450.
    1993   Clovis-Folsom Geochronology and Climatic Change. In: *From Kostenki to Clovis: Upper Palaeolithic-Paleo-Indian Adaptations*, edited by O. Soffer and N. D. Praslov, pp. 219-236. Plenum Press, New York.

Haynes, Gary C.
    1992   *Mammoths, Mastodons, and Elephants*. Cambridge University Press, Cambridge.

Huelsbeck, David
    1988a   The Surplus Economy of the Central Northwest Coast. In: Prehistoric Economies of the Pacific Northwest Coast, edited by B. L. Isaac, pp. 149-178. *Research in Economic Anthropology*, Supplement 3.
    1988b   Economics of Whaling at the Ozette Site, Olympic Peninsula, WA. Paper presented at the 53rd Annual Meeting of the Society for American Archaeology, Phoenix.

Klein, Richard G.
    1987   Reconstructing How Early People Exploited Animals: Problems and Prospects. In: *The Evolution of Human Hunting*, edited by M. H. Nitecki and D. V. Nitecki. Plenum Press, New York.

Krupnik, Igor W.
    1993a   *Arctic Adaptations: Native Whalers and Reindeer Herders of Northern Eurasia*. University Press of New England, Hanover, NH.
    1993b   Prehistoric Eskimo Whaling in the Arctic: Slaughter of Calves or Fortuitous Ecology? *Arctic Anthropology* 30(1):1-12.

Kurten, Bjorn, and Elaine Anderson
    1980   *Pleistocene Mammals of North America*. Columbia University Press, New York.

Lanata, Jose Luis
    1990   Humans and Terrestrial and Sea Mammals at Peninsula Mitre, Tierra del Fuego. In: *Hunters of the Recent Past*, edited by L. B. Davis and B. O. K. Reeves, pp. 400-406. Unwin Hyman, Boston.

Lupo, Karen D.
    1994   Butchering Marks and Carcass Acquisition Strategies: Distinguishing Hunting from Scavenging in Archaeological Contexts. *Journal of Archaeological Science* 21:827-837.

Lyman, Lee
    1992   *Prehistory of the Oregon Coast*. Academic Press, San Diego.

Martin, Paul S.
    1967   Prehistoric Overkill. In: *Pleistocene Extinctions: The Search for a Cause*, edited by P. S. Martin and H. E. Wright, Jr., pp. 75-120. Yale University Press, New Haven.
    1984   Prehistoric Overkill: The Global Model. In: *Quaternary Extinctions: A Prehistoric Revolution*, edited by P. S. Martin and R. G. Klein. University of Arizona Press, Tucson.

Martin, Paul S. and John E. Guilday
    1967A   Bestiary for Pleistocene Biologists. In: *Pleistocene Extinctions: The Search for a Cause*, edited by P. S. Martin and H. E. Wright, Jr., pp. 75-120. Yale University Press, New Haven.

McCartney, Allen P.
    1980   The Nature of Thule Eskimo Whale Use. *Arctic* 33(3):517-541.

McCartney, Allen P. and E. D. Mitchell
    1988   Thule Eskimo Bowhead Whale Selection on Somerset Island, Arctic Canada. Paper presented at the 53rd Annual Meeting of the Society for American Archaeology, Phoenix.

McCartney, Allen P. and James M. Savelle
    1985   Thule Eskimo Whaling in the Central Canadian Arctic. *Arctic Anthropology* 22(1):37-58.

Morlan, Richard E. and Jacques Cinq-Mars
    1982   Ancient Beringians: Human Occupation in the Late Pleistocene of Southern Alaska and the Yukon Territory. In: *Paleoecology of Beringia*, edited by D. M. Hopkins et al., pp. 353-382. Academic Press, San Diego.

O'Connell, James F., Kristen R. Hawkes, and Nicholas G. Blurton Jones
    1988   Hadza Hunting, Butchering, and Bone Transport and Their Archaeological Implications. *Journal of Anthropological Research* 44(3):113-161.
    1990   Reanalysis of Large Mammal Body Part Transport Among the Hadza. *Journal of Archaeological Science* 17(4):301-316.

Owen-Smith, R.N.
    1988   *Megaherbivores: The Influence of Very Large Body Size on Ecology*. Cambridge University Press, Cambridge.

Potts, Richard
    1984   Hominid Hunters? Problems in Identifying the Earliest Hunter-gatherers. In: *Hominid Evolution and Community Ecology,* edited by R. Foley, pp. 129-166. Academic Press, New York.

Saunders, Jeffrey J.
    1980   A Model for Man-mammoth Relationships in Late Pleistocene North America. *Canadian Journal of Anthropology* 1:87-98.

Savelle, James M., and Allen P. McCartney
    1988   Geographical and Temporal Variation in Thule Eskimo Subsistence Economies: A Model. *Research in Economic Anthropology* 10:21-72.
    1991   Thule Eskimo Subsistence and Bowhead Whale Procurement. In: *Human Predators and Prey Mortality*, edited by M. C. Stiner, pp. 201-216. Westview Press, Boulder.

Schiavini, Adrian C. M.
    1986   Una Aproximacion a la Predacion de los Aborigenes Prehistoricos del Canal Beagle sobre los Pinnipedos. *Actas de Reuniao de Trablho de Especialistas em Mamiferos Aquaticos da America do Sul* 2:81-82.
    1993   Los Lobos Marinos Como Recurso Para Cazadores-recolectores Marinos: El Caso de Tierra del Fuego. *Latin American Antiquity* 4(4):346-366.

Shipman, Patricia
    1983   Early Hominid Lifestyle: Hunting and Gathering or Foraging and Scavenging? *British Archaeological Reports* 163:31-49.
    1986   Scavenging or Hunting in Early Hominids: Theoretical Framework and Tests. *American Anthropologist* 88:27-44.

Smith, Eric Alden
    1991   *Inujjuamiut Foraging Strategies*. Aldine de Gruyter, Hawthorne, NY.

Stanford, Dennis J.
    1976   The Walakpa Site. *Smithsonian Contributions to Anthropology* No. 20.

Veniaminov, Ivan
    1984   *Notes on the Islands of the Unalashka District*. Translated by L. Black and R. H. Geoghegan. Limestone Press, Kingston, ON. (1840)

Winterhalder, Bruce
- 1981 Optimal Foraging Strategies and Hunter-Gatherer Research in Anthropology: Theory and Models. In: *Hunter-Gatherer Foraging Strategies: Ethnographic and Archaeological Analyses*, edited by B. Winterhalder and E. A. Smith, pp. 13-35. University of Chicago Press, Chicago.

Yesner, David R.
- 1981 Archaeological Applications of Optimal Foraging Theory: Harvest Strategies of Aleut Hunter-Gatherers. In: *Hunter-Gatherer Foraging Strategies: Ethnographic and Archaeological Analyses*, edited by B. Winterhalder and E. A. Smith, pp. 148-170. University of Chicago Press, Chicago.
- 1989 Effects of Prehistoric Aleut Exploitation on Sea-mammal Populations. *Arctic Anthropology* 25(1):28-43.
- 1991 Fuegians and Other Hunter-Gatherers of the Subantarctic region: "Cultural Devolution" Reconsidered. In: Hunter-Gatherer Demography: Past and Present, edited by B. Meehan and N. White, pp. 1-22. *Oceania Monographs* No. 24, Sydney.
- 1992 Evolution of Subsistence in the Kachemak Tradition: Evaluating the North Pacific Maritime Stability Model. *Arctic Anthropology* 29(2):167-181.
- 1993 Assessing the Impact of Early European Contact on Maritime Hunter-Gatherers of the Subarctic/Subantarctic Regions. In: *Culture and Environment: A Fragile Coexistence*, edited by R. W. Jamieson et al., pp. 61-76. Chacmool, Calgary.
- 1994 Seasonality and Resource "Stress" among Hunter-Gatherers: Archaeological Signatures. In: *Key Issues in Hunter-Gatherer Research*, edited by E. S. Burch, Jr., and L. J. Ellanna, pp. 151-167. Berg, New York.

Yesner, David R., Charles E. Holmes, and Kristine J. Crossen
- 1993 Investigating the Earliest Alaskans: The Broken Mammoth Archaeological Project. *Arctic Research of the United States* 6:1-6.

Siberian Eskimos as Whalers and Warriors

Hans-Georg Bandi
Swiss-Liechtenstein Foundation for Archaeological Research Abroad
POB 79, CH-3000
Bern 15, Switzerland

Abstract. *Bering Strait Eskimos of the Punuk period (c. A.D. 500-1000) developed baleen whaling into an important subsistence industry as warfare became increasingly common with the increase in population density, settlement size, territoriality, and competition over Asian-American trade. Weaponry, armor, burial patterns, and settlement patterns of St. Lawrence Island Yupik Eskimos are discussed in this paper as evidence for such conflict. Unlike whaling, which developed in the Bering Sea region, the bow and arrow complex and the armor complex used in battles were developed in interior northeastern Asia and then brought to the coast. Whaling crew structure and sociopolitical organization of large coastal communities provided a framework within which intersocietal warfare could thrive.*

INTRODUCTION

Warfare is a cultural dimension that requires discussion in the context of native whaling in the Western Arctic and Subarctic. The well-known Danish explorer and Eskimologist, Kaj Birket-Smith, once made the general observation that it is naive to emphasize the peacefulness of the Eskimos, as did Fridtjof Nansen for the Greenlanders (Birket-Smith 1959:49). Burch and Correll (1972:33) point out that among North Alaskan Eskimos "Warfare, feuding, and a more or less continuous state of mutual enmity and armed conflict were just as characteristic of inter-regional relations as were the more peaceful sorts of relations ..." Other authors such as Nelson (1899), Ray (1975), Burch (1974, 1988a), and Fienup-Riordan (1990) treat 19th century Eskimo warfare in detail, especially in northwestern and western Alaska. This paper briefly describes warfare among Siberian Eskimos and especially among St. Lawrence Island Eskimos, where warfare appeared rather early and became important long before this region came under the control of Russian colonists and later Euroamericans. It appears that the bellicose tendencies found among the Yupik Eskimos of Chukotka and St. Lawrence Island developed with whaling. It is possible and even probable that the early activities of Siberian Eskimo warriors initiated warfare in other parts of Alaska. This paper primarily

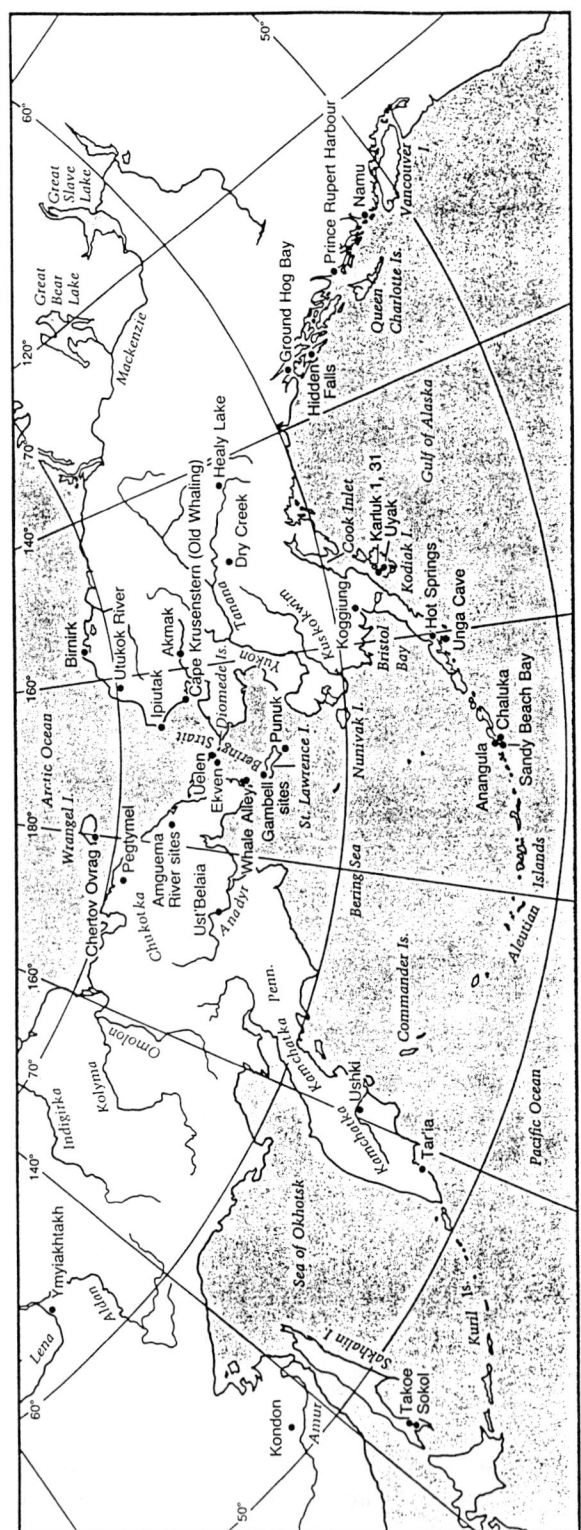

Figure 1. Map of the Bering Sea region (after Arutiunov and Fitzhugh 1988).

treats the results of archaeological research, but some attention is also given to ethnographic and ethnohistoric data.

PUNUK WHALERS

The investigations of Henry B. Collins and Otto W. Geist on St. Lawrence Island (Fig. 1) in the late 1920s and early 1930s provided much information about the succession and dating of Okvik, Old Bering Sea, and Punuk cultures in the Bering Strait region and their relationship to other prehistoric Eskimo cultures of the North American Arctic (Collins 1937; Geist and Rainey 1936). Later research carried out on Chukchi Peninsula by archaeologists of the former Soviet Union complemented archaeological knowledge about the cultural sequence between c. 500 B.C. and A.D. 1500 (Rudenko 1961). Although there is no doubt that walrus hunting has always been of great subsistence importance for these populations, they also practiced whaling. Especially during and following the Punuk period (c. A.D. 500-1000), whaling became an increasingly central focus of their subsistence. But, it must be stressed that not only did whaling distinguish Punuk culture from the preceding Bering Strait cultures but warfare did as well. Whereas whaling certainly developed in the Bering Sea region in an Eskimo cultural context, warfare and its necessary equipment seems to have reached the Bering Strait region from Northeast Asia. On the other hand, some influences from aggressive Northwest Coast Indians in southeastern Alaska cannot be totally excluded. These two cultural complexes — whaling and warfare — combined or developed together in the northern Bering Sea coastal zone to create large, productive, and aggressive societies, compared to those of the preceding Okvik-Old Bering Sea period.

BURIAL CUSTOMS

The increased whaling of the Punuk period is also reflected in burial customs. During the early excavations on St. Lawrence Island, practically no graves were discovered. Collins (1937:246) reported that:

> ... neither at Gambell nor elsewhere on St. Lawrence Island have I ever found burials which from the accompanying grave offerings could be identified as of Punuk or Old Bering Sea age. If, as seems probable, the prehistoric St. Lawrence Eskimos followed the usual practice of simply placing the body among the rocks, either on a flat surface or in a natural crevice, it is easy to understand why no skeletal remains of any antiquity have been found, since in this method of burial the bones very soon become scattered and broken.

Some years later, S. A. Arutiunov, M. G. Levin, and D. A. Sergeev (1964) found important graveyards at the Uelen and Ekven sites on northeastern Chukchi Peninsula. These Asian discoveries made the existence of similar and contemporary cemeteries on St. Lawrence Island very probable. A team of Swiss archaeologists operating under a research program which I organized between 1967 and 1973 succeeded in discovering a total of 149 graves on the western part of this large island. A total of 98 graves were found in the area of Gambell, 11 at Dovelavik Bay near Gambell, and 40 at Kitnepaluk, the latter located about 20 km south of Gambell (Bandi 1984, 1987). Compared to the Siberian cemeteries of Uelen and Ekven, where most of the burials belonged to the Okvik or Old Bering Sea cultures, the St. Lawrence graves had more baleen whale bones used in grave construction (Fig. 2). Especially near Gambell, most of the burials were framed and/or covered by mandibles, ribs, and skulls. This framing is particularly present in Punuk graves. The prevalence of large whale bones in graves and the fact that the heavy whale bones had been carried for a long distance from the beach to the graveyards seem to indicate that whales and whaling must have been important to the Punuk people. They certainly might be called whalers. These prehistoric graves are hidden because there are no signs or marks left on the surface to indicate their presence.

The men buried in these graves were not only whalers but were warriors as well. This fact is unmistakably shown by one of the first burials which I found in the summer of 1967, when our team traced the Siberian coastal burial customs to St. Lawrence (Bandi and Bürgi 1971/72). We found the skeleton of a man who must have died at age 35-40 that was placed in a carefully built grave with large stones located at both ends. The skeleton was well covered by whale mandibles, some ribs, scapulae, and skull pieces, as well as by some stones (Figs. 3-6). In the chest area, we found 13 ivory and antler arrowheads. Some of them were entirely preserved, some were broken, one was still sticking in a vertebra, and another was embedded in the right scapula. Two more broken arrowheads were found, one in the opening of the nose and one between the tibia and the femur of the left leg. Finally, a small basalt spearhead or knife blade was found behind the right clavicle. It is probable that the man was killed by the projectile which hit him in the face. The other arrows seem to have entered his body from behind while he was running or after he had already collapsed. The stone point would have injured his back.

At first, we thought that this was an example of human sacrifice, similar to that expected in the killing of a slave or a prisoner of war with many arrows. Siberian natives

Figure 2. Punuk period burial located near Gambell, St. Lawrence Island. It is framed by whale mandibles; two additional mandibles covered this skeleton of an adult man (C-14 date: A.D. 850±70).

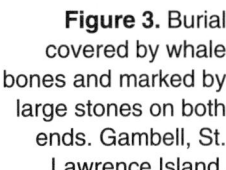
Figure 3. Burial covered by whale bones and marked by large stones on both ends. Gambell, St. Lawrence Island.

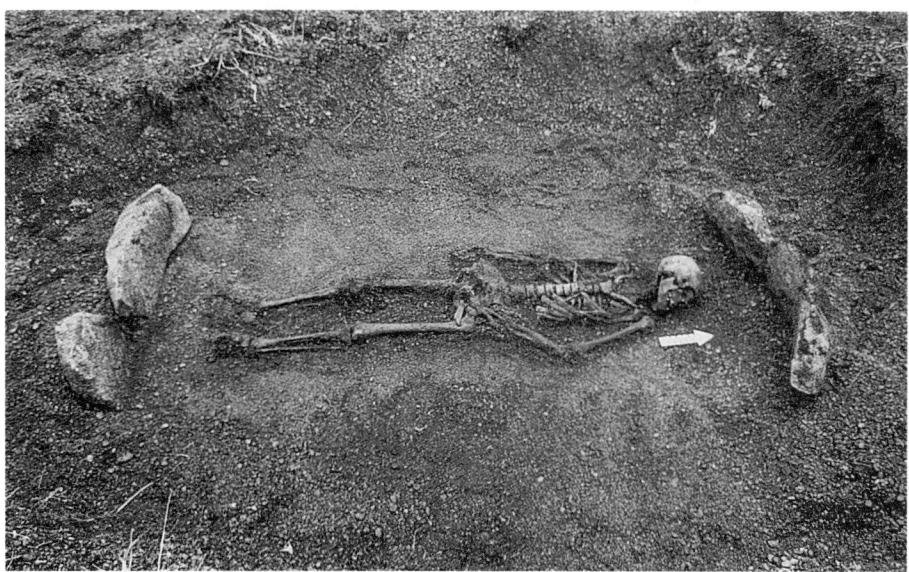

Figure 4. Same burial as shown in Fig. 3 but opened. The skeleton is of a man aged 35-40 years. A total of 15 arrowheads and a basalt point were found, most around the chest.

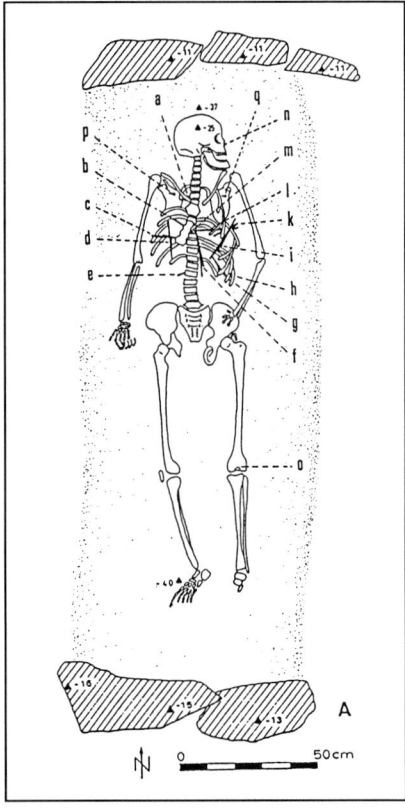

Figure 5. Drawing of the Eskimo skeleton shown in Fig. 4. The letters mark the locations of the arrowheads and basalt point.

are reported to have killed slaves before leaving their villages on raids. But information recorded by Nelson (1899:329) is probably more to the point. Nelson was told by an old native that "Sometimes ... a man would be shot [in battle] so full of arrows that his body would bristle with them, and, falling, be held almost free from the ground by their number." It is likely that this Punuk warrior was the victim of such an attack and was buried afterwards by his family. It is interesting to note that two of the arrowheads in this person's body were made of antler, probably reindeer antler. Because no caribou or reindeer were known on St. Lawrence Island before the introduction of domesticated reindeer by missionaries in the early 20th century, it is likely that the two Punuk arrowheads originated on the Siberian or the Alaskan mainland.

PUNUK BATTLE EQUIPMENT

Let us turn now to the evidence for warfare in this part of the Bering Sea region. As Collins has emphasized, new cultural elements appeared on St. Lawrence Island once Punuk culture became dominant in the mid-first millennium A.D. Together with the increasing importance of the bow, probably of the baleen reinforced type, Punuk arrowheads appear "so distinct from those of Old Bering Sea that we would seem justified in regarding them as later importations" (Collins 1937:323). Most of them are triangular in cross-section, and have a single, prominent barb. Apart from blunt arrowheads for bird hunting, bows and arrows could never have been of great value in hunting on St. Lawrence. Harpoons were used on sea mammals, caribou did not exist on the island, and polar bears were usually attacked with spears and knives, once they were surrounded by dogs. Therefore, it is probable that the importance of bows and arrows during the Punuk period is due to warfare, although people from St. Lawrence occasionally may have crossed over to Chukchi Peninsula to hunt large terrestrial game. Also, bone and ivory wrist guards used to protect the arms of archers are an element which Collins (1937:325) emphasized as "characteristic of the Punuk stage." Sinew twisters used to rig compound bows and relatively abundant metal appeared during Punuk times (Collins 1961:14; Ackerman 1984:112-113). Rudenko (1961:153) points out that the "sudden appearance" of Punuk

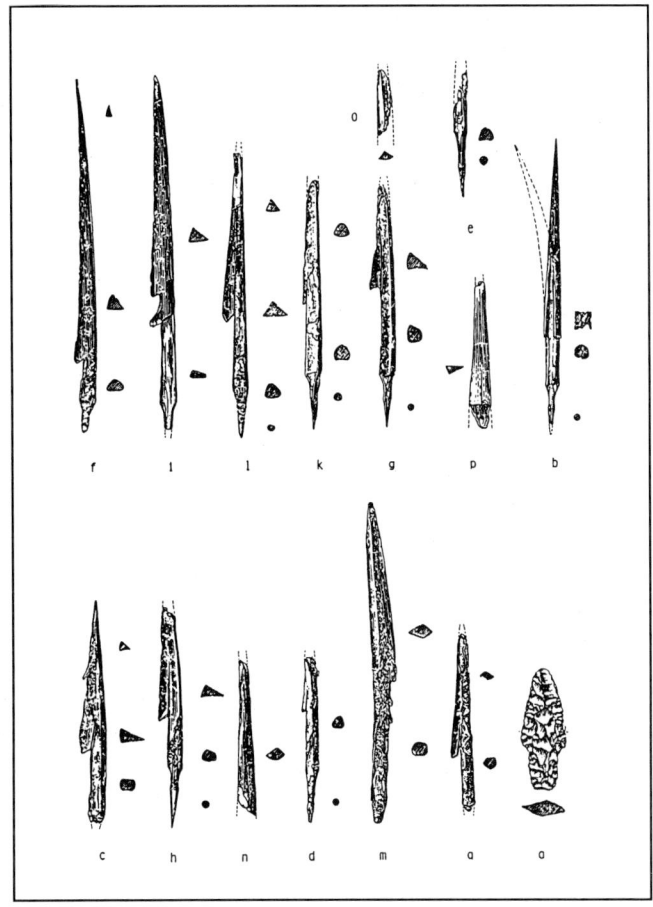

Figure 6.
Arrowheads and basalt point (lower right) from the burial shown in Figs. 3-5.

traits in the Bering Strait region "may indicate either an intrusion of hostile groups into northeastern Siberia or the gradual northward diffusion of weapons and improved means of warfare."

Burch (1974:5) calls attention to the fact that hunting tools and skills in using them are easily transferred to the tools and skills of battle. Bows and arrows, spears, and clubs can be used on people as well as game animals. Although bows and arrows on St. Lawrence support an interpretation of warfare because caribou, the common bow and arrow prey, were lacking, the presence of body armor leaves no doubt that warfare was commonplace.

ARMOR

Collins (1937:325) says that "plate armor is another characteristic St. Lawrence element that makes its appearance during the Punuk stage. The plates are all made of bone and in form are identical with those used on St. Lawrence Island in the nineteenth century." Corresponding information comes from Otto Geist who found many armor plates, especially in his excavations at the huge site of Kukulik, near the modern village of Savoonga on the north coast of the island (Geist and Rainey 1936). Because armor plate continued to be used until the end of the 19th and perhaps into the early 20th centuries, we have ethnographic specimens available as well as excavated ones (Fig. 7).

Nelson (1899:330, Pl. XCII) reports that "[d]uring the wars formerly waged among the people living on the coasts and islands of Bering strait, there was in common use a kind of armor made of imbricated plates of walrus ivory fastened together with sealskin cords" (see an example illustrated in Burch 1988b:37). Nelson also illustrates "a nearly complete set of this body armor, which was obtained on the Diomede Islands" towards the end of the 19th century. The Sheldon Jackson Museum at Sitka has a well preserved armor set collected from Cape Prince of Wales on Seward Peninsula. Adolf Etholen, who served the Russian-American Company during the 1820s-1840s, collected a plate armor set from St. Lawrence or St. Matthew island (Varjola 1990:271-271, Pl. 448). Another set of plate armor, which is reminiscent of the bulletproof vests of today, is found among the collections of the National Museum of Natural History, Smithsonian Institution. For comparison, the reader may refer to a similar rod armor from a prehistoric Aleutian Island burial cave, a Tlingit slat armor *cuirasse*, and a Tlingit painted hide armor tunic ornamented with carved bone "shark's teeth" and Chinese coins illustrated by Burch (1988a:230, Figs. 306 & 308). Dall (1878: Pl. 6) illustrates wood slat armor found at an eastern Aleutian burial cave on Kagamil Island.

Some years ago, Curator William Holm showed me parts of two sets of armor and two helmets, all made of iron plates, among the collections of the Thomas Burke Memorial Museum, University of Washington. They had been purchased from the Alaska Fur Company in 1952, and were identified as originating from the Siberian Koryak. Burch (1988a: Fig. 307) mentions that "[t]he forging and inlaying of iron was an art that spread to the Koryak perhaps via the Evenk but ultimately from the Chinese or Japanese." Also of interest is the fact that one of these helmets is decorated with French coins of the 18th century. They are riveted on the iron plates, and their stamp, LUD XVI D G FR ET NAV REX (read by the numismatist Dr. B. Kapossy, Bern, as: LUDovicus XVIth Dei Gratia Franciae ET NAVarrae REX), shows that they were issued during the reign of this unlucky king, who was executed in Paris on January 21, 1793, during the French

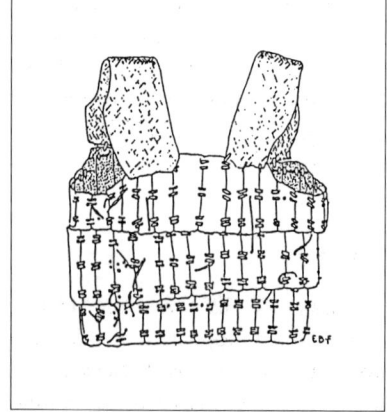

Figure 7. Drawing of plate armor.

Revolution. This gives a *terminus post quem* for the use of these coins by a Koryak blacksmith, although they might have been brought to Siberia by European (French?) whalers sometime later.

The Koryak were not the only native whalers in the Bering Sea who used metal plates for their armor. The University of Alaska Museum in Fairbanks has in its St. Lawrence Island collection five complexes of armor plates of a brass-like material, all excavated or purchased by Otto Geist. Three came from the Kukulik site mentioned above, including two that were found in "modern" meat caches. The exact provenience of the fourth is not known. The fifth might be identified by the following story in Geist's biography, written by Charles J. Keim:

> As Otto had pieced together the story of the great famine and sickness of 1879-80 he had learned that possibly the richest man at Kukulik had owned a brass armor. This man ... had perished in the famine. Further investigation disclosed that the rich man's name was Seechoohak. Otto obtained ... the armor for the College Museum. Seechoohak had got the brass from the white whalers, cut it into strips and plates, then fashioned these into an armor ... Otto doubted whether the Eskimos ever wore the armor, except for defensive purposes. The Siberians, especially the Chukchi, were aggressive. The islanders were a more peaceful people. When Otto arrived on the island they still were talking about the aggressiveness two visiting Siberians had displayed toward one of the early missionaries who attempted to strip and search them for whisky (Keim 1969:195-196; Geist and Rainey 1936).

The metal plates originating from Geist's researches on St. Lawrence have the same size and form as the precontact bone plates known from excavations. They are also fixed together by rawhide cords. An analysis of the metal by Joseph Riederer of the Doerner Institute in Munich shows that it is an alloy which was unknown until the second half of the 19th century, when it was then used in shipbuilding (Bandi 1974/75).

Collins thoroughly studied the origin of the plate armor which appeared on St. Lawrence early in the Punuk period (after A.D. 500). He believed that the use of this equipment by Eskimos, Chukchi, and Koryak warriors was peripheral to a larger center of use. He put the center somewhere in the vast area between China and Northeast Asia. The fact that the Su-Shen, situated on the border between Korea and China, had plate armor by the 3rd century A.D. might be of importance in establishing the chronology of later armor use in the Bering Sea region (after A.D. 500-700). With regard to the rod, slat, and hide armors used by Northwest Coast Indians in Alaska, Collins (1937:325-333) was unsure whether they were derived from Bering Sea plate armor or represented an independent invention.

Reminiscences of enemy attacks coming from the sea were still present in 1967 when Alaska celebrated the centennial of its purchase by the United States from Tsarist Russian. The inhabitants of Gambell built old-fashioned dwellings in the form of a semisubterranean winter house and a summer house made of driftwood and walrus skins. Furthermore, they prepared a performance representing an attack of warriors, arriving by *umiak*, upon villagers wearing fur coats as might have occurred in earlier times.

VanStone (1983:20) reports that similar remembrances can be viewed through recent engravings of Chukchi and Eskimo body armor from the Chukchi Peninsula. The same is true of some ivory figurines from Chukotka as well as examples from St. Lawrence

174 *Hans-Georg Bandi*

Island. In 1972, the young carver Larry Aningayou from Gambell sold me such a sculpture, measuring 11 cm high. He told me that he had been informed by old men about archery equipment used on raiding journeys in former times. His sculpture illustrates exactly a second type of armor known from St. Lawrence Island and the Chukchi Peninsula. It is called telescoping banded armor or simple band armor (Fig. 8). Several museums such as the National Museum of Natural History (Smithsonian Institution), the Field Museum of Natural History, and the Ethnographical Museum in St. Petersburg own such sets of armor (the latter set was obtained in 1840 on St. Lawrence Island). This armor consists of two parts, an upper shield made of bleached sealskin and wood that protects the upper part of the back, the occiput, and the arms without reducing mobility too much, and a lower part that consists of sealskin hoops which are linked and fixed telescopically. As Burch (1988a:227, Fig. 302) points out, a warrior equipped with such armor "could turn his back to a hail of incoming arrows and be protected by the upper shield … the

Figure 8.
Drawing of band armor.

collapsible lower body armor ... could be tied up around the waist to free the legs for running." Collins (1937:326) states that the Siberian Eskimos, including the St. Lawrence Islanders and also the Chukchi, used plate armor as well as band armor.

In contrast to plate armor, no band armor has been found in an archaeological context. However, the fact that the University of Alaska Museum has an ivory figure found at the Kukulik site that seems to depict band armor (VanStone 1983:21, Fig. 11) raises the possibility that this armor type also existed during the Punuk period. I know of no information as to where this kind of armor originated. The fact that it had also been used by the Itelmen of Kamchatka Peninsula suggests that it, too, originated in Asia.

BERING STRAIT WARFARE

Nelson (1899:327-330) began studying northern Bering Sea warfare among whalers who were armed and equipped as described above. Additional investigations by such authors as Birket-Smith (1959), Burch (1974, 1988a), Malaurie (1974), VanStone (1983), and Fienup-Riordan (1990) have added significantly to our knowledge of armed conflicts between Bering Sea Eskimo societies. By summarizing these reports of warfare for this part of the Arctic, we have the following picture. One of the reasons for hostilities might have been the proximity of different and competing ethnic and linguistic groups. Another reason might be the incursion of aggressive and militarily organized societies spreading into the Bering Strait region from other parts of Asia. In referring to blood vengeance among Alaskan natives, Birket-Smith (1959:151) states that the "contact with the North Pacific cultures has resulted in the adoption both of the advantages and the drawbacks of higher civilization." Burch (1988a:229) further states that "The Chukchi and Eskimos in particular were aggressive people by disposition and effective as fighters. Their men constantly prepared for combat through vigorous physical exercise, weaponry training and drill, and practice in dodging missiles." Citing information from the 18th century, he adds that "hostility seems to have prevailed over trade as the dominant theme of international affairs" during this period. It is quite possible that the role of *umialiqs* or boat captains was not only to organize successful whaling, but also to organize hostile activities directed against other villages. To plunder them might have been more profitable than to trade with them.

Nelson makes several references to "an almost constant intertribal warfare" between Eskimos on the Alaskan shore of the Bering Sea. He mentions that:

> In ancient times the Eskimos of Bering strait were constantly at war with one another, the people of Diomede islands being leagued with the Eskimo of the Siberian shore against the combined forces of those on King Island and the American shore from near the head of Kotzebue sound to Cape Prince of Wales and Port Clarence. An old man from Sledge Island told me that formerly it was customary among the people of the Siberian coast to kill at sight any Eskimo from the American shore who might have been driven by storm across the strait, either in umiaks or on the ice (Nelson 1899:330).

Jean Malaurie (1974:8) provides a graphic description of the activities of fighting whalers in the Bering Sea area (translated from the French):

> The atrocity of battles, the cruelties of massacres and the tortures can be — at last partially — explained by the fact that is was practically impossible to embark a considerable

> number of prisoners on an umiak, where they could have been dangerous ... The conquerors humiliated the defeated by a sodomitic sexual act. And before the head of a victim was cut off, one urinated on his face. The eyes were torn out or the eyelids were sewn together. The tongue was wrapped with a string. Heart and liver were fed to dogs or eaten by a warrior. Rebellious or old women were speared at the vagina, defeated enemies were castrated. After returning to their own village sexual or head trophies were exhibited.

Burch confirms the severe treatment of women captured in battle. They "were sometimes taken prisoner, but usually this was for only a brief period; eventually they were tortured — often horribly — before being put to death" (Burch 1988a:231). The raiding journeys made overland or by boat (in one or more equipped *umiaks*) must have been motivated in part by a search for prestige and for securing booty.

As mentioned above, Burch notes that the Chukchi as well as Eskimos constantly prepared themselves for combat through vigorous physical exercise. One example of such exercise came from so-called jumping stones, which can be seen at many places on St. Lawrence Island. Collins lists several such places, and reports that similar training places also are found in Siberia. He describes three rows of these jumping stones on top of Mount Chibukak (Mount Sevuokok) near Gambell.

> The principal one is something over 100 yards long and is evidently very old, as the stones are deeply embedded in the ground and covered over with moss and lichens. The ground all around has been cleared of stones, some of which have been placed in two piles about five feet high, on either side of the row at about the center. The line of stones curves somewhat but extends in a general north and south direction. The stones are closely spaced, some being hardly more than a foot apart (Collins 1937:355).

Collins was told by St. Lawrence Eskimos that these stone rows had been used in former times "by young men who were training to be runners or strong men. While carrying a heavy log, they would jump over and between the stones" (Collins 1937:355). I received the same information from Gambell informants about the purpose of such boulder rows, which are now no longer used. But we could observe a memento of hard training for men during our visit to Gambell. When celebrating July 4th there in 1972, there was a competition of running a long course over difficult terrain. When one of my collaborators came in second at this little island Olympiad, his prestige, as well as that of the whole Swiss group, increased immediately! Of course, in former times such training might also have been intended to make the men strong for hunting.

LOOKOUTS AND DEFENSIVE SITE LOCATIONS

There is other evidence of internecine activity found on St. Lawrence. This concerns dwelling sites at places which are difficult to reach and/or where one can see for a great distance out to sea. Nelson (1899:327) reports that prior to the arrival of the Russians, Bering Sea Eskimo "villages were built on high points, where defense was more easily made against an attacking party and from which a lookout was kept almost constantly." Birket-Smith (1959:151-152), who probably was aware of this statement of Nelson's, said that "The settlements ... were often placed in a position of defence by being built upon steep slopes with an open outlook, or on spits of land from which the inhabitants could flee in boats away from an approaching enemy."

Between 1967 and 1973, we made corresponding observations on St. Lawrence's west coast between Boxer Bay and Kongkok Bay as part of our research program. The distance between the two bays measures 12 km in a straight line. Near Boxer Bay, the cliff reaches an elevation of about 500 m above sea level. Near Kongkok Bay, the cliffs diminish to about 150 m above sea level. On a reconnaissance trip in 1967, we first discovered a considerable number of house ruins along the cliffs north of Kongkok Bay. The height of the cliffs above the beach was not great. The house ruins are often situated near the edge of the cliff, some a little deeper on terraces or in hollows on the slopes (Fig. 9). They are rather small and have a more or less circular outline. In most cases, the roofs are missing but the walls, constructed of large stones, are preserved to a height of about 50 cm or more above the natural ground level. The floors of these dwellings are usually dug below ground level. In a few cases, small houses or shelters have roofs constructed exclusively of stone, and are similar to *trulli* in Mediterranean countries (small, round, single-room houses with stone roofs; Fig. 10). In the area adjacent to Boxer Bay, in the direction of Kongkok Bay, we found about 30 small house groups or single house ruins, but there must have been more. Due to lack of time and because we did not find middens connected with these house ruins, we do not know their age. Some small potsherds associated with the houses lead us to believe that these dwellings are not very old. They probably belong to the late Eskimo period on St. Lawrence or sometime during the 18th or 19th centuries.

Figure 9. House ruin at Boxer Bay, St. Lawrence Island.

Figure 10. *Trulli*-like shelter on top of a cliff at Boxer Bay, St. Lawrence Island.

Near some of these house ruins, we noted upright stones or slabs, some 50 cm to 1 m high, in rocky areas. In some locales, we found only a few of these upright stones, whereas in other locations we found a considerable number (Fig. 11). From their upright position, it would appear that people erected them.

Following the coast to the north by *umiak*, we stopped half-way between Taveeluk Point and Situuluk Bay. On a terrace about 60 m above sea level, we observed a group of 20 or more house ruins and related features nearby. These occurred in small groups. From Ivekan Mountain, we could see the Kongkok Bay area, where a considerable number of house ruins are located, but we had no opportunity to reach them on our 1967 journey.

In 1973, we returned to Kongkok Bay, located about 27 km south of Gambell. There we found more than 50 house ruins similar to those we had seen in 1967 in the Boxer Bay area, although all of them are in the tundra and not on rocky terrain. Some had been constructed close to beaches, but most were situated on slopes, which are increasingly steep towards the inner part of the Kongkok Basin. These features are not visible from the shore. They are dispersed in small groups or larger clusters (Fig. 12). We located them as much as 1 km from the shore and up to about 650 m above sea level. As the native authorities did not give us permission to dig in these house ruins, we could only map them. These investigations were carried out under bad weather conditions, including dense fog, and it is quite possible that there are other house ruins in this area that we did not identify.

Our Eskimo helpers gave us two different explanations for these constructions in the Boxer Bay-Kongkok Bay area. Some believed that they had been shelters used by people

Figure 11. Upright stone, one of many found near Boxer Bay, St. Lawrence Island.

hunting birds or looking for eggs and nestlings. The other explanation was that they were hiding places and lookouts in case of attack from the sea. Although bird hunting and egg collecting could easily be practiced on the slopes and cliffs, this interpretation is not particularly persuasive. Hunters and their families could no doubt have traveled by boat, and it would have been logical for them to camp under upturned *umiaks* or in tents during short stops. It would have been much easier to make round trips to the cliffs for hunting birds than to construct shelters there. Furthermore, bird hunting was done in spring or summer, and there would have been no reason to construct stone houses dug into the ground like winter dwellings.

The need for hiding and protection against attacking enemies is much greater. Also, the clustering of this large number of house ruins in the Kongkok Basin would not support the interpretation of bird hunting shelters. On the other hand, sea mammal hunters must have had a special reason for not building their settlements on the shore, where they would have been in proximity to seals, walrus, or whales. Having their living quarters away from shore would give these people a greater chance of being hidden from the sea or having

Figure 12. House ruins in the Kongkok Basin, St. Lawrence Island.

an escape route if a dangerous enemy approached the beach. The little *trulli*-like shelters on the top of the cliffs, similar to those found near Boxer Bay, might have been lookouts from which aggressors could be identified long before they reached the shore. These lookouts may also have been used for whale-watching. An explanation which we received from an Eskimo friend concerning the upright stones near some dwelling places supports this interpretation. He suggested that skins may have been fixed to such stones in former times, which floated in the wind to give approaching enemies the impression that the number of defenders was greater than was actually the case.

WHALERS AS WARRIORS

We see that a northeastern Asian warfare pattern probably developed among Siberian Yupik Eskimos during the Punuk period or c. A.D. 500-1000. Whaling developed simultaneously in this area. Whaling might be the key, or at least one of the keys, to understanding why these Eskimos were both specialized sea mammal hunters and accomplished warriors. As we know from Birket-Smith (1959:145), the Eskimos in former times accepted leadership only in special cases, such as from a "strong man" in Alaskan villages or from the *isumataq* ("he who thinks ... for the others") among the Central Eskimo. Raids, as described above, certainly demanded some kind of leadership. The same is true of whaling. The increase of this most dangerous variant of sea mammal hunting during the Punuk period may have led to a perfection of navigation techniques, improved *umiaks*, disciplined crews, and boat captains with strict military-like authority. Even today on St. Lawrence Island, a boat captain, who on shore might be a reserved man, has absolute command on the *umiak*, during the hunt, and in the distribution of captured whales, although he has to respect traditions. It is obvious that successful attacks on other villages, whether carried out on land or by boat, depended upon good leadership. The strong authority of a leader was needed not only for the attack proper, but also for the treatment of captives, for the distribution of loot, and, in case of defeat, for leading a successful retreat (see Burch 1974:6 for a discussion of discipline and command patterns of North Alaskan warfare).

Punuk period villages are larger and more numerous than are those of the preceding Old Bering Sea occupation, suggesting that regional population density was greater and communities were more complexly organized. Sociopolitical organization, if similar to the ethnohistoric period, would have been based on large, extended families headed by *umialiqs* or production leaders who served as "big men." Maritime subsistence, thought to have been based primarily on walrus and baleen whales, relatively high population density, needs for intersocietal alliances, control of transcontinental trade, and new weaponry/armor technology from distant Asian peoples, seems to have combined in the Bering Strait region to provide a cultural environment for persistent, large-scale armed conflict (see Burch and Correll 1972). Thus, it was during the Punuk period that whaling and warfare developed as an interrelated pattern that was not interrupted until a millennium later, when Euroamericans arrived in the Bering Strait region.

DISCUSSION

On St. Lawrence Island, there is archaeological as well as ethnographic and ethnohistoric evidence that confrontations with enemies occurred over a long period. There is little doubt that there were hostilities between aggressors arriving by boat and prehistoric inhabitants of the island. Such raids not only increased the attackers' prestige, but also offered the chance of obtaining booty in the form of walrus meat, hides, ivory, baleen, and possibly slaves. It is quite probable that at least some of the aggressors reaching St. Lawrence Island were Chukchi from the Siberian coast. As pointed out above, Geist claimed that Siberians, especially the Chukchi, were aggressive, but that the St. Lawrence Islanders were more peaceful (Keim 1969:196). Burch (1988a:229) claims that both the Chukchi and the Eskimo were aggressive peoples. The degree of enmity may have varied with the degree of cultural-linguistic distinctiveness, with greater hostility exhibited between major groups such as the Chukchi, Siberian Yupik Eskimos, and mainland Alaskan Inupik Eskimos.

There is also the possibility of hostilities between the St. Lawrence Islanders and other Bering Sea Eskimos. VanStone (1983:16), referring to a report by W. F. Doty, mentions that "bitter feuds are said to have existed between the St. Lawrence Islanders and the people of Mys Chaplino" near Provideniya. Furthermore, there are stories among St. Lawrence Islanders about raids from mainland Alaska or from adjacent islands such as King Island, south of Cape Prince of Wales. These groups are said to have crossed Bering Strait and followed the coast of Chukchi Peninsula, from where they might have reached St. Lawrence Island. We also have to consider the possibility that, during the Punuk period, Siberian Eskimos, including those from St. Lawrence, made hostile trips in the opposite direction and, thus, initiated warlike activities on mainland Alaska. This hypothesis is supported by the fact that the Old Bering Sea culture had reached the Diomede Islands, and its influence can be shown in some regions of mainland Alaska. The same has been documented for Punuk culture, traces of which are found in Alaska from Cape Prince of Wales on Seward Peninsula as far north as Point Hope and Point Barrow. In any case, there is no doubt that Punuk period Eskimo whalers in the Bering Sea region engaged in war-like or raiding activities until the beginning of the 20th century.

REFERENCES

Ackerman, Robert E.
   1984   Prehistory of the Asian Eskimo Zone. In: *Handbook of North American Indians*, Vol. 5, *Arctic*, edited by D. Damas, pp. 106-118. Smithsonian Institution, Washington, D.C.

Arutiunov, S. A. and William W. Fitzhugh
   1988   Prehistory of Siberia and the Bering Sea. In: *Crossroads of Continents*, edited by W. Fitzhugh and A. Crowell, pp. 117-129. Smithsonian Institution Press, Washington, D.C.

Arutiunov, S. A., M. G. Levin, and D. A. Sergeev
   1964   Ancient Cemeteries of the Chukchi Peninsula. *Arctic Anthropology* 2(1):143-154.

Bandi, Hans-Georg
   1974/75   Metallene Lamellenpanzer der Eskimos auf der St. Lorenz Insel, Alaska. *Folk* 16-17:83-95.

Bandi, Hans-Georg (editor)
   1984   St. Lorenz Insel-Studien, Vol. 1, Allgemeine Einführung und Gräberfunde bei Gambell am Nordwestkap der St. Lorenz Insel, Alaska (in collaboration with E. Anliker-Bosshard and A. B. Hofmann-Wyss). *Academia Helvetica* 5/1, Berne.
   1987   St. Lorenz Insel-Studien, Vol. 2, A. B. Hofmann-Wyss, Prähistorische Eskimo-Gräber an der Dovelavik Bay und bei Kitnepaluk im Westen der St. Lorenz Insel, Alaska. *Academia Helvetica* 5/2, Berne.

Bandi, Hans-Georg and J. Bürgi
   1971/72   Gräber der Punuk-Kultur bei Gambell auf der St. Lorenz 72 Insel, Alaska. *Jahrbuch des Bernischen Historischen Museums* 51-52:41-116.

Birket-Smith, Kaj
   1959   *The Eskimos*. Methuen and Co., London.

Burch, Ernest S., Jr.
   1974   Eskimo Warfare in Northwest Alaska. *Anthropological Papers of the University of Alaska* 16(2):1-14.
   1988a   War and Trade. In: *Crossroads of Continents*, edited by W. Fitzhugh and A. Crowell, pp. 227-240. Smithsonian Institution Press, Washington, D.C.
   1988b   *The Eskimos*. University of Oklahoma Press, Norman.

Burch, Ernest S., Jr., and Thomas C. Correll
   1972   Alliance and Conflict: Inter-Regional Relations in North Alaska. In: Alliance in Eskimo Society, edited by L. Guemple, pp. 17-39. *Proceedings of the American Ethnological Society*, University of Washington Press, Seattle.

Collins, Henry B., Jr.
   1937   Archeology of St. Lawrence Island, Alaska. *Smithsonian Miscellaneous Collections* 96(1).
   1961   Eskimo Cultures. *Encyclopedia of World Art*, Vol. 5, Cols. 1-28. McGraw Hill, New York.

Dall, William H.
 1878 On the Remains of Later Pre-Historic Man Obtained from Caves in the Catherina Archipelago, Alaska Territory, and Especially from the Caves of the Aleutian Islands. *Smithsonian Contributions to Knowledge* 22(318).

Fienup-Riordan, Ann
 1990 Yup'ik Warfare and the Myth of the Peaceful Eskimo. In: *Eskimo Essays: Yup'ik Lives and How We See Them*, pp. 146-166. Rutgers University Press, New Brunswick, NJ.

Geist, Otto W. and Froelich G. Rainey
 1936 Archaeological Excavations at Kukulik, St. Lawrence Island, Alaska. *University of Alaska Miscellaneous Publication* 2. U.S. Government Printing Office, Washington, D.C.

Keim, Charles J.
 1969 *Aghvook, White Eskimo: Otto Geist and Alaskan Archaeology*. University of Alaska Press, College.

Malaurie, Jean N.
 1974 Raids et esclavage dans les sociétés autochtones du detroit de Behring. *Inter-Nord* 13-14:3-29.

Nelson, Edward W.
 1899 The Eskimo About Bering Strait. *18th Annual Report of the Bureau of American Ethnology for the Years 1896-1897*. Washington, D.C.

Ray, Dorothy J.
 1975 *The Eskimo of Bering Strait, 1650-1898*. University of Washington Press, Seattle.

Rudenko, Sergei I.
 1961 The Ancient Culture of the Bering Sea and the Eskimo Problem. Translated by Paul Tolstoy. *Arctic Institute of North America, Anthropology of the North, Translations from Russian Sources* 1. University of Toronto Press, Toronto.

VanStone, James W.
 1983 Protective Hide Body Armor of the Historic Chukchi and Siberian Eskimos. *Etudes/Inuit/Studies* 7(2):3-24.

Varjola, Pirjo
 1990 *The Etholen Collection: The Ethnographic Alaskan Collection of Adolf Etholen and His Contemporaries in the National Museum of Finland*. National Board of Antiquities of Finland, Helsinki.

Whaling Surplus, Trade, War, and the Integration of Prehistoric Northern and Northwestern Alaskan Economies, A.D. 1200-1826

Glenn W. Sheehan
Department of Anthropology
Bryn Mawr College
Bryn Mawr, PA 19010

Abstract. *Prehistoric northern and northwestern Alaskan economies were integrated to a high degree of interdependency through trade during the period A.D. 1200-1826. Supporting data for this assertion are in the literature but have remained unemphasized. McCartney's (1991:39) comment on metal use applies to trade in general: "the sociocultural dimensions ... have all but been ignored." People carried on voluminous trade, exhibited significant differences in access to basic raw materials from one group to another, and engaged in massive and deadly warfare whose underlying causes are best interpreted as economic ones. Access differences were most marked between inland caribou hunters and coastal whalers, but there were also differences in access to necessities between the various whaling villages themselves and between all of the Alaskan groups and the adjacent Canadian groups. Groups traded as a means of long-term survival. The whaling surplus generated by the coastal whaling villages underwrote these integrated economies.*

INTRODUCTION

The Eskimo nomads of the interior and the sedentary Eskimo whalers of the northern and northwestern Alaskan coast formed a highly integrated regional economy and culture during the prehistoric period of c. A.D. 1200-1826 and into the historic period up to the late 19th century (Fig. 1). They exhibited a regional economy and a regional division of labor that supported sedentism and whale hunting on the coast and maintained populations of nomadic caribou hunters in the interior. The whaling surplus, distributed through a native trade network, was the material basis for this integrated regional structure. Previously, Larsen and Rainey (1948:24ff; Larsen 1973), elaborated upon by Spencer (1959), posited an "ecological dichotomy" between the two groups. Burch (1981:62), in rejecting an inland/coastal dichotomy in favor of an inland/coastal continuum approaches

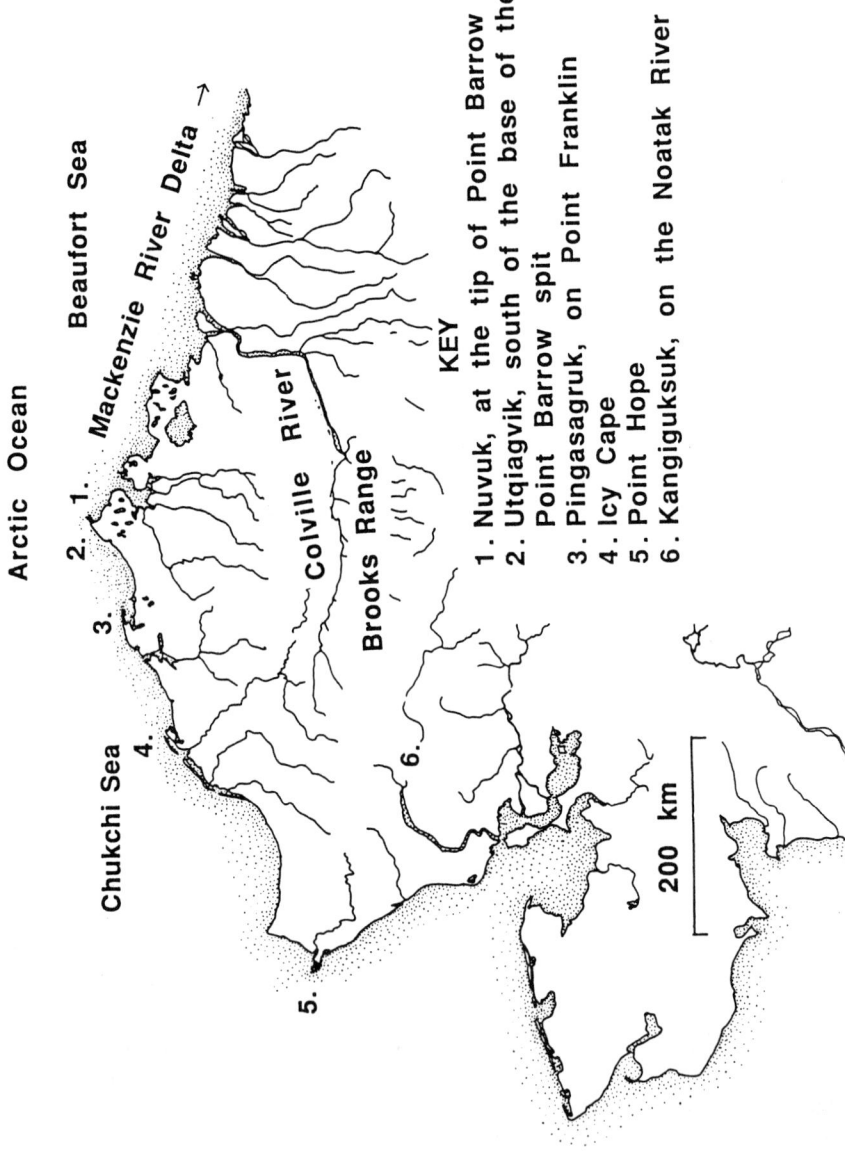

Figure 1. Locations of North Alaskan villages and archaeological sites mentioned in the text.

the position presented here more closely than previous commentators. Neither the term "dichotomy" nor "continuum" is appropriately descriptive for the regional situation, which I argue was one of material and social interdependency.

Trade enabled both sets of groups to schedule their hunts to maximize the harvest of their respective major game species, caribou and bowhead whale. The interior caribou hunters owed their long-term survival to their close relations with the coastal villagers. The whalers depended upon supplies from the inlanders so they could focus their efforts upon the whale hunt.

The development of trade volume over time, the growth of populations, and the prosecution of warfare were interconnected. The growth of whaling village populations led to the development of small outlying settlements in order to take advantage of resources in their vicinity, but these local offshoots, whose members moved back to the villages for spring and fall whaling, remained part of the village economies and they also depended on trade goods from the interior. Growing populations strained the trade network during the prehistoric period. Warfare intensified and eventually became devastating to human populations, while at the same time relieving some of the stress on the trading networks.

Trade from the coast was primarily in sea mammal oil. Trade from the interior was mainly caribou skins and other caribou products. Each side received items that were necessary for survival. For the inlanders, sea mammal oil provided a buffer against possible catastrophic failures of the caribou harvest. For whalers, caribou skins were the only suitable raw material from which to fashion arctic clothing needed for the hunt and for traveling. Exotic goods not critical for survival were also traded, such as jadite used for some blades and marked Siberian reindeer skins used for fancy parkas. The volume of trade in exotics was relatively small, and its possible reinforcing effects upon the status of some individuals was outweighed by the bulk trade in items necessary for physical survival.

In addition to their ties with inlanders, coastal communities also had close ties with each other. These ties served to even out irregularly distributed resources (i.e., everything but whale products) in normal years, and possibly provided the villagers with access to whale products in those unusual years when an individual village's whale harvest was inadequate or failed completely. Village-to-village ties also served to strengthen the emerging social order, wherein a few among the many whaling captains came to play a preeminent role in village political and economic life. Messenger feasts, multiple trading partners, and, ultimately, the mobilization of villagers for warfare were all centered upon these *umialit* (whaling captains; s. *umialik*).

The degree to which the various coastal Eskimo groups were integrated is illustrated archaeologically by the remarkable similarity of their prehistoric material culture and midden materials, despite the differences in their local resource bases. These were mainly differences of access to various resources, hunted and otherwise (e.g., caribou, walrus, bearded seal, cryptocrystalline lithics, slate, soapstone, prepared wood planks). As our archaeological data base grows through ongoing research, we should be able to test these hypotheses of material uniformity in a quantitative way.

The integration of coastal and inland groups is most dramatically illustrated by the responses of both groups to the historic contact-induced decimation of the coastal populations, starting after the middle of the 19th century. As coastal whalers died, the inland people came to the coastal villages and joined the whaling crews, replacing those who had died and, thereby, maintaining the whale hunting that had sustained their economies. Ultimately, almost the entire coastal population was replaced by inlanders in this fashion. Stefansson (1913:66) noted before 1910 that the Cape Smythe/Barrow community itself was thriving, but "only four persons are now living who are considered by the Eskimo themselves to belong to the Cape Smythe tribe" and no more than 21 had one parent from Barrow.

TRADE AND WARFARE ON A LARGE SCALE

The coastal distance between Point Hope and Point Barrow is about 400 mi. Many people of Nuvuk (Point Barrow) knew the coastline in detail for 600 mi, from Point Hope to the Colville River (Dr. Simpson in Bockstoce 1988:2:542), and some went as far east as the Mackenzie River to trade. Maguire, who overwintered twice in the early 1850s, believed that probably all adults were familiar with the coast between Point Barrow and Point Hope (Bockstoce 1988:1:184). This was the geographic scale of trade and war. Their scope was comparably large, in terms of war casualties and in terms of trade volume.

McCartney (1991), Hickey (1979), and others make the case for geographically extensive prehistoric trade among Eskimo groups, although Morrison (1991:242) characterizes Alaskan-Canadian trade in soapstone as "meagre and sporadic" until the early historic period. Based on Maguire's journals (Bockstoce 1988) and archaeological excavations (Hall 1990), it is reasonable to assume that protohistoric and early historic native trade was massive in volume, both in numbers of people involved and in quantities of material moved. Burch (1974, 1981, 1988) argues persuasively for large-scale warfare among protohistoric and, by extension, prehistoric groups in northern and northwestern Alaska. However, trade's significance and relation, if any, to warfare is undocumented. The economic basis of Eskimo warfare has not been addressed previously. This paper suggests some connections and answers.

Language and material culture were held in common among the Eskimo whaling villages in northern and northwestern Alaska at the time of contact in the 1820s and through the historic period (Rasmussen 1927). Ethnohistoric research (e.g., Burch 1980, 1981 et seq.) and archaeology (e.g., Hall 1990; Sheehan et al. 1991; Stanford 1976; Bockstoce 1976, 1977, 1979; Ford 1959) reveal a time depth to this uniformity without attributing causality. Perhaps whaling groups were similar through time, simply because they started out that way or because their environments allowed few alternatives. Were their arctic environments as similar as their cultural ones? Archaeology at Point Franklin (Sheehan et al. 1991) and Barrow (Hall 1990) and ethnohistoric sources (Bockstoce 1988) suggest small environmental differences may have made significant differences for people in their access to resources. I suggest that the efforts of whaling groups to compensate for their dissimilarities through trade of locally available resources played an important part in supporting their common culture. Trade, in this view, facilitated a geographic cultural continuity over time, and served as a mechanism that at times ensured people's physical survival. Trade's basis, therefore, was not focused upon wants or status, as much as it was focused upon material requirements for daily life.

Acquisition of traded goods influenced scheduling, allowing groups to concentrate their time on specialized seasonal hunting activities. Trade allowed people to reside in locations whose suite of resources was insufficient to meet all of their needs, but where at least one of the resources could be taken in surplus. Morrison (1991:245), describing the Kangiryuarmiut of Prince Albert Sound, notes that "[t]he opportunities for acquiring non-food resources may have attracted a larger population than could be sustained in the long term." For whaling villages, whales were part of the surplus and a main part of the diet, while the local availability of other resources varied among the villages (Sheehan 1985). As growing populations outstripped village catchments, trade helped make up any shortfalls. The intense and highly murderous warfare documented by Burch (1981) for the protohistoric period is best explained as a result of population growth outstripping trading capacity.

Nomadic Thule folk of c. A.D. 800-1200 were open-ocean whalers (McGhee 1984), an adaptation that became impossible when the overall climate cooled and ice covered the oceans even in seasonal warm periods. Sedentary villages grew at the few places where whaling could continue by means of hunting through spring ice-leads. The period for hunting whales, formerly continuous from spring through the summer and into fall, was now concentrated in spring and a few weeks in fall. As the villages were settled between c. A.D. 1200-1400, scheduling became critical to the whalers' success.

Punuk folk preceded those of the Thule culture in the Bering Strait region, and Birnirk preceded Thule in North Alaska, and both Punuk and Birnirk were apparently coastal cultures. The interior was virtually uninhabited when the whaling villages were first established. For instance, according to the latest interpretation of radiocarbon dates, "inland Ipiutak is partly contemporaneous with terminal Birnirk and the beginning of Thule," and Birnirk has a radiocarbon range of 1300-1000 B.P. (Gerlach and Mason 1992:63). The only other candidate for this time period, Punuk, has newly calibrated dates ranging from 1300 to 800 B.P. (Gerlach and Mason 1992:64).

It has been argued that prehistoric Eskimo populations tended to grow over time, at least from the Birnirk period through the Thule period and up to protohistoric times in northern and northwestern Alaska (Sheehan 1992). McCartney (1991:33-34) shows that early Thule groups throughout the Arctic were connected interregionally through trading in easily transported valuables. He suggests that interregional trade was supported by a more dense network of regional systems for more utilitarian commodities. This dense network of exchange made up for the shortfalls left when individual hunters, small groups, and whole villages were unable to maintain a complete logistic inventory. Almost every element of Eskimo material culture was necessary for survival, and few substitutions were suitable in raw materials or in end products. Shortfalls in any types of resources were potential threats to people's lives and lifeways.

Early trade, when population levels were low, may have included relatively fewer traders as a proportion of the population and relatively more exotic goods as a proportion of total trade volume. Smaller populations would have been better able to supply their non-whale material needs within their local catchments, thus placing less of a premium on bulky utilitarian goods. Before villages were settled c. A.D. 1200, nomadic whalers were more free to move for purposes of acquiring locally scarce products instead of trading for them, since almost wherever they moved along the coast they could conduct open-water whaling. However, when sedentary whaling villages were established between c. A.D. 1200-1400, ice-lead whaling was confined to a few geographic points.

As village populations grew, local catchments could not supply enough game-derived raw materials for clothing, covers, containers, hunting gear, tools, and even construction materials to keep pace with demand. Simultaneously, increased population resulted in increased crew launchings and greater whaling efficiency. Food harvests went up absolutely and probably relatively, while raw materials went down in amounts. Large-scale warfare, rather than contributing to social complexity or preceding large-scale trade, probably grew out of, or was intensified by, trade restrictions, whether the limits on trade were logistical or political.

Even though trade may have benefitted some people disproportionately (Sheehan 1987), it was not confined to an elite, based on the volume and numbers of people involved in native-to-native trade, as seen in the early historic period. Nelson wrote that in 1881, "[n]ear Cape Lisburne we met nine umiaks containing about one hundred people from Point Hope, who were on their way to the vicinity of Point Barrow to trade" (Nelson 1983:231). From 75% to 82% of the inland population may have traveled to coastal trade fairs each summer (Hall 1971:67, agreeing with Foote 1965:256). Bockstoce (1988:1:241 footnote 4), using Maguire's data, calculates that in July, 1853, about 50% of the population of the whaling village of Nuvuk on Point Barrow traveled east to the Colville trade fair. Additional people traveled west to trade, and still others attended the nearby fair at Birnirk at the base of the Point Barrow spit. With the majority of the population and heavy loads of goods involved, trade was not primarily in exotic or scarce items. Rather, "exchange of regional goods ... took place to level out needed resources from unevenly distributed sources ..." (McCartney 1991:35).

Nuvuk people moved their trade goods, mostly blubber and oil, on sledges as they left the village. One *umialik*'s trading party was made up of six sledges accompanied by four *umiaks* (skin whaling boats), although the typical family might pull a single heavily loaded sledge of trade goods (Maguire in Bockstoce 1988:1:238, 2:408).

In these Eskimo societies, a large trade volume in non-exotic items, controlled mainly by a few of the many whaling captains, is a phenomenon implying and contributing to social complexity, while differential possession of rare or status items seems not to have been especially significant. At least in northern and northwestern Alaska, continued growth in trade can be seen as continued successful accommodation to logistical problems raised by growing populations. Eskimos lived by hunting, and hunters are typically seen as self-sufficient. Therefore, the only way to establish that increased numbers of hunters would necessarily require an increased volume of trade is to show that they were not self-sufficient and in what respects they were not self-sufficient.

Once the whaling villages were established, enough resources for everyone may have been harvested in northern and northwestern Alaska and adjacent Canada, but it was not harvested equally. Trade equalized imbalances, at least to the carrying capacity of the trade network. I believe that to the extent imbalances were not resolved by trade, warfare was promoted. Whaling provided villagers with a food surplus, and the more whale hunters there were, the more the surplus grew. Whales were the only relatively unlimited resource that could be harvested from any village location, so eventually an increase in the number of hunters, while still yielding more food from whales, would have led to a relative decrease in the harvests of the limited non-whale resources.

An economic effect of war, at least for villagers, was to reduce the demand for trade by reducing the population and, therefore, the need for non-whale materials from outside the village catchment. Positive proof for this position is not available, but this much is known: large-scale warfare eventually killed hundreds of people, causing significant decline, for example, in the Point Hope population. Following this murderous climax around A.D. 1800 (Burch 1981), indigenous warfare diminished, finally disappearing by the mid-19th century. During that same period, the indigenous trade network continued to flourish and to move massive amounts of commodities. War and regional population peaked together, with trade as a constant underpinning that preceded the peak and continued after significant population reductions and long after war disappeared.

Examples of war behavior in North Alaska include: the burning of the Utqiagvik Mound 34 *qargi* (s., *qargich* pl., ceremonial center) at Barrow, the terminal deposits of which were capped with large amounts of body armor at c. A.D. 1400s-1500s (Sheehan 1990); the construction of at least one escape tunnel at Utqiagvik, the only proposed explanation for which is war-related (Sheehan 1990); war as a predominant phenomenon of c. 1600-1800 (Burch 1988:234); the abandonment of Pingasagruk village at Point Franklin in the early 1700s (Sheehan 1992; Sheehan et al. 1991); the post-1800 Point Barrow bone pile with over 50 skulls attributed to battle casualties (Rasmussen 1927, Neg. 120809; shown in Burch 1988:230, Fig. 305); the Point Hope wars, culminating in the early 1800s, including one battle with over 200 killed, that included the use of defensive works that are documented by Burch (1981) and Kashevarov (VanStone 1977:54, 76 footnote 78); attacks on explorers starting in the 1820s; and, finally, the mobilization of the Nuvuk village (Point Barrow) for an abortive attack on Captain Maguire's command in the 1850s (Bockstoce 1988).

WHALING CAPTAINS

There are social organizational concomitants of a dependency upon trade, and villages that are not self-sufficient and are dependent upon trade have a higher or more complex level of social organization than do self-sufficient groups. This is especially true in this case, where the material basis of trade, the whaling surplus, was already unequally distributed within the village even before any of it was traded outside. VanStone (1962) and Worl (1980) document the whaling captain's disproportionate share of each whale. With village life dependent upon trade, and trade dependent upon the whaling surplus, and the surplus in the hands of the *umialit*, increased social complexity was a necessity simply to organize the logistics.

Whaling was primarily an organized hunting task and the hunt leader received a disproportionate share of the harvest, as was traditional in other types of communal hunting. Whaling may or may not have produced a surplus among the early nomadic whalers, but they had no ability to capitalize on it anyway, due to their nomadic lifestyle (McGhee 1984:371). When whaling produced a surplus in the sedentary villages, whaling captains were direct beneficiaries. Village life concentrated people together, as well as resources, and for longer times than among their nomadic predecessors. This extended the duration of interactions between *umialit* as leaders and redistributors and their co-villagers.

The whale cult provided ideological justification for the *umialik's* role, and helped promote it by its emphasis on the *qargi,* the pivotal communal institution. *Umialit* were able to incorporate community followers into their extended families through liberal use of fictitious kin ties and by additions to whaling crews. To the extent social incorporation of familial outsiders (other villagers) was accepted, an *umialik's* actions on behalf of the community paralleled the case of a primary hunter acting for his dependents. Warfare, a phenomenon of increased social complexity, may have been carried out in keeping with the traditional obligations of family heads to dependents.

Individuals and groups benefitted from trade in more than material ways. Whaling captains owned the bulk of each whale and, hence, they owned the bulk of trade goods. They were able to translate their surplus into prestige items, such as marked reindeer skins from Siberia, a material benefit with social overtones. Probably more importantly, they were able to multiply and maintain their political and social alliances beyond their local base. *Umialit* had a potential to create about 10 times the numbers of alliances with other individuals than could non-*umialit* (Hennigh 1983, reviewed below). Not all *umialit* were equal, and the leading *umialit* exerted control over their compatriots' trading activities (Charles Brower n.d.) and over warfare.

Cultural continuity in the archaeological record and cross-cultural similarities between whalers and their contemporary arctic neighbors should not obscure the social complexity required by village life and exemplified in the *umialik's* role. Hennigh (1983) shows 59 potential one-, two- and three-way interrelations between the role of *umialik* and other individuals, but a much smaller number between shamans and other people, and very few for typical villagers. Contrary to Hennigh's (1983:26) modern informants' beliefs, but consistent with Captain Maguire's observations almost 150 years ago (Bockstoce 1988:1:292-294; 2:317), some of the most important *umialit* were also shamans. This emphasizes the extreme disparity between the most important of the *umialit* (Hennigh's "rich men") and typical villagers. This combination of whaling captain's and shaman's roles further increased an individual's potential number of interrelations.

Hennigh (1983:23) argues "the rules for alliance combination were part of the larger social structure which existed in the minds of the North Alaskan Eskimos," based on a structured pattern of potential alliance combinations. Rules for alliance structuring depended upon role position. The number of potential alliances, also role dependent, are "[e]vidence of enormously disproportionate opportunities for rich men to combine alliances ..." (Hennigh 1983:30).

There is no evidence of village rich men who were not whaling captains, but not all whaling captains were rich, thus all potential alliances were not necessarily activated by all *umialit*. Hennigh (1983:30) concludes that "[a]pparently, the social system was one which encouraged concentration of power in the hands of a few, and a scattering of alliances among the rest of the people."

Hennigh (1983:24) analyzed 10 named social positions for alliance potential. He glossed the positions as consanguine, affine, name partner, wealth-exchange partner, spouse-exchange partner, shaman, rich man, amulet partner, chief or organizer, and servant or helper. Interestingly, he included the *umialik's* position as rich man, but excluded the position and relations of whaling captain from his analysis (this was not by error, but for convenience). Some of the most intense interpersonal alliances were formed within whaling crews (see e.g., Spencer 1984:320; Maguire in Bockstoce 1988:1:106), suggesting that Hennigh understates the alliance potential of *umialit*. He also omits adoption partnerships. Data reviewed in Sheehan (1992) suggest orphan adoptions were mainly or uniquely found among *umialit*, as were cases of multiple wives.

Hennigh's results indicate more potential interpersonal alliances for rich men, and that their positions enabled them to forge stronger alliances. In most alliances, an *umialik* could count upon identifying two or even three different connections with the same person. In only one alliance might a rich man have only one connection to a person, when he is allied with another rich man.

> In every other case, the rich man must be in some additional alliance position. Such an individual has 16 possible opportunities to combine two types of alliance with someone, and 42 opportunities to combine three types of alliance.
>
> A shaman also has only one opportunity to be in a single alliance, in this case, with another shaman. A shaman has 10 opportunities to form two types of alliance, and eight opportunities to form three types of alliance simultaneously with another individual. An "ordinary" person, who is not a servant, has five possible ways to be in a single type of alliance, and eight possible ways to be in two types of alliance, where the second type is that of name partner. However, in North Alaska, personal names are distributed at random, and so in a "typical" situation, an individual would not have any possible means of forming more than a single type of alliance with someone (Hennigh 1983:28-30).

Hennigh (1983:29, Figs. 1 & 2) works out potential alliances in two matrix diagrams. The results are reproduced in Table 1:

Removing namesake alliances dramatically reduces the potential for an ordinary person by more than two-thirds, from 13 to 4, but discounting name alliances for a rich man does not proportionately reduce his potential alliances, which drop by less than one-third, from 59 to 42. The rich man in this configuration has a potential of more than 10 times the alliances of a typical man, 42 to 4. As Hennigh (1983:23) points out, the qualitative (i.e., multiple connections to the same person) and quantitative (by a factor of

Table 1. Hennigh's logical possibilities of multiple alliances (from Hennigh 1983:29, Fig. 3; Sheehan 1992:297, Tables 5 & 4).

Combinations	*Rich Man*	*Shaman (not rich)*	*Ordinary Person*	
			A	& B
One	1	1	5	4
Two	16	10	8	0
Three	42	8	0	0
Total	59	19	13	4

A = not a servant
B = not a servant or namesake

10) differences in alliance potential between the *umialik* and the ordinary person open a window on Eskimo social structure.

Religious aspects of the *umialik's* role are not included in the above computations. Communal ceremonies for the entire village, presided over by the most important of the *umialit*, created additional individual ties between them and each person in the village. Warfare was organized by a few of the *umialit*, but included all able-bodied men in a village, even those belonging to other *qargich* (Maguire in Bockstoce 1988:1:103, 177, 253). Thus, warfare provided an additional tie between each villager and the most important *umialit*. Hennigh's analysis, limited to individual ties, becomes more comprehensive when viewed in conjunction with community-wide activities.

SUBSISTENCE AND ITS RELATIONSHIP TO TRADE AND WAR

Cohen (1977) argues that human groups tend to create a hierarchy of values, and under stress to their food supply they attempt to preserve higher order values (e.g., type of food and manner of acquisition) unchanged by sacrificing lower order values (e.g., residential location). In other words, people will move if they believe they can then successfully replicate their previous food-getting activities. Cohen argues that subsistence activities tend to be very high order; a group will attempt to preserve subsistence pursuits intact above almost anything else, showing a conservatism when it comes to what they eat and how they obtain it. For example, Cohen's interpretation of the evidence suggests that big game hunters do not become broad spectrum hunter-gatherers unless big game hunting becomes locally impossible and they cannot move to new hunting grounds. Stress for whalers would have included anything that threatened the whaling schedule. For inlanders, stress was more direct: no caribou, no food. Following Cohen, people would have accentuated warfare, with all its attendant destruction, before they would have allowed their whale hunting schedules to be disrupted.

Cohen's data set does not include the Arctic, so arctic groups can serve as an independent test of his theory. In fact, arctic prehistory and early history seem to conform to Cohen's thesis (Sheehan 1992), with coastal villagers' attending to whaling over everything else, and inland folk attending to caribou hunting until they were *in extremis* as a result of declining trade with the coast in the late historic period at the end of the 19th century. This lack of trade resulted from a massive die-off of their coastal trading partners, during which time the inlanders moved to the coast and maintained the whaling focus there, joining and sustaining established but decimated whaling crews (Charles Brower n.d.; Burch 1981:17-18; Stefansson 1913:66-67). This suggests a widespread native perception, extending to inlanders, that whaling per se was the prime necessity for continuing traditional life, rather than generalized hunting access to other coastal resources that could be harvested almost anywhere. By forsaking their territory and their nomadism in order to whale, the inlanders demonstrated that their ultimate subsistence base was whaling.

During this period of upheaval, as the combined inland/coastal population plummeted, social complexity decreased: warfare disappeared and feuding remained as the most complex social expression of hostility, the native trading network fell apart, the communal whale cult disappeared, and eventually shamanism disappeared and only a personal religion of amulets, songs, names, and aversive strategies remained (Sheehan 1985 et seq.).

INLAND CARIBOU HUNTING SUPPORTED BY SEA MAMMAL OIL

There were few alternatives to local preferred subsistence pursuits in the Arctic. The larger populations grew, the fewer places on the coast that were unoccupied, so that at the time of maximum population there was literally no place to go that was not already spoken for (Burch 1981). Some number of villagers could have survived without hunting whales, but the food available from non-whale sources was not sufficient for the entire population at its maximum (Sheehan 1985). In this case, the preferred way of life eventually became the only feasible manner of maintaining life.

Group survival, facilitated by a largely successful attempt to avoid changes to the preferred local "way of life," was the ultimate impetus for trade between native whalers on the coast, inland caribou hunters, and native groups further to the east. Inlanders' subsistence focus upon caribou was very risky. Caribou populations can drop severely over a short time, and variations in caribou travel routes can be misjudged by hunters. In either case, these differences can lead to a failed harvest. One failed harvest might destroy a hunting group. Similarly, harvest timing was critical. The food value of caribou changes radically through the seasons of the year. A large harvest of meat with little fat can fill a hunter's belly and still starve him. Stefansson (1913:140) experienced the problems of an all-meat no-fat diet, which resulted in "symptoms that ... are practically those of starvation."

Recent archaeological research by Douglas Anderson (pers. comm., 1994) indicates that human population levels in the interior did not rise and fall as caribou populations rose and fell, and that while human population distributions seem to have changed as caribou populations fluctuated, these distributional changes were within home territories and did not result in changes in territorial population densities for people. Consumption of oil traded from the coast was one strategy that helped the interior people cope with these caribou shifts.

As Oswalt (1967:133-134) observes: "Those Eskimos who depended on caribou meat as the mainstay in their diet ... had to obtain additional fat for a balanced diet". Rasmussen (Ostermann 1952:140) also documents the need for fat. The inland groups who traded with the whaling villages were the first and only inlanders who did not disappear during prehistory. The main difference between the interior Eskimo from c. A.D. 1200/1400 to 1826 and the preceding interior occupations that disappeared is that the inland/coastal trade network was fully developed and available to support the later people.

Caribou hunters needed a backup system to ensure food for those possibly rare occasions of harvest failure, and to ensure fat for supplementing lean meat for out-of-season harvests. That backup system was trade with coastal groups for sea mammal oil. Backup mechanisms such as trading partners and trade fairs might be critical only once a generation, but constant maintenance through use was required to keep the trade network in place. Thus, long-term survival for inland groups depended upon continuing trade. Eskimos seem not to have received anything essential in trade from the inland Indians, but the Indians received "seal oil and other sea products" (Gordon 1906:75), and the full extent of Indian/Eskimo trade remains to be documented. Oil was available for an occasional critical role in survival, because it was in constant use for a variety of purposes: heating, lighting, preserving food, as a condiment, and as a food itself.

For inlanders and coastal people, caribou were also more than food. A family of seven probably needed between 57 and 66 skins for its own use each year (Foote cited in Spiess 1979:167) for such uses as clothing, sleeping mats, and tents. "Caribou were essential to Inuit. They provided food and, most important, the best fur for Arctic clothing" (Bruemmer 1995:61). Inlanders had to hunt a much larger number in order to procure trade stock as well, requiring them to schedule a great deal of time for hunting caribou. Oswalt notes that:

> If trading contacts with northern coastal Eskimos were essential for the inland peoples to survive, the reverse was also true. Above all else, coastal Inuit required caribou skins as parka materials ... Thus, the one inland product which could be supplied by caribou hunters was an essential need of the coastal peoples ... The overall trade complex indicates that in the north self-sufficient inland Eskimos did not exist and that coastal Eskimos in turn could maintain themselves without trade only if the caribou necessary for clothing skins happened to be available locally (Oswalt 1967:134-135).

In cases where timing was everything, failure to intercept the migrating caribou herd required an immediate decision: risk everything to try to catch up, when another failure might mean starvation, or forego the caribou harvest completely and move to the coast to try other resources for that year. As Burch (1981:61-62) notes, by the protohistoric period at the beginning of the 19th century, practically the entire coastline from below Point Hope to east of Point Barrow was under the control of the various whaling villages. Inlanders trying to occupy coastal areas would either have had to fight their way there or make arrangements with the territorial claimants. But, if they were assured of the availability of trade goods, especially oil, even if their harvest failed, then they could safely choose to hunt the caribou.

Food was the main value of the trade for inlanders. This position is supported by an example derived from Hall's (1971) data, showing that an intensively studied prehistoric inland household must have relied upon sea mammal oil to provide the margin of survival during their occupation of Kangiguksuk, located on the Noatak River, south of the Brooks Range. Hall (1971) excavated and analyzed a prehistoric interior occupation at Kangiguksuk dating to the late 16th century A.D. Spiess (1979:165-168) reanalyzed Hall's data and agreed in the essentials with Hall's methods and conclusions. Hall excavated almost the entire midden associated with this isolated household, including well-preserved faunal remains that represent continuous, year-round occupation. Hall's work at Kangiguksuk provides the basis for further insight into the necessary part that access to sea mammal oil may have played in maintaining a continuous inland occupation.

Using Foote's calculations of 57-66 caribou skins required each year to support a family of seven, the single family at Kangiguksuk would have needed an average of 70 caribou per year (Hall 1971:62, Table 10) or 279 over four years of occupation, leaving them with anywhere from four to 13 surplus skins to trade each year. This excludes any caribou hunted solely for skins, whose remains may not have entered the residential archaeological record. Most ethnographic accounts indicate caribou hunters, including

Indians, tended to kill as many caribou as possible, whether or not they could consume or even transport the meat (see e.g., Burch 1991:442). This suggests that the number of skins each inland family might have had available to trade was probably higher than the Kangiguksuk archaeological data indicate. Rasmussen (Ostermann 1952:140) recorded that inlanders traveled to the coast "to procure blubber, to buy blubber bags." If a whaling family needed 66 skins and traded for all of them, they would need about 13 seal pokes for trading, according to Rasmussen's calculated rate of five good caribou skins for "a sealskin full of blubber." Dr. Simpson (Bockstoce 1988:2:537) recorded that *umiaks* leaving Nuvuk to trade were usually so full of trade goods that men had to ride in kayaks to make room, while the women paddled the *umiaks*. A traveling *umiak* (as opposed to a smaller whaling *umiak*) would easily hold more than a dozen pokes.

Hall (1971:64) calculates that the Kangiguksuk family consisted of seven people in an analytical procedure and with a result that Spiess (1979:165) terms "ingenious." Hall (1971:60-61) uses faunal counts to determine a length of occupation based on available protein and carbohydrates. Spiess (1979:166-167) determines length of occupation based on protein and fat, arriving at essentially the same length of stay, 1500 days versus Hall's 1368 days.

Spiess' figures show the family could have eaten a calorically sufficient and balanced diet (fat and protein) for 1003 days at a group rate of 2 kg fat and 2 kg meat per day, after which they would have exhausted their caribou-derived fat supply. By consuming meat alone, they would have required 21 kg per day, so the remaining meat would have lasted another 465 days, with miscellaneous food adding 32 days (total of 1500 days or roughly four years). This proposed diet relies almost solely on caribou. As Hall (1971:53) observes, this family had a good four years. Even so, "the Kangiguksuk people extracted all possible fat from the caribou bones," and all caribou long bones found at the site were cracked for marrow removal (Hall 1971:59).

The reason for meticulous attention to fat is that without fat, meat consumed by the family increases from 2 kg to 21 kg per day (Spiess 1979:167), a factor of 1000%. Stefansson related from his own experience that:

> Now with a diet of lean meat everything was different. We had an abundance of it as yet and we would boil up huge quantities and stuff ourselves with it. We ate so much that our stomachs were actually distended much beyond their usual size — so much so that it was distinctly noticeable even outside of one's clothes. But with all this gorging we felt continually hungry. Simultaneously we felt like unto bursting and also as if we had not had enough to eat (Stefansson 1913:140-141).

If any fat were lost or consumed for heating and lighting at Kangiguksuk, the length of stay would decline considerably. If 2 kg of fat were diverted to heat and light, another 19 kg of meat were necessary for eating. If even 100 kg of the total 2007 kg of fat were redirected or lost, total length of occupation would drop by 45 days. Calculations are shown below:

Hall's basic data are 2007 kg fat, 11,782 kg meat:

at 2 kg fat and 2 kg meat/day, supply lasts 1003 days plus 9776 kg meat remains
at 21 kg meat/day, meat lasts another 465 days for total of 1468 days

remove 100 kg fat, new total 1907 kg fat with same 11,782 kg meat
at 2 kg fat and 2 kg meat/day, supply lasts 953 days plus 9,875 kg meat remains
at 21 kg meat/day, meat lasts 470 days for total 1423 days (45 days less)

remove 200 kg fat, new total 1807 kg fat and same 11,782 kg meat
at 2 kg fat and 2 kg meat/day, supply lasts 903 days plus 9,975 kg meat remains
at 21 kg meat/day, meat lasts 475 days for total 1378 days (90 days less)

As the calculations indicate, the redirection of each 5% of available fat from human consumption through loss or other uses results in a potential reduction of 45 days or over six weeks of occupation at the site.

Even a few days can make the difference between survival and death for groups who depend upon moving herds and other seasonally available resources. Seal oil makes the ideal insurance policy, because it is edible and does not deteriorate quickly (in addition to being a food, it is used as a preservative for other foods such as caribou meat, birds, and berries, which are dropped into entire sealskins that have been converted into containers called pokes). Hall does not report finding distal seal elements, the most likely remnants of sealskin pokes used to transport and store oil, but such ephemeral evidence might never have been deposited if the skins were reused for other purposes and taken offsite. Hall (1971:25) did find a cache that was probably a storage place for oil in a poke.

This particular site lies on the northern boundary of the tree line, and most or all domestic fuel was willow or spruce rather than oil (Hall 1971:47). People living just a little farther north would not have had the option of burning wood throughout the year.

The coastal whalers generally seem to have given more than they received when trading with inlanders, even in normal years (Charles Brower n.d.). Whalers must have been willing to extend "credit" to their trading partners. There were a variety of reasons for whalers to maintain a trade imbalance, but the result was that inlanders whose harvest had failed and who had little to offer in any particular year could still expect to receive a survival ration from their trading partners. This is similar in many ways to the relationship between traders, trappers, and hunters in the historic period as described by Thomas Paneahtak Brower (n.d.).

WHALE HUNTING SUPPORTED BY INLAND CARIBOU HUNTERS

What circumstances would allow coastal whalers to maintain a trade relationship that favored the inlanders with supplies, even if the inlanders had little to give in any particular year? First, coastal people had a very large surplus of a commodity that inlanders most needed, sea mammal oil. Prior to the depredations of the Yankee whalers, native whalers could be relatively confident of renewing the surplus each year. Therefore, holding onto oil would serve no purpose, once it was transported to the trading grounds. Second, trade enabled inlanders to continue hunting caribou when they might otherwise temporarily be forced to drop the hunt in favor of less rewarding but perhaps more certain pursuits. Coastal villagers' focus on whales could only be maintained by devoting a great deal of time to whaling and its preparations, and that meant procuring some caribou skins through trade. By adopting a long-term view of the trading relationship, the coastal people could more securely bind their inland trading partners and ensure their own long-term interests.

For an *umialik* with trading partners from more than one inland group, there often may have been at least one out-of-balance relationship. Significant status also was derived from having and supporting multiple trading partners, another factor that probably inclined *umialit* toward generosity.

Disproportionate giving in trade by whalers can be understood as a reasonable insurance tactic, and perhaps also as a way of reinforcing status similar to that seen among Northwest Coast groups. They usually had more than enough oil and meat for themselves from whaling, and could count on annual renewal of the harvest, but inlanders could not always count on caribou. Coastal villagers could also hope to obtain whale products from other villages if they did have a bad year. To avoid distractions from whaling, making sure of inland trading partners was advantageous to the whalers, whether they gave a great deal for each skin or whether they gave only on promise of a return the following year.

Nineteen-plus weeks per year were devoted to whaling (Sheehan 1992:302 Table 5-5), as follows: fall whaling took over two weeks and the associated festivals took another two weeks, winter festivals took over two weeks and whaling preparations took another four weeks, spring whaling took eight weeks, followed by over a week for *nalukataq* (another festival) and a short break. An additional 17 weeks each year were devoted to trading and travel to and from trade fairs.

A subsistence focus upon whales apparently was not risky in terms of food and fuel (blubber and meat), since the hunt seemed to steadily provide bowhead and beluga whales and walrus until Yankee whalers disturbed and then almost eliminated these resources (Bockstoce 1986:101).

But arctic coastal Eskimos did not live by food and fuel alone. Time invested in whaling meant they probably would get more than enough food and fuel for survival, but the necessary scheduling precluded or at least reduced other hunting opportunities. What whales provide, they provide in abundance, but there are other (mostly hunted) raw materials critical to arctic survival, including caribou, walrus, bearded seal, and smaller seals, various lithics for knapped and ground tools, soapstone for lamps, and wood for house construction and probably wood for manufacture of bows and arrow and kayak and *umiak* frames. Time devoted to whaling interferes with their accumulation. In addition to scheduling conflicts brought on by whaling, the increasingly large populations of coastal whaling villages simply placed too many people into circumscribed areas. While locally available non-whale resources remained static, the number of people and their needs for those resources climbed. The whalers settled on those few points of land where whaling was consistently productive, and those points were not necessarily the best places for maximizing the harvest of other game species.

The logistics of whaling were undermined when local village populations could no longer support themselves solely from land and sea resources nearby. To hunt away from the village at the wrong time could interfere with whaling. As other, non-whale game that was vital to survival became relatively less available per capita and with more people trying to capitalize on relatively inelastic local stocks of game, trade came to play an increasingly important role in complementing the pursuit of whaling.

As pointed out above, caribou products, especially skins, were critical to the whalers. Caribou also were the most time-consuming game for whalers to hunt themselves without interfering with whaling. For instance, Kashevarov (VanStone 1977:42) observed that fall whaling did not take place every year, but Maguire (Bockstoce 1988:1:76, 281) documented that when fall whaling was undertaken, whalers were on the coast at the same time the big caribou herds migrated in the interior. Bruemmer (1995:61) points out that "[t]he Inuit made every effort to kill caribou for winter clothing in August and early September, when the fur was of optimum quality."

Caribou skins made the best insulating garments, because of their hollow hairs. Substitutes were not as effective, and could cause hunters to "lose their edge" against the cold arctic environment, leading to death. For the same reason, skins had to be replaced annually. A used skin was not an adequate substitute for a new, clean, and intact one. Caribou skins also served as bedding in winter houses, where no clothing was worn. Other important caribou products included sinew for sewing and for bowstrings and backings on compound hunting bows, antler for tools and parts of composite tools (arrowheads, harpoon points, knife handles, etc.), and antler for armor slats which were sewn together into protective garments for war.

While many people from each village participated in trade, the *umialit* had the most goods to trade and maintained multiple trading partnerships. Some *umialit*, those few among the whaling captains who were also owners of *qargich*, exerted a level of control over all the trade within their villages. They occupied positions that allowed them to look to the long term in their trade relations, and that may have permitted them to enforce short-term sacrifice upon their fellow villagers (but not necessarily on themselves; Burch 1980:268). These sacrifices may have included inducing co-villagers to extend favorable terms to inlanders, or arbitrary reductions in their support of co-villagers when they transferred goods to favored trading partners.

The most powerful *umialit* could also impel co-villagers to undertake warfare, contrary to Burch's view (1974). For example, some *umialit* were able to control the entire mass of villagers. O-mi-ga-loon, Angunisua and Erk-sing-era organized an attack on Captain Maguire's *Plover* in the mid-19th century by *all* of the able-bodied men in the village of Nuvuk, and O-mi-ga-loon also called off the attack (Bockstoce 1988:1:103, 177, 253). These three men were allied *umialit* from one *qargi*, suggesting the possibility that one of the two Nuvuk village *qargich* was more important than the other.

While war may not have been conceptualized (seen emically) as an economic crusade, its etic results had significant economic impacts by cutting off trade to enemies, by clearing coastal areas of "foreign" villagers and opening these areas to exploitation, and by reducing populations on both sides, thereby reducing pressure on the possibly overloaded trade networks. It is worth considering the possibility that the virtual Armageddon in precontact times described by Burch (1974, 1981, 1988), wherein Point Hope suffered a significant population decline, was brought forth by the inability of the various trade networks to satisfy the logistic problems that, in turn, were caused by the creation of large protohistoric arctic populations in localized areas.

TRADE NETWORKS AND THEIR INTERRUPTION

Caribou products and oil merely topped the list of things for which people traded. Various groups apparently had relatively unequal access to a variety of game and materials. Among the whaling villages, walrus and *ugruk* (bearded seal) were not equally available to hunters (Sheehan 1992). Other items were available mainly through long-distance trade, such as metal (e.g., McCartney 1991), soapstone lamps, and wooden construction planks (Bockstoce 1988:1:239).

For most villages, the main source for one or more of these trade goods lay on the far side of at least one other village. In other words, the potential for trade disruption and/or "extortion" was present. There are numerous cases in the New and Old Worlds of associations between real or potential middlemen and the prosecution of warfare. One of the most spectacular examples concerns the Neutrals, Hurons, and Iroquois in the mid-17th century (Trigger 1976). Although Iroquoian war seems to have reached its highest pitch due to the introduction of trade goods during contact, the pattern of prehistoric wars seem to have grown out of similar indigenous trade issues.

There is some evidence to suggest that at least one whaling village, Pingasagruk (c. A.D. 1400-1700) on Point Franklin, was abandoned before Euroamerican contact due to economically driven warfare (Sheehan et al. 1991; Sheehan 1992). We have learned that Pingasagruk was comparable in size to its prehistoric contemporaries; it was abandoned at c. A.D. 1700 and was resettled generations later, during the 1870s in the historic period. Among potential reasons for abandonment, sociopolitical reactions to economic problems appear to be most likely. If trade in unevenly distributed resources between villages, and trade in oil and skins between inlanders and villagers could not keep up with growing human populations, the strain may have led to a new "solution" to that uneven access: warfare to control trade and/or to eliminate competing groups. War or loss of access to trade would have made life untenable for the people of Pingasagruk, also making it impossible for them to reestablish their community nearby, as was done by whaling groups known to have relocated for other reasons such as the Eskimos of Nuvuk (Maguire in Bockstoce 1988:2:378-379), which was eroding, and Point Hope, which had been inundated.

The material culture of the whaling villages, as represented archaeologically, has always suggested deep uniformities. Recently, it has been argued that these uniformities result not from uniform distribution of and access to raw materials (mainly, but not only, game) across the arctic seas and landscape, but from archaeological masking, attributable to a remarkably effective trade network that leveled out dissimilar access to critical resources (Sheehan et al. 1991; Sheehan 1992). Maguire's journals (Bockstoce 1988) suggest several material resources and game animals that were not equally available to the whaling villages. Walrus was abundantly available at Pingasagruk, but not at Utqiagvik or Nuvuk; this point is also emphasized by Charles Brower (n.d.) and Tom Brower (n.d.), was assert that *ugruk* was also more abundant at Point Franklin and Icy Cape. Soapstone lamps and prepared driftwood planking from the Canadian trade had to pass through the territory and perhaps the hands of the Utqiagvik and Nuvuk villages before reaching Pingasagruk or Icy Cape. The most important house floor planks were the ones into which the *katak* (floor level entryway) was cut, since they had to bear the load of people hoisting themselves up from the tunnel to the house floor. These were the largest and sturdiest planks in the floor, and the first to be recycled into new houses when old ones were abandoned. The trade from Canada in wood planking probably focused on these and other unique construction elements that would have been difficult to obtain from coastal driftwood near the whaling villages. As for walrus, today Point Franklin is the effective terminus of the annual walrus migration, although the extent is sometimes reported as "the Barrow vicinity" (Stoker 1984:a:64). Many observations in Maguire's journals (Bockstoce 1988) affirm the same distribution for the mid-19th century.

Our current field project at Pingasagruk is investigating this lack of uniformity in accessible raw materials, including research into biome gradations or changes from one village location to the next. Masked in the archaeological record is the material reflection of McCartney's (1991:38) "cultural homogeneity achieved through constant interaction." This leveling through trade may have promoted distribution of "valuable" trade items over long distances and "commodities" or bulk trade over relatively shorter distances, to follow McCartney's (1991:34, 38) phraseology, but in the case of whaling village *umialit,* they controlled most of both kinds of trade.

Trade in these Eskimo hunting societies was a necessity for their long-term survival. Trade was not based on rare or elite-oriented goods, but on utilitarian bulk commodities, all or almost all of which the various groups were capable of acquiring independently, but only at the cost of disrupting their annual schedules. Warfare arose from, and was driven by, the same factors that supported trade. Late prehistoric whale hunters and their contemporary interior groups traded as a means of survival, and their trade was underwritten by the whaling surplus. By maintaining and expanding upon a long-existing trading network, people incidentally maintained rank and obtained exotic goods, but these were by-products of the bulk commodities trade in necessities.

Coastal villagers required caribou skins, but could not consistently acquire all they needed by hunting without diverting effort from the whale hunt. Inlanders required sea mammal oil for heating, lighting, preserving birds and meat, and food. Caribou hunting, notoriously variable in its success, had to fail for only one season for interior populations to starve. Caribou hunters, even in relatively good seasons, required supplemental fat in the form of sea mammal blubber and oil. In bad seasons, oil and friendly relations with coastal whale hunters were their "life insurance policies."

Constant maintenance of trading networks was required to keep the networks available for times of necessity. Each village had the potential to serve as middlemen between some resources and some other villages. In the role of middlemen, they could facilitate or disrupt trade passing through their territories. Wooden planks and soapstone from Canada had to pass Nuvuk to reach Utqiagvik, then Pingasagruk to reach Icy Cape and, ultimately, to Point Hope. If the trade went cross-country and reached Point Hope first, products would then follow the reverse sequence. Trade between the coastal villages may not have been as important to the villages as was their trade with the inlanders, but it may have required as much attention from them. The Messenger Feast may have been one sign of traders' concern for intervillage relations, since the feasts were organized by the most important of the whaling captains, who also were the biggest traders.

CONCLUSION

It is postulated that regional northern and northwestern Alaskan Eskimo population increased from the time the villages were settled at c. A.D. 1200-1400, and that local hunting and trade could not keep pace. As a result, warfare was greatly intensified between c. A.D. 1600-1800, the period Burch (1981) documents as the height of regional warfare. Warfare declined first through precontact population losses from warfare itself and then through additional losses following European contact in 1826, when diseases killed more people and the decline of the whale stock left others to starve.

During the period of maximum population, when Point Hope society reached up to 1300 people and Nuvuk/Utqiagvik was over 700, the already complex social relations of the Eskimo whalers were entering a new phase. Villages exerted control over larger territories whose boundaries eventually met (Burch 1981), and where smaller villages like Pingasagruk (300+ people) might be eliminated or absorbed. If warfare had continued unabated through the 19th century (i.e., if there had been no contact disruptions), this new phase of social relations might have led to an entirely new level of social complexity. Without contact, would the eventual result have been a single regional power, at Point Hope or at the Nuvuk/Utqiagvik nexus, or would warfare have led to stabilization at a lower level of complexity by continuing to reduce population levels? In fact, dislocations and deaths associated first with war and then with contact reduced the population, eliminating the economic basis of warfare, so that resources available in the vicinity of each village more nearly matched the diminished demands of the survivors. Or, without contact, would war have ceased to be a continuation of politics and trade by other means, and "when it came in a 'true' form ... [would it have] proved to be a termination first of politics, then of culture, ultimately almost of life itself" (Keegan 1993:28)?

Acknowledgments. The National Science Foundation (OPP-9321112) is currently supporting a three-year arctic coast field project through Bryn Mawr College and the University of Indianapolis that includes the participation of the author, Gregory R. Reinhardt, Anne M. Jensen, John Bockstoce, and Robert L. Chufo, in cooperation with the North Slope Borough's Commission on Inupiat History, Language, and Culture. Earthwatch is joining the 1995 field season. This project is investigating some of the issues raised in this paper. Previous and current support for North Alaskan prehistoric whaling research has also come from the villages of Wainwright and Barrow, the Research Foundation of the State University of New York, the U.S. Fish and Wildlife Service, and the National Science Foundation (BNS 8214594). The comments of Allen McCartney and anonymous reviewers added greatly to the clarity of this paper.

REFERENCES

Bockstoce, John
- 1976 On the Development of Whaling in the Western Thule Culture. *Folk* 18:41-46.
- 1977 Eskimos of Northwest Alaska in the Early Nineteenth Century. *University of Oxford, Pitt Rivers Museum, Monograph Series* No. 1.
- 1979 The Archaeology of Cape Nome, Alaska. *University of Pennsylvania, University Museum Monograph* 38.
- 1986 *Whales, Ice, and Men: The History of Whaling in the Western Arctic.* University of Washington Press, Seattle.

Bockstoce, John (editor)
- 1988 *The Journal of Rochfort Maguire, 1852-1854: Two Years at Point Barrow, Alaska, Aboard HMS* Plover *in the Search for Sir John Franklin,* 2 Vols. Works Issued by the Hakluyt Society, Second Series No. 169. The Hakluyt Society, London.

Brower, Charles
   n.d.   MS. on file, unabridged source of *Fifty Years Below Zero* (1942). Dartmouth College Library Archives, Hanover.

Brower, Thomas Paneahtak
   n.d.   MS. on file, life history prepared with the assistance of Glenn W. Sheehan. Department of Anthropology, Bryn Mawr College, Bryn Mawr.

Bruemmer, Fred
   1995   Sentinels of Stone. *Natural History* 104:1:56-63.

Burch, Ernest S., Jr.
   1974   Eskimo Warfare in Northwest Alaska. *Anthropological Papers of the University of Alaska* 16(2):1-14.
   1980   Traditional Eskimo Societies in Northwest Alaska. In: Alaska Native Culture and History, edited by Y. Kotani and W. B. Workman, pp. 253-304. *Senri Ethnological Studies* No. 4. National Museum of Ethnology, Osaka.
   1981   *The Traditional Eskimo Hunters of Point Hope, Alaska: 1800-1875*. The North Slope Borough, Barrow.
   1988   War and Trade. In: *Crossroads of Continents: Cultures of Siberia and Alaska*, edited by W. W. Fitzhugh and A. Crowell, pp. 227-240. Smithsonian Institution Press, Washington, D.C.
   1991   Herd Following Reconsidered. *Current Anthropology* 32(4):439-444.

Cohen, Mark N.
   1977   *The Food Crisis in Prehistory: Overpopulation and the Origins of Agriculture*. Yale University Press, New Haven.

Foote, Don C.
   1965   *Exploration and Resource Utilization in Northwestern Arctic Alaska Before 1855*. Unpublished Ph.D. thesis, Department of Geography, McGill University. MS. on file, Archives, Elmer E. Rasmuson Library, University of Alaska Fairbanks.

Ford, James A.
   1959   Eskimo Prehistory in the Vicinity of Point Barrow, Alaska. *Anthropological Papers of the American Museum of Natural History* 47:1.

Gerlach, Craig and Owen K. Mason
   1992   Calibrated Radiocarbon Dates and Cultural Interaction in the Western Arctic. *Arctic Anthropology* 29(1):54-81.

Gordon, G. B.
   1906   Notes on the Western Eskimo: First Paper. *Transactions of the Department of Archaeology, Free Museum of Science and Art* II(1): 69-101 (plus plates). University Museum, Philadelphia.

Hall, Edwin S., Jr.
   1971   Kangiguksuk: A Cultural Reconstruction of a Sixteenth Century Eskimo Site in Northern Alaska. *Arctic Anthropology* 8(1):1-101.

Hall, Edwin S., Jr. (editor)
   1990   *The Utqiagvik Excavations*, 3 Vols. The North Slope Borough Commission on Inupiat History, Language, and Culture, Barrow.

Hennigh, Lawrence
   1983   North Alaskan Eskimo Alliance Structure. *Arctic Anthropology* 20(1):23-32.

Hickey, Clifford G.
   1979   The Historic Beringian Trade Network: Its Nature and Origins. In: Thule Eskimo Culture: An Anthropological Retrospective, edited by A. P. McCartney, pp. 411-434. *National Museum of Man, Mercury Series, Archaeological Survey of Canada Paper* No. 88.

Keegan, John
   1993   *A History of Warfare.* Alfred A. Knopf, New York.

Larsen, Helge
   1973   The Tareormiut and the Nunamiut of Northern Alaska: A Comparison Between Their Economy, Settlement Pattern, and Social Structure. In: *Circumpolar Problems: Habitat, Economy, and Social Relations in the Arctic,* edited by G. Berg, pp. 119-126. Pergamon Press, Oxford.

Larsen, Helge and Froelich Rainey
   1948   Ipiutak and the Arctic Whale Hunting Culture. *Anthropological Papers of the American Museum of Natural History* 42.

McCartney, Allen P.
   1991   Canadian Arctic Trade Metal: Reflections of Prehistoric to Historic Social Networks. In: Metals in Society: Theory Beyond Analysis, edited by R. M. Ehrenreich. *MASCA Research Papers in Science and Archaeology* 8(11):27-43.

McGhee, Robert
   1984   Thule Prehistory of Canada. In: *Handbook of North American Indians,* Vol. 5, *Arctic,* edited by David Damas, pp. 369-376. Smithsonian Institution, Washington, D.C.

Morrison, David
   1991   The Copper Inuit Soapstone Trade. *Arctic* 44(3):239-246.

Nelson, Edward W.
   1899   The Eskimo About Bering Strait. *18th Annual Report of the Bureau of American Ethnology for the Years 1896-1897.* (1983)

Ostermann, H. (editor)
   1952   The Alaskan Eskimos as Described in the Posthumous Notes of Dr. Knud Rasmussen. *Report of the Fifth Thule Expedition, 1921-24* 10(3). Copenhagen.

Oswalt, Wendell
   1967   *Alaskan Eskimos.* Chandler Publishing Company, San Francisco.

Rasmussen, Knud
   1927   *Across Arctic America: Narrative of the Fifth Thule Expedition.* G. P. Putnam's Sons, New York.

Sheehan, Glenn W.
   1985   Whaling as an Organizing Focus in Northwestern Alaskan Eskimo Societies. In: *Prehistoric Hunter-Gatherers: The Emergence of Cultural Complexity,* edited by T. D. Price and J. A. Brown, pp. 123-154. Academic Press, New York.
   1987   Accentuation of Inequality in Eskimo Whaling Societies During Early Contact. Paper presented to the Annual Meeting of the Society for American Archaeology, Toronto.
   1990   Excavations at Mound 34. In: *The Utqiagvik Excavations,* edited by E. S. Hall, Jr., 2:181-325, 337-353. The North Slope Borough Commission on Inupiat History, Language, and Culture, Barrow.
   1992   *Proto-Historic Social Organization of the Coastal Whaling Communities of North and Northwest Alaska.* Unpublished Ph.D. dissertation, Department of Anthropology, Bryn Mawr College.

Sheehan, Glenn W., Gregory A. Reinhardt, and Anne M. Jensen
- 1991 *Pingasagruk: A Prehistoric Whaling Village on Point Franklin, Alaska,* 2 Vols. SJS Archaeological Services, Inc.; reports prepared for U.S. Fish and Wildlife Service, Anchorage. Contract 14-16-0007-86-6612, pp. 1-199.

Spencer, Robert
- 1959 The North Alaskan Eskimo: A Study in Ecology and Society. *Bureau of American Ethnology Bulletin* 171.
- 1984 North Alaska Eskimo: Introduction. In: *Handbook of North American Indians,* Vol. 5, *Arctic,* edited by D. Damas, pp. 278-284. Smithsonian Institution, Washington, D.C.

Spiess, Arthur E.
- 1979 *Reindeer and Caribou Hunters: An Archaeological Study.* Academic Press, New York.

Stanford, Dennis J.
- 1976 The Walakpa Site, Alaska: Its Place in the Birnirk and Thule Cultures. *Smithsonian Contributions to Anthropology* No. 20.

Stefansson, Vilhjalmur
- 1913 *My Life with the Eskimo.* The MacMillan Company, New York.

Stoker, Sam W.
- 1984 Subsistence Harvest Estimates and Faunal Resource Potential at Whaling Villages in Northwestern Alaska. In: *Subsistence Study of Alaska Eskimo Whaling Villages,* Stephen Braund, Helen Armstrong, and Sam W. Stoker, Appendix A. 1983. Alaska Consultants, Inc., with Stephen Braund & Associates. Prepared for U.S. Department of the Interior, Anchorage.

Trigger, Bruce G.
- 1976 *The Children of Aataentsic: A History of the Huron People to 1660.* McGill-Queens University Press, Montreal.

VanStone, James W.
- 1962 *Point Hope: An Eskimo Village in Transition.* University of Washington Press, Seattle.

VanStone, James W. (editor)
- 1977 A. F. Kashevarov's Coastal Explorations in Northwest Alaska, 1838. *Fieldiana: Anthropology* 69.

Worl, Rosita
- 1980 The North Slope Inupiat Whaling Complex. In: Alaska Native Culture and History, edited by Y. Kotani and W. B. Workman, pp. 305-320. *Senri Ethnological Studies* No. 4. National Museum of Ethnology, Osaka.

And Then There Were None: The "Disappearance" of the Qargi *in Northern Alaska*

Mary Ann Larson
Oral History Program
Alaska and Polar Regions Department
Elmer E. Rasmuson Library
University of Alaska Fairbanks
Fairbanks, AK 99775

Abstract. *The activity of whaling penetrated nearly every aspect of coastal Inupiat life around the turn of this century. Economics, social organization and regulation, and ceremonialism were all intertwined with the pursuit of these "largest animals." The one place where all of these facets came together in the most obvious fashion was in the* qargi *or ceremonial house. By 1910, most of the* qargi *structures in Inupiat whaling villages had been abandoned or demolished, but that does not necessarily indicate that the institution disappeared with the buildings. This distinction is important for archaeological analyses, where the presence or absence of an institution tends to be tied to the structural forms associated with it. This paper focuses on the "disappearance" of both the* qargi *structure and institution and how archaeologists might be able to distinguish between the two.*

INTRODUCTION

From Beechey's (1831:260) first mention of an Alaskan *qargi*, explorers and ethnographers have noted the existence of *qariyit* and have analyzed their place in northern social organization. Archaeologists, however, have not focused on the *qargi* to the same extent that their cultural anthropological colleagues have. Although many archaeologists have paid passing attention to the *qargi*, most have been content to identify the physical remains of the old *qargi* buildings. Very little emphasis has been placed on analyzing its actual role in North Slope communities, where it served as both a structure and an institution. This latter point is particularly important, as the structure of the *qargi* became separated from the institution of the *qargi* when the ceremonial houses were demolished in the early years of the 20th century. If archaeologists look only to the presence of the physical structure to indicate the existence of the *qargi* institution, they will be missing the mark, as the institution lived on in many and various forms long after the permanent structures had disappeared.

This paper focuses on Inupiaq whaling villages of the northwestern and northern Alaskan coast, specifically the settlements at Cape Prince of Wales, Tikigaq/Point Hope, Utkiagviq/Barrow, and Nuvuk/Point Barrow. This study addresses the role of the *qargi* in these communities, the destruction of *qargi* structures, and the subsequent retention of the *qargi* institution when the buildings were no longer extant. Considering the centrality of the *qargi* to whaling, this information will have a bearing on much of the archaeological analysis pertaining to northern whaling communities.

The sources utilized for this research are diverse and require some explanation. I have cited archival collections, published works, "grey" literature, and existing oral history collections. I have also consulted elders and other residents of Barrow, Fairbanks, Kotzebue, Point Hope, and Wales during research undertaken between May, 1991, and November, 1993. These are the "interviews" noted by number throughout the text. They are referenced numerically in order to maintain the anonymity of consultants, while at the same time illustrating the range of individuals cited. Because of its historical association with shamans, the *qargi* can be a difficult subject for some people. Therefore, although everyone involved in these discussions signed release forms, I am hesitant to publicly tie particular pieces of information to certain individuals.

The *qargi* has manifested itself in a number of different forms in Alaska. Based on dialect, language, or regional preference, this structural type has been referred to variously as a *kazigi* (Giddings 1961; Anderson 1988), *qalgi* (Burch 1981), *karigi* (Spencer 1959), or *kashim* (Nelson 1899), and it boasts (by the author's count) over 72 orthographies. For the purposes of this study, I am adopting the orthography of the North Slope Borough Inupiat History, Language, and Culture Commission (singular form *qargi*; plural form *qariyit*).

The *qargi* has had a number of functions and physical forms, varying according to region and cultural group. Because of many researchers' familiarity with Yupik *qariyit*, the *qargi* is often stereotypically thought of as a "men's house." Indeed, in southwestern Alaska, the Yupik version of the *qargi* has at least partially served the purpose of a "men's house," because it served as a dormitory, sweatbath, and workshop for the men of the community (Nelson 1899:286-287). It was never utilized exclusively by men, however (Larson 1989). Women entered the *qargi* throughout the day to bring food, and they also took part in the ceremonial life of the community which was centered in the *qargi* (Nelson 1899:287). The emphasis of this paper is on the Inupiat *qargi*, which differed somewhat in function. It had at least four different structural manifestations, the permanent and temporary coastal forms and the permanent and temporary interior forms (for a more complete discussion of types, see Larson 1989). This study focuses on the *qariyit* in the coastal communities, and more particularly on those in the larger whaling villages. In these settlements, there were usually multiple permanent *qariyit* whose members often comprised rival factions in the community. At various points in time, Cape Prince of Wales had two to four *qariyit* (Trollope 1855:863 cited in Ray 1983; Anderson and Eells 1935:68), Utqiagvik/Barrow had three (Murdoch 1892:79; Spencer 1959:183), Nuvuk/Point Barrow had two or three (Simpson 1855:259; Murdoch 1892:79; Spencer 1959:183), and Tikigaq/Point Hope had either six or seven depending upon the source cited (Spencer 1959:186; Rainey n.d.b; hereafter, archival manuscripts are identified by authors' initials followed by a document box number).

The permanent *qariyit* were usually built much like normal dwellings except on a larger scale (Ahngasak n.d.). They were semisubterranean sod structures supported by whale bones or driftwood, and they often had subterranean entrance tunnels, although that was not always the case (Rainey n.d.b). They differed from most of the houses in the community in that they had benches along all four walls but had no sleeping platforms (Larson 1989:37, 42). Wooden floors often marked dance or ceremonial staging areas within the structure, while the ceilings were adorned with animal carvings and the inflated bladders of the whales which had been killed by members of the qargi (Rainey n.d.b).

The coastal whaling settlements also used a number of temporary forms of the *qargi* even when permanent forms were in place. Snow houses, upturned *umiat*, and tents were all considered by consultants on the North Slope to be *qariyit* proper. Snow houses tended to be used for work on whaling equipment (Interview 1; Rainey n.d.b) while the *umiak* windbreaks were used as *qargi* settings during the annual Nalukatuq celebrations following the whaling season (During these festivities, food and presents were given away by the crew or *qargi* hosting the event, and events such as the blanket toss were held.) Large skin tents were put into service for very large gatherings or in coastal locations with no permanent *qariyit* (Interview 1; Foote n.d.a).

There is also a form of the *qargi* referred to as a *qargiruk* or a "small *qargi*." These were often structurally the same as various temporary forms, but made to a smaller scale. One elder described the difference as follows: "That's a celebrating house, a *qargi*. *Qargiruk* is a smaller eating place. *Qargi* is a large, *qargiruk* is a small." This same elder also talked about a *qargiruk* as a place for "lunch" (as opposed to dinner or feasts; Interview 1). After the destruction of the permanent *qargi* structures, the *qargiruk* were very important in allowing the people of the North Slope to continue certain *qargi* traditions. This will be discussed in more detail elsewhere in this paper.

FUNCTIONS

To understand the role of the *qargi* in a community, it is first necessary to understand the numerous functions that it served both structurally and institutionally at the turn of this century. These functions fall into six basic categories: (a) whaling crews and preparation, (b) feasts and celebrations, (c) shamanic performances, (d) games and competitions, (e) social regulation, and (f) miscellaneous functions. I have tried to group them according to how my consultants discussed them and how topics were mentioned in previous interviews. It should be noted that none of these categories is exclusive of any of the others. I have organized them in this manner out of a need to simplify, but they are all inextricably intertwined, and this breakdown into topics should not be taken to suggest that these functions are separate from one another.

Whaling Crews and Preparation

In the larger coastal settlements, one of the primary functions of the *qargi* was to serve as a center for whaling preparations and crew organization. Each whaling crew captain or *umialik* (plural, *umialgich*) was affiliated with a particular *qargi* and the crew members were, in turn, members of that same qargi. *Qargi* and crew membership had much to do with the redistributive system, something which crosscut many of the categories but was ultimately a major function of the *qargi*. When whales were caught, the various blubber and meat cuts were distributed along crew lines first, then according to *qargi* affiliation, and then to the community as a whole, but crew and *qargi* membership determined which cuts an individual would receive. Food was also distributed over the course of the year in conjunction with various whaling activities (e.g., work on the *umiat*, Nalukatuq celebrations, and Flipper Feasts), and crew members especially could count on their *umialgich* supplying them with food year-round (see e.g., Rainey n.d.b).

A considerable amount of work was necessary to prepare for the whaling season. Traditionally, it was considered essential for the boats or *umiat* to have new, clean skin coverings each year. The whales appreciated clean, white boats and gave themselves to the crews who showed respect for them in this manner (North Slope Borough 1992). Therefore, one of the major tasks that had to take place each year was the recovering of the *umiat* (Interview 1). This could take place either in the permanent *qargi* structures or in temporary snow houses, although there are indications that in some villages a man could not prepare his whaling gear inside the permanent *qargi* unless he had caught at least five whales (Rainey n.d.b; Foote n.d.b). This regulation probably was meant to relieve overcrowding in the *qargi* during whaling preparations, as there were often six to nine boats per *qargi* (Rainey n.d.b) and it would be impossible to recover that many *umiat* at one time in such a limited space.

The snow houses were usually built in February or March, when other preparations (such as sewing new *umiak* covers or repairing gear) were also beginning. Crew members would first build or repair the *umiat* and then they would recover them once the frames were solid (Interview 1). Some elders remember that as children they were given food at the snow house *qariyit* while this work was ongoing, so the *umialik* was apparently responsible for distributing food to people other than his crew at this time of year (Rainey n.d.b).

After the initial preparations were made, there was a time of "sitting-in," when the *umialgich*, their *angatkot* (shamans; singular, *angatkok*), and their crews would sit in the *qargi* for a number of days, singing the songs associated with that *qargi*. These songs were meant to bring them closer to the whale. After the sitting-in time, other ceremonies were conducted prior to the beginning of the whaling season, including predictions by *angatkot* and the burning of the *poroks* (animal figures; Rainey n.d.b).

During this season of preparation, instructional activities occurred in the *qargi* as well, and the social organization was cemented through various activities. For example, it was common at this time of year for elders to remind the younger *umialgich* and crew members of the proper whaling procedures (Rainey n.d.b), something which became more pronounced during the periods of rapid change resulting from contact with Euroamerican whalers. The elders needed to remind the young of the traditional ways and the *qargi* was the forum. Traditional social organization was also reinforced in the *qargi*, both through seating arrangements and through the display of whale bladders on the ceiling. Although the younger *umialgich* might have been the ones catching the most whales, the older *umialgich* had the central seats in the *qargi*, the places of honor. If the younger *umialgich* needed to be reminded of why, they only needed to look to the ceiling, where the bladders of the whales caught by the older *umialgich* were strung (Rainey n.d.b).

Social organization was not the only function of the *qargi* in this respect. It served as a form of social regulation of whaling as well. In the *qariyit* of some villages, for example, there were lessons on game control. One source mentions that *umialgich* were limited to three whales per season, a regulation which was discussed in the setting of the *qargi* (Andrews n.d.).

During the whaling season, the crews were out on the ice, so there was not much *qargi* activity until a crew caught a whale, and then preparations for Nalukatuq would begin. Although Nalukatuq was (and still is) held outside rather than in the *qargi*, at Point Hope at least each crew celebrated at the Nalukatuq area affiliated with its *qargi*. Nalukatuq celebrations were highly competitive, with each crew and *qargi* trying to hold their celebrations first and each attempting to outdo the others in terms of generosity (Rainey n.d. b). At Nalukatuq, gifts were given to the crew by the captain's wife (Rainey n.d.a), and food was distributed to the community as a whole (again, with the precise cuts determined by one's crew and qargi affiliation). This redistribution was taken to the extreme for *umialgich* catching their first whale. Instead of being able to keep part of the whale for himself, "[a] captain's first whale is all given away" (Rainey n.d.b). At Point Hope, a captain was also expected to give away all of his whaling gear and equipment upon catching his first whale (Burch, pers. comm.). This was in keeping with the traditional image of an *umialik* as a generous person.

Celebrations and Feasts

Generosity was one of the tenets exhibited through the annual cycle of celebrations and feasts. Not only were these events generally held in the permanent *qargi* structures, but they were often organized by particular *qariyit*, which is one of the reasons one elder referred to the *qargi* as "the celebrating house" (Interview 2). Depending upon the scale of the celebration in question, redistribution of food and goods at these events could range from a local to a regional level.

Point Hope is one of the few places where feasts between local *qariyit* were documented prior to the destruction of the permanent structures. One of Rainey's consultants in the 1930s talked at length about the trade that went on between Ungasiksikaaq *qargi* and Qagmaqtuuq *qargi* at annual feasts (Rainey n.d.b). These celebrations featured exchanges between established trading partners in the opposite *qargi*, and they later mutated into a Christmastime feast after the introduction of Christianity.

Similar redistributive patterns occurred on a larger scale during regional Kivgiq feasts (also known as "Inviting-in" or "Messenger" feasts; see Spencer 1959:210-228). On these occasions, a runner would invite entire villages to a celebration to be hosted by a specific *qargi*. According to one elder, these might be held whenever a particular community had a good year or had otherwise managed to save food for such an event. People would travel for days to attend Kivgiq, where they would trade with established partners and participate in intervillage and/or inter*qargi* competitions of skill and strength. Their contacts at these occasions were partially structured by their *qargi* affiliations, which gave strangers from other communities a way of categorizing each other (Rainey n.d.b).

Other, less formal celebrations and dances would also occur between villages and *qariyit* over the course of the year. At all of these, large and small, formal and informal, food was distributed to the community as a whole, with an emphasis on feeding the poor, thus fulfilling the distributive function of the *qargi* (Rainey n.d.b,c).

Shamanic Performances

When missionaries first arrived on the North Slope, they objected to the *qargi* because it was the domain of the *angatkok* or shaman. For many missionaries, dancing was not an obstacle to Christianity nor was feasting, but the ceremonies and seances conducted by the *angatkot* were viewed as the work of the Devil.

Every whaling crew had an affiliated *angatkok*, so shamanic performances and predictions featured heavily in the preparations for whaling season. The *angatkot* were also involved in the ceremonialism associated with many of the celebrations and feasts, but there were occasions when they were called upon to do their work in the *qargi* independent of other events. Ceremonies were held for healings and for shamans to make songs, and in times when food was lacking, the *qariyit* would be specially opened so that the shamans could call the game back to the community (Rainey n.d.b).

Other descriptions of *angatkok* performances also appear in the literature and in references made by consultants, but in many cases the exact reason for these performances is not as clear as in the instances noted above (e.g., Interview 2; Rainey n.d.b). Generally, these seances probably took place as a means of bolstering the status of the *angatkok*, the *qargi*, and the community as a whole, particularly because many of these performances appear to be done in direct or indirect competition with other *angatkot*. Since the services of *angatkot* were secured and compensated by a particular *qargi*, it was in their best interests to make themselves and their *qargi* look as powerful as possible.

Games and Competitions

One of the ways that individuals gained status within *qariyit* was through various games of strength and skill. Running and jumping competitions were ways of displaying strength to opponents (Andrews n.d.), and these often took place at Nalukatuq as well as at the inter*qargi* or regional feasts and celebrations. For example, one of the better known competitions on the North Slope was the race to the host *qargi* that was held at the beginning of every Kivgiq feast. The winner secured both status for himself and for his *qargi*, for which his victory won the right to host the next Kivgiq (Rainey n.d.b; Interview 3).

Again, the *qargi* also served as a means of categorizing strangers, although in this case through competition rather than through established *qargi* affiliation. One consultant related that when an unknown individual came into a community, the "boss" of a *qargi* (i.e., the dominant *umialik*) or other people in the village would start inviting that person to the *qargi* to do what this elder called "testing it, how they strong" (Interview 1). By putting an individual to various competitive tests in the *qargi*, a community could evaluate the strength or intelligence of the newcomer, thereby estimating the threat he might pose. In the old days, whether or not a totally unknown person was treated in a friendly or a hostile manner might depend on how well he performed tests of skill and strength in the *qargi*.

Social Regulation

The *qargi* also served as the "courthouse" and council room in the larger coastal villages, as it was the place where justice was dispensed and community decisions made. One elder from Barrow related the story of a woman who gossiped too much. The council decided that her spreading of gossip had begun to damage the community, and they disciplined her by cutting her mouth open even further on both sides, allowing her to "open her mouth" even wider to talk about others (Interview 3). Besides being a punishment and judgment on the woman, her disciplining served as a constantly visible reminder to the community of the evils of gossip. Other stories of *qargi* discipline include tales of women who refused to marry and were subsequently punished for their disobedience by being subjected to severe pain and humiliation (Foote n.d.a).

The *qargi* also served as the "courthouse" for happier occasions, however, because it was in the *qargi* that a marriage was signified. No special ceremonies were performed, but it was announced to the community that a woman had married a man when she brought him fish or meat at his *qargi* (Mozee n.d.).

Miscellaneous Functions

One of the standard forums for communal cultural transmission was the *qargi*, where stories, songs, and traditions were passed from one generation to another. This obviously took place during the normal work, celebrations, and feasts, but there were also times set aside for other pursuits. In some cases, storytelling, joking, or song-singing would take place informally at the end of the day, but there were also specific times of the year dedicated to it (Rainey n.d.b). One of Don Foote's consultants described one such occasion, *oui-veer-aazi*, as the time before the hunting ice came when people stopped regular work for work in the *qargi* and for storytelling (Foote n.d.a).

Among all the *qargi*'s functions, perhaps the most mundane but most common was that of workshop. It has already been mentioned that work in preparation for the whaling season took place in the *qargi*, but so did everyday tasks throughout the other months that the *qargi* was open (generally early October through January and again in the spring; Rainey n.d.b; Foote n.d.a,b).

In Wales, the *qargi* served the additional purpose of sweatbath, which would have been unusual for an Inupiat qargi. Generally speaking, the sweatbath was only associated with *qariyit* in the Yupik areas. This function at Wales may have been a recent aberration, however, because no mention is made of sweatbathing in Wales until 1935 (Ross 1958:197).

SPATIAL EFFECTS

The *qargi* was not merely the focal point of a community's social life, however. It was also a spatially organizing force in residential patterning. In the larger Inupiat villages having more than one *qargi*, neighborhoods appeared in association with the *qargi* structures. House clusters were mentioned in passing by early explorers and ethnographers, and they have been discussed by more recent researchers as well. Ernest Burch has conducted one of the most in-depth studies of neighborhoods in Inupiat communities, and although he attributes the formation of these clusters to kinship ties (Burch 1975:22-24), there is also a correlation between *qargi* affiliation and these neighborhoods. Norman

Chance, basing his claims on Burch's work, notes clusters of houses into neighborhoods, particularly relative to population centers such as Point Hope and Barrow (Chance 1990:21). Spencer, although differing with Burch on the nature of *qargi* affiliation, still agrees that "through the karigi, a focus was given to the physical arrangement of the village" (Spencer 1959:51).

THE "DISAPPEARANCE" OF THE *QARGI*

The *qargi*, like all structures and institutions, existed within a particular historical context. Although we have archaeologically-known remains that have been labeled *qariyit*, we cannot be certain that institution was the same as the one we now recognize. This fluidity of institutions as they operate within historical frameworks is something we would do well to remember as researchers. Accepting this, it is important to note that the *qargi* under consideration in this paper is the one which existed from the time of the first extensive contact with Euroamericans (around the 1870s) to the present day. There obviously have been changes in the role of the *qargi* during this period, but many of these have been recorded through oral histories and various documents which allow us to trace alterations in the use of the *qargi*.

Loss of the *Qargi* Structures

The influx of missionaries into coastal areas was at least partially responsible for the abandonment and/or destruction of the *qargi* buildings. Dr. John Driggs founded the St. Thomas Mission at Point Hope in 1890, the same year in which Harrison Thornton and William Lopp established a mission school at Cape Prince of Wales (Episcopal Church n.d.; Thornton 1976:xiii). The Yukon Presbytery started the Utqeagvik Presbyterian Church at Barrow nine years later (Yukon Presbytery n.d.).

Many of the missionaries associated the *qargi* structures only with shamanic activities and "pagan" ceremonialism, and considered them places of evil that needed to be eradicated (Larson 1992). Some missionary concern over the *qariyit* may have been based on early ethnographic reports, which would have been the major source of available background information on the North. Dall, for example, tied shamanism to the *qariyit* (which he called *casines*) in his explanation of the origins of the word "Eskimo," giving the impression that shamanism was an integral part of the Inupiat identity. "The Indians call the Innuit and Eskimo "Uskeemi" or sorcerers. Kaguskeemi is the Innuit name for the *casines* in which their shamans perform their superstitious rites. From this root comes the word "Eskimo" " (Dall 1884:144). Murdoch's 1892 report of the Point Barrow Expedition outlined a number of ceremonies held in the *qargi*, including a detailed description of a *qargi* meeting designed to drive out *tungak* spirits (Murdoch 1892). Simpson described shamanistic *qargi* ceremonies as well in his 1855 report of information collected while wintering near Nuvuk with the ship Plover (Bockstoce 1988).

Compounded with its automatic affiliation with shamans, perhaps another difficulty concerning the *qargi* was that many nonspecialists at the turn of the century viewed all Arctic Eskimos as being alike. Missionaries trying to ready themselves for this field of service would have had more access to information from the Eastern Arctic in the late 1890s, including literature on the natives of Greenland. In western Greenland, there was an institution called a *qassi* (a word similar to the Alaskan *qargi* or *kazigi*), and this structure served as a "festival house." The *qassi* was defined in an 1804 Greenlandic dictionary as "a little house without windows 'where both sexes used one another indiscriminately'" (Kleivan 1984:601). A footnote in Murdoch's (1892:80) 1892 notes one definition of the Greenlandic term *kagsee* as "a brothel." If the missionaries had any knowledge of the ill repute of this more eastern institution, added implications of immorality may have been added to the Alaskan *qargi*. This disregards the fact that the buildings were also used for community feasts and celebrations, social regulation, and redistributive purposes.

Of the settlements under discussion here, Barrow was the first to lose its *qariyit*. In the winter of 1900-1901, the structures were not only abandoned but were demolished and used for firewood (Brower n.d.:573). In this case, the missionaries disapproved of both the shamanic associations and the perceived behavior of young people in the *qargi*. Charles Brower disagreed not only with the action, but with the reasoning. "One of the biggest mistakes, I think, was in tearing down the dance houses for fuel. While this put an end to young people freely congregating there at night and sometimes, no doubt, doing things they shouldn't, it didn't improve matters to take away this common family rendezvous" (Brower 1965:232).

The exact date is not known for the abandonment of the one *qargi* that remained in Nuvuk in 1895, but based on Brower's texts it would seem to have occurred at approximately the same time as the closing of the *qariyit* in Barrow (Spencer 1959:49). The *qariyit* structures at Point Hope remained intact during Dr. Driggs' tenure there, but they were deserted during or just prior to 1910, after the Reverend Augustus Hoare had been the resident missionary for three years. The details of their abandonment are unclear.

The permanent *qargi* at Cape Prince of Wales was the longest-lived of the group by far, lasting into the early 1960s (Interview 4). By that time, however, it was being used only occasionally for sweatbathing by the older men of the village, and had only been utilized for that purpose for decades (Ross 1958:116; Interview 4). Actually, despite the presence of the *qargi* at Wales, some of the traditional *qargi* functions took place in other buildings starting sometime after the mid-1930s.

Structural Substitutes

In spite of the fact that the permanent *qargi* buildings disappeared in most North Slope communities before 1910, the institution lived on as the activities were carried out in various other structures. This is important from the standpoint of archaeological interpretation, because it means that we cannot necessarily equate the abandonment of a structural type with the abandonment of the institution housed within it.

The permanent and temporary structures in which *qargi* functions were held were very similar from one community to the next. In many instances, villages utilized the existing temporary forms of *qariyit* for the functions traditionally handled by the permanent structures. As was mentioned previously, *umiat* were often recovered in snow houses if there was not room for all the boats in the *qargi* or if an *umialik* did not yet enjoy the privilege of bringing his boat into the permanent *qargi* for repairs. After the abandonment of the old *qargi* buildings, however, almost all *umialgich* worked on their boats in snow houses. Various sources reference this behavior in Barrow (Interviews 1 and 5), Wainwright (Andrews n.d.), and Point Hope (Rainey n.d.b). Overturned *umiat* were also used, as they were in previous years, as windbreaks during Nalukatuq celebrations and general gatherings following the whaling season at Barrow (Interview 1) and at Point Hope (Foote n.d.a).

The *qargiruk*, or "small *qargi*," began to serve the function of the *qargi* proper, generally taking the form of a skin tent. Dances at Point Hope often took place in these more temporary structures (Interview 2). The *qargiruk* had the advantage of being portable, and one consultant from Barrow mentioned that it was utilized at fish camps over the summer when people were out of view of the missionaries (Interview 1). In this way, a number of the *qargi* ceremonies were kept alive in spite of a disapproving missionary presence in the community. Much of the ceremonialism associated with whaling ceased to be practiced around this time, however, because people were in town and in proximity to the missionaries during the season that most preparations of that type would have taken place.

In Wainwright, following the demise of the *qargi* building, the school was a primary location for "Eskimo dances"(Andrews n.d.; Mozee n.d.), games and competitions, preparation for whaling season (Andrews n.d.), and everyday tasks. William Van Valin's journal entry for April 8, 1914, reads in part, "Much and divers (sic) work going on in the schoolroom. Making and mending sleds — snowshoes washing filingsaws — making knife handles — ivory pulleys — mending igloo tent poles ..." (Van Valin n.d.). Schoolrooms were also utilized in Point Hope for dances and council meetings (Rainey n.d.a) and in Barrow for *umiak* recovering (Van Valin n.d.). In Wales, the school became the meeting place for the men (Ross 1958:116).

Although schoolhouses became the primary location for *qargi* activities, other large, spacious structures were utilized as well. Browning Hall at Point Hope was built to be a community gathering place, and it certainly served that purpose. Traditional *qargi* shows, complete with animated whales and marmots, took place there prior to whaling season, and other performances and ceremonies using mechanized animals were held at Browning Hall throughout the course of the year (Rainey n.d.b). People also used the space to hold dances and to recover *umiat* (Foote n.d.a,b). Point Hope had a number of other options as well, including the Armory and the store for dances (Foote n.d.a; Rainey n.d.b) and the Community Hall for Flipper Feasts, which were an integral part of the redistributive system affiliated with whaling communities (Foote n.d.a).

Although "Eskimo dances" were held on a regular basis in schoolrooms or other public venues in both Barrow and Point Hope, there were times when people would have dances in private homes (Rainey n.d.b; Interview 1). In Barrow around the 1930s, there was a visual code to allow people to know in which house a dance would be held. The day of the dance, a kerosene lamp with a burlap sack over it would be placed in front of the house, and then people would gather there later in the evening (Interview 1). The consultant relating this information could not say whether this had been done to avoid missionary attendance or if it was simply a way of announcing a dance.

Loss of the *Qargi* as an Institution

Of the communities discussed in this paper, only Point Hope still retains a *qargi* institution in a form resembling that of the turn-of-the-century. In Point Hope, individuals still maintain *qargi* affiliations (Qagmaqtuuq and Unasiksikaaq), the *qariyit* still host inter-*gargi* feasts, and there is still a fierce rivalry between the two factions of the community. Whaling crew captains have strong *qargi* ties and Unasiksikaaq and Qagmaqtuuq still maintain ownership of *qargi* songs. At the old village site 2 mi to the west of the present community, there were two Nalukatuq sites, one for each *qargi*. When the village was moved in 1976, two new Nalukatuq areas were established at the new town site according to the old traditions. The Unasiksikaaq area was placed closer to the "Point," and the Qagmaqtuuq site was placed further from the "Point" (Interview 6). The institution of the *qargi*, particularly as it pertains to whaling, is still very much alive in Point Hope in many ways.

At Barrow and Wales, the story is very different. Wales, although it retained the *qargi* structure longer than its counterparts did, did not retain the institution, at least in any recognizable form. Although sweatbathing continued in the village until the early 1960s, this was only a small part of the *qargi's* role, and, as was mentioned previously, that particular role may be a fairly recent one.

The community of Barrow does not maintain *qargi* affiliations, and much of the ceremonialism and other aspects of the *qargi* related to whaling has gone by the wayside in past decades. Recently, however, there has been a resurgence in interest in some of the feasts and celebrations and in the idea of the *qargi* as a community center for intergenerational contact. Much of this has resulted from efforts by the Inupiat History, Language, and Culture Commission as well as by other individuals in the community, and this move has resulted in the reinstitution of Kivgiq (Messenger feasts) on the North Slope.

CONCLUSIONS

The question remains, where does this leave us as researchers, and, more specifically, where does this leave archaeologists? It is obvious that we must be careful of the assumptions that we make about institutions and the structures that house them. First, the example of Point Hope illustrates that we cannot equate the disappearance of a structural form with the disappearance of the institution housed in that structure. Second, and conversely, the case at Wales shows that the existence of a building type does not necessarily indicate maintenance of an associated institution. And, Barrow serves to remind us that institutions, or facets of them, can be reborn, with or without accompanying structures. For archaeologists studying past whaling communities, however, particularly in relation to societal interactions, it is crucial to know whether or not the *qargi* as an institution was present. So, how to do it?

The conclusion we must ultimately reach is that we cannot make assumptions about the existence of social institutions based simply on the presence or absence of building types alone. There needs to be some other corroborating thread of information that assists us in drawing conclusions of this type.

In the case of the *qargi*, there are two possibilities. First, since we do not know where all of the *qargi* activities were held following the demise of the permanent structures, we cannot necessarily search for structural remains. Instead, we need to look for the specialized artifacts associated with the *qargi*. These will vary from location to location, based on the specific functions of the *qargi* in different communities. For example, one might expect to find a woven grass respirator in association with sweatbaths in Wales, but not in association with *qargi* structural replacements in Point Hope or Barrow, where sweatbathing is not known to have been a *qargi* activity. Where animated or mechanized animals were part of *qargi* shows, such artifacts could indicate a continued existence of an institution apart from a particular structure. So, too, with paraphernalia related to whaling ceremonies such as the "sitting-in time" or to different feasts and dances.

The second possibility would be to look for other visible effects of the *qargi*, such as the organization of neighborhoods or house clusters, or the maintenance of specific ceremonial areas, such as the two Nalukatuq sites at Point Hope. Determining the nature of neighborhoods is a complex endeavor, but if the information to do such an analysis is available, it could serve to corroborate the existence of the *qargi* institution.

Archaeologists have often been wary about delving into the realm of social institutions, and with good reason. The material record can be sketchy, and, depending up the assumptions we make, it can be misleading. We must be careful that we are not too simplistic in our analyses. There needs to be information that corroborates the initial assessment, whether it be artifactual data, spatial patterning, or some other visible aspect of the material record, so that we do not let the jury decide the verdict based only on a single line of circumstantial evidence.

Acknowledgments. Portions of this research were made possible by grants from the Geist Fund at the University of Alaska Fairbanks and the Alaska Humanities Forum and by nonmonetary support from the North Slope Borough Inupiat History, Language, and Culture Commission. I would like to thank Karen Brewster and Connie Oomittuk, who provided me with housing and other support when I was in Barrow and Point Hope, and Douglas Anderson, Craig Gerlach, William Schneider, and Rose Speranza, who have contributed comments and encouragement throughout this research. I would also like to express my appreciation to Ernest Burch for his generosity in sharing with me the results of his research in the Point Hope area. Above all, however, I am grateful to those many individuals in Barrow, Point Hope, Wales, Kotzebue, and Fairbanks who patiently answered my questions and instructed me in my ignorance. Berna Brower, Levi Griest, Jana Harcharek, Terza Hopson, Bertha Leavitt, and Arthur Neakok of Barrow, Dinah Frankson, Leon Kinneavauk, Carol Omnik, Kirk and Rosemary Oviok, and Alice Weber of Point Hope, Ralph Anungazuk of Wales, Rachel Craig of Kotzebue, and Ron Senungetuk of Fairbanks all advised me during my research. I owe a great debt to them all, and I hope that I have used the information that they gave me in a responsible and appropriate way.

REFERENCES

Ahngasak, R.
    n.d.   Oral history interview on file at the Oral History Collection, Alaska and Polar Regions Department, Elmer E. Rasmuson Library, University of Alaska Fairbanks.

Anderson, D. D.
    1988   Onion Portage: The Archaeology of a Stratified Site from the Kobuk River, Northwest Alaska. *Anthropological Papers of the University of Alaska* 22(1-2).

Anderson, H. D. and W. Eells
    1935   *Alaska Natives: A Survey of Their Sociological and Educational Status.* Stanford University Press, Palo Alto.

Andrews, C.
    n.d.   Archival collection on file at the Alaska and Polar Regions Archives, Elmer E. Rasmuson Library, University of Alaska Fairbanks. Box 4 of the collection.

Beechey, F. W.
   1831   *Narrative of a Voyage to the Pacific and Beering's Strait to Co-operate with the Polar Expeditions Performed in His Majesty's Ship Blossom ... in the Years 1825, 26, 27, 28.* Colburn and Bentley, London.

Bockstoce, J. (editor)
   1988   *The Journal of Rochfort Maguire, 1852-1854.* Hakluyt Society, London.

Brower, C.
   n.d.   The Northernmost American: An Autobiography. MS. on file, Naval Arctic Research Laboratory, Barrow.
   1965   *Fifty Years Below Zero.* Dodd, Mead and Co., New York. (1942)

Burch, E., Jr.
   1975   *Eskimo Kinsmen: Changing Family Relationships in Northwest Alaska.* West Publishing Co., New York.
   1981   *The Traditional Eskimo Hunters of Point Hope, Alaska: 1800-1875.* North Slope Borough, Barrow.

Chance, N.
   1990   *The Inupiat and Arctic Alaska: An Ethnography of Development.* Holt, Rinehart, and Winston, Fort Worth.

Dall, W.
   1884   On Masks, Labrets, and Certain Aboriginal Customs, With an Inquiry Into the Meaning of Their Geographical Distribution. *3rd Annual Report of the Bureau of American Ethnology for the Years 1881-1882.* Washington, D.C.

Episcopal Church
   n.d.   Archival collection on file at the Alaska and Polar Regions Archives, Elmer E. Rasmuson Library, University of Alaska Fairbanks.

Foote, D. C.
   n.d.a   Archival collection on file at the Alaska and Polar Regions Archives, Elmer E. Rasmuson Library, University of Alaska Fairbanks. Box 5 of the collection.
   n.d.b   Archival collection on file at the Alaska and Polar Regions Archives, Elmer E. Rasmuson Library, University of Alaska Fairbanks. Box 53 of the collection.

Giddings, J. L.
   1961   Kobuk River People. *University of Alaska Studies of Northern People* No.1.

Kleivan, I.
   1984   West Greenland Before 1950. In: *Handbook of North American Indians*, Vol. 5, *Arctic*, edited by David Damas, pp. 595-621. Smithsonian Institution, Washington, D.C.

Larson, M.
   1989   An Interpretation of the Band 3, Level 2 Choris House at the Site of Onion Portage, Alaska. MA paper, Department of Anthropology, Brown University, Providence.
   1992   I Saw Jesus and the Whale Fly By. Paper presented at the 1992 Annual Meeting of the Alaska Anthropological Association, Fairbanks.

Mozee, B.
   n.d.   Archival collection on file at the Alaska and Polar Regions Archives, Elmer E. Rasmuson Library, University of Alaska Fairbanks. Box 1 of the collection.

Murdoch, J.
   1892   Ethnological Results of the Point Barrow Expedition. *9th Annual Report of the Bureau of American Ethnology*, pp. 19-441. Government Printing Office, Washington, D.C.

Nelson, E.
   1899   The Eskimo About Bering Strait. *18th Annual Report of the Bureau of American Ethnology*, Government Printing Office, Washington, D.C.

North Slope Borough
   1992   *Uiniq: The Opeen [Open] Lead*. North Slope Borough, Barrow.

Rainey, F.
   n.d.a   Archival collection on file at the Alaska and Polar Regions Archives, Elmer E. Rasmuson Library, University of Alaska Fairbanks. Box 1 of the collection.
   n.d.b   Archival collection on file at the Alaska and Polar Regions Archives, Elmer E. Rasmuson Library, University of Alaska Fairbanks. Box 2 of the collection.
   n.d.c   Archival collection on file at the Alaska and Polar Regions Archives, Elmer E. Rasmuson Library, University of Alaska Fairbanks. Box 5 of the collection.

Ray, D. J.
   1983   *Ethnohistory in the Arctic: The Bering Strait Eskimo*. Limestone Press, Kingston, ON.

Ross, F.
   1958   *The Eskimo Community House*. Unpublished MA thesis, Department of Anthropology, Stanford University, Palo Alto.

Simpson, J.
   1855   Observations on the Western Esquimaux and the Country They Inhabit. *Great Britain, Parliament, House of Commons, Sessional Papers, Accounts and Papers, 1854-55* 35(1898).

Spencer, R.
   1959   The North Alaskan Eskimo: A Study in Ecology and Society. *Bureau of American Ethnology Bulletin* 171.

Thornton, H.
   1976   *Among the Eskimos of Wales, Alaska 1890-93*. AMS Press, New York.

Trollope, H.
   1855   Proceedings of Her Majesty's Discovery Ship "Rattlesnake". *Great Britain, Parliament, House of Commons, Sessional Papers, Accounts and Papers, 1854-55* 35(1898).

Van Valin, W.
   n.d.   Archival collection on file at the Alaska and Polar Regions Archives, Elmer E. Rasmuson Library, University of Alaska Fairbanks. Box 1 of the collection.

Yukon Presbytery
   n.d.   Yukon Presbytery Minutes. MS. on file, Elmer E. Rasmuson Library, University of Alaska Fairbanks.

Also, the collections of the following individuals or groups in the Rasmuson Library, Alaska and Polar Regions Archives include documents, field notes, journals, and letters of the following persons: Clarence Andrews (CA), the Episcopal Church (EC), Don C. Foote (DCF), Ben Mozee (BM), Froelich Rainey (FR), and William Van Valin (WVV).

Paul Silook's Legacy: The Ethnohistory of Whaling on St. Lawrence Island

Carol Zane Jolles
Department of Anthropology, DH-05
University of Washington
Seattle, WA 98195

Abstract. *A considerable body of archival data exists concerning whaling practices on St. Lawrence Island, Alaska, in the early 20th century. Numerous researchers, including Alexander Leighton, Dorothea Leighton, and Charles Hughes, have utilized this information. What is seldom discussed, however, is that much of the data relies on the contributions of a single indigenous consultant, Paul Silook. In this article, I discuss the influence of Paul Silook's work on our understanding of the ethnohistory of St. Lawrence Island whaling traditions, and its implications in terms of ethnographic representation. The article also discusses the role of the "other" in ethnographic and ethnohistorical research as well as the influence of indigenous consultants on the interpretation of culture.*

INTRODUCTION

In the summer of 1940, Paul Silook (Fig. 1), a Yupik resident of Gambell, St. Lawrence Island, Alaska, recorded his life story for anthropologist Alexander Leighton. Silook's recitation was unusual both for its length and because, as it later turned out, it would be his last important collaboration with researchers. His homeland, St. Lawrence Island, is the largest of the Bering Sea islands and sits at the center of what William Fitzhugh and Aron Crowell (1988) have called the "crossroads of continents." Its pivotal location at the entry into Bering Strait proper caught the attention of explorers, traders, and whalers from the beginning of the 18th century onward. By the early 20th century, researchers had begun to excavate its ancient shoreline sites, hoping to track the evolutionary history of human entry into the so-called New World and to track the history of northern indigenous peoples whose populations reached from the present Russian Far East to the eastern shores of Greenland. The archaeological remains, including human and sea mammal bones, uncovered on the island by Otto Geist (Geist and Rainey 1936), Henry Collins (1937), Hans Georg Bandi (1969), J. Louis Giddings, James Ford, Ales Hrdlicka (1930), and others revealed a society which had, for some 2000 years, made its living from the careful exploitation of marine mammal resources.

Figure 1. Photograph of Paul Silook and his wife, Margaret (Dorothea Leighton, Collection, acc. #84-31-22N, Archives, Alaska and Polar Regions Department, Elmer E. Rasmuson Library, University of Alaska Fairbanks).

For the last several hundred years at least, a major component of the island residents' economic and social systems has been whaling, especially bowhead whaling. Even the earliest underground dwellings, sketched and described by Edward Nelson (1899) and later excavated by Henry Collins, Otto Geist, and J. L. Giddings, among others, included the massive crania of the bowhead as structural elements, while mandibles comprised the uprights and curved roof supports. Not surprisingly, then, when cultural anthropologists began to document the ethnohistorical and ethnographic record for the island, whaling practices, both cultural and economic, figured significantly. Paul Silook, whose own family history tied him to the archaeological site at Kukuleq on the northern shore of the island and to Ungazik or Chaplino on the Siberian shore, was intimately involved in each of these documentation projects.

THEORETICAL BACKGROUND

This article focuses on Paul Silook's remarkable career as an indigenous St. Lawrence Island historian and self-educated and trained archaeological assistant. I suggest that energetic and influential individuals such as Silook influence the shape of the historical understanding of their communities. Silook's knowledge and the degree of recognition which his contributions have received are the central theme here. Together, these highlight the theoretical and pragmatic problems of ethnographic representation, the "other," and the nature of ethnographic authority which continue to hold Native and non-Native researchers alike in their interwoven grip.

Ethnographic research on St. Lawrence Island began with the visit of Edward Nelson and his companion, John Muir, in June, 1881. Nelson collected substantial archaeological and human skeletal materials for the fledgling Smithsonian Institution. The two were followed by Waldemar Bogoras who worked under the auspices of the Jesup North Pacific Expedition. Bogoras' monumental study of the Chukchi and his compilation of traditional stories (*ungipaghaate*)[1] in a smaller companion volume, *Eskimos of Siberia*, contain numerous references to St. Lawrence Island. His data, collected from 1887 to 1903, are concerned with the full range of northern coastal life, but deal with St. Lawrence Island only as an adjunct to the Siberian coast. By 1912, however, physical anthropologist Ales Hrdlicka and his student, Riley Moore, began a combined social and physical study of St. Lawrence Island residents. It is this research, which relied heavily on the willing cooperation of Silook as local consultant, which serves as the starting point for this discussion.

The principal concern, already noted, is the theoretical engagement with process and representation through which the complex relationship between ethnographer and consultant and the textual representation of conversations from "the field" are sorted out. For the ethnographic enterprise, it has to do with the notion voiced by Vincent Crapanzano that "... speech and writing ... are themselves interpretations" (Crapanzano 1986:52). It also focusses on the problem which has gained so much currency in the 1980s and 1990s: defining and redefining the "other." Crapanzano has discussed this in terms of interpretation, ethnographic authority, and the boundaries assumed to operate among the ethnographer, the ethnographer's readership, and the subjects of ethnographic texts. Particularly applicable is his comment on the historicity of ethnographic texts:

> ... ethnography is historically determined by the moment of the ethnographer's encounter with whomever he is studying ... (Crapanzano 1986:51)

At the same time, he notes that the researcher's ability to render the observed, that which is potentially culturally "foreign" into the familiar, while at the same time convincing readers of the authenticity of observation and explication is also relevant. George Marcus and Michael Fischer (1986) and others have referred to this as the "current crisis of representation," one most often played out in ethnographic texts. We are asked to consider the interesting question of the degree to which any ethnographer's major scientific contributions are the products of personality and an informed imagination. As a corollary, to what degree such contributions are informed by a "... largely unique research experience to which only [the individual researcher] has practical access in the academic community ..." (Marcus and Fischer 1986:21).

For Native American researchers, and particularly northern researchers, the questions of "the other" and of ethnographic authority are closely articulated with important ethical and social questions of indigenous representation and acknowledgment in contemporary texts. For Native historians on St. Lawrence Island, such as Anders Apassingok

---

1. Spelling of Yupik words comes from Badten et al. (1987).

and his colleague, Willis Walunga, who have themselves engaged in extensive life history and oral narrative research (Apassingok et al. 1985, 1987, 1989; Walunga 1987), ethnographic authority is conceptualized in terms of giving voice to the wisdom of local elders. Thus,

> ... the wisdom can be used in the formal education of [their] grandchildren ... the words ... are more than just facts or history. Behind the words [of elders] is the heartbeat of a people. We [elder historians] want this heartbeat to live on in our children (Apassingok, in Apassingok et al. 1985:xv-xvi).

Partly in recognition of the concerns of elder historians such as Apassingok and Walunga and partly to grapple with the issue of "... how to represent cultural experience ... anthropology's claims to provide authoritative interpretations of culture are being challenged from both inside and outside the discipline" (Cruikshank 1990:1). Thus, Native life history narrative texts, once utilized primarily for the "flavor" of personalized human experience which they could add, now receive greater attention because they provide alternative windows into indigenous histories. Some notable illustrations include studies edited by Margaret Blackman (1989), Julie Cruikshank (1979, 1990), Robin Ridington (1988), and William Schneider (1986). Additionally, northern anthropologists and historians have begun to give special attention to local voices to serve as the basis for historical works. An important example is Ann Fienup-Riordan's *Boundaries and Passages: Rule and Ritual in Yup'ik Eskimo Oral Tradition,* which attempts, in the author's words, to "... address the organization of ethnographic detail by the anthropologist and the place of that detail in the lives of the people the anthropologist represents ..." (Fienup-Riordan 1994:xiii). According to Fienup-Riordan:

> The result is not an ethnography ... but a history in which I have used the oral texts as an archive, citing the orators as authors (Fienup-Riordan 1994:xv).

Silook, who died in 1946,[2] represents a different facet of the problem of ethnographic authority, interpretation, and representation through his legacy of recorded data. As with any ethnographer, we must consider him in historical context. However, as "silent author" of ethnographic texts, his role is somewhat unique. As I will try to demonstrate, much of our contemporary knowledge of traditional Yupik whaling practices on St. Lawrence Island in the early 20th century can be traced directly to Silook's critical role as "key informant" or Native consultant or Native local expert.[3] Silook was one of several men and women living in Gambell during the first half of the century who worked with researchers. His contributions are distinguished by sheer volume and by his singular interest and devotion to the research process itself. Although any number of local individuals worked as consultants (informants) and as workmen, Silook became a research assistant to the scholars who inevitably came to rely on his assistance. While

---

2. This is an approximate date. Penapak, Silook's oldest daughter, was unable to give an exact death date.
3. These changing terms suggest the evolution of attitudes toward indigenous people manifested among anthropologists over the last 100 years.

each of the researchers mentioned above,[4] along with cultural anthropologists Alexander Leighton, Dorothea Leighton, Charles Hughes, myself, and presumably many others, are deeply indebted to the extensive work of this one remarkable indigenous scholar, he has received little if any public recognition.[5] Silook produced several hundred pages that document local cultural practices in his community. Of particular interest here are his contributions to the ethnohistory of marine mammal hunting practices, especially St. Lawrence Island whaling traditions.

Even if we put aside the interesting question of Silook's viewpoint as an indigenous scholar, the representative character of his information is an important concern, because so many researchers have relied on Silook's work. At the center of that concern is Silook's clan or *ramket* affiliation. The existence of clans on St. Lawrence Island is a contested issue among anthropologists (see e.g., Hughes 1958, 1960, 1984; Jolles 1991; Schweitzer 1989, 1992). Regardless of how scholars label them, however, *ramket*, sub-clan segment, and lineage membership was an accepted element of Sivuqaq (St. Lawrence Island) social organization and concomitant daily practice in Silook's time. It is still an important element in contemporary St. Lawrence Island society. Membership was, and continues to be, defined patrilineally. While clans themselves derived, most probably, from extended family settlement sites, these same sites are now associated with *ramket* membership and with specific cultural practices, including *ramket*-specific names, hunting traditions, and specialized whaling traditions and practices reserved to themselves. Traditions of other *ramke* (plural) "belong(ed)" to "those people over there," an often repeated phrase in the present village of Gambell, and therefore, either in theory or in actual practice, remained outside the realm of an individual's own *ramket*-defined knowledge and experience.

This quality of membership is particularly important when one considers how much of contemporary ethnohistorical research has depended on Silook as the ultimate resource. Silook's primary *ramket* association was as a member of Aymramka, a *ramket* or clan-type unit associated most immediately with Siberian Yupik immigration to the island from Chaplino in the late 1800s. Through his mother, he retained some identification with Kukuleq (Kukulghet).[6] The practices which distinguished Aymramka from other large *ramke* units like Pugughlmiit (people of Southwest Cape), while they might be utilized politically, do not actually distinguish between "Siberian" and St. Lawrence Islander. Rather, they represent micro characteristics of lineage-based and more general practices of each individual *ramket*. These differences are often minor in the sense that they refer to slightly different ceremonial foods or practices which grew out of specific encounters of individual members of a single lineage with a particular animal. Nevertheless, they

---

4. Nelson, of course, came to the island before Silook was born, while Bogoras never spent time on the island, collecting his data from Siberian Yupik relatives on the Russian shore.
5. I do not mean to suggest that researchers have not utilized other consultants in the field; however, Silook's recorded data is unparalleled in its breadth. His willingness to describe processes inherently sensitive also sets him apart from most consultants of his time.
6. The Kukleghmiit were persons from the well-known archaeological site, Kukulik, excavated by Geist, Rainey, and Collins, among others. Kukuleq was abandoned at the time of depopulation (1878-1880).

take on a special character in demanding allegiance to one's *ramket* unit. In the present context, issues of interpretation and ethnographic authority, embedded in Silook's cultural representation of particular lineally-based practices, integrally affect present understanding of the island's whaling traditions.

SILOOK AND THE ETHNOHISTORY OF ST. LAWRENCE ISLAND WHALING TRADITIONS

According to his autobiography, recorded in 1940 (Leighton 1982), Paul Silook was probably born in September, 1892, some 12 years after severe famine and disease had reduced the island's population by as much as 1000-1500 persons.[7] At the time of his birth, the small remaining island population had begun to concentrate at the contemporary village of Gambell or Sivuqaq, located at the island's northwestern tip (Northwest Cape), 38 mi from the Chukchi Peninsula and the shores of the former Soviet Union. It was here that the first school had been built under the direction of Sheldon Jackson in 1891. By 1894, the first school teacher-missionaries, Vene Gambell and his wife, Margaret, had taken up residency. Thus, Silook was born into a society in which major changes had already been effected. Primary among these were the reorganization of social life following the depopulation and the introduction and/or imposition of English literacy and Christianity upon the traditional Sivuqaghhmiit (people of Sivuqaq) world. When Silook began his education, these forces shaped both his world view and his future livelihood. He was strongly influenced by school teachers and missionaries, and expressed in his own words:

> In school I was promoted from my schoolmates, from my grades, in one year. And when I can read a little, then my thought was, my idea yielded to white man's way. Then I am in third grade, good speller and can work all the works of that grade. Then I skipped fourth grade and already took up fifthgrade (Silook 1940:1).

As he reflected on the early stages of his growing-up years in June, 1940, Silook commented to cultural anthropologist Alexander Leighton: "I love to write and read ..." (Silook 1940:13). For Leighton, Silook both wrote and related his autobiography, cited above. Aware of his desire to keep on writing in a world where supplies were few, Leighton "... gave him a package of note book paper" (Silook 1940:13).

In 1894, when the first school teachers began their work, St. Lawrence Island had a population of approximately 220 persons, residing mainly in Gambell and at Southwest Cape (Pugughileq). One of the first tasks of teachers was to communicate with their constituents, identifying, if they could, translators who could assist both in the school-

---

7. Edward Nelson (1899: 269) described the depopulation at the present site of Gambell and elsewhere on the island as follows:

    In July I landed at a place on the northern shore where two houses were standing, in which, wrapped in their fur blankets on the sleeping platforms, lay about 25 dead bodies of adults, and upon the ground and outside were a few others. Some miles to the eastward, along the coast, was another village, where there were 200 dead people. In a large house were found about 15 bodies placed one upon another like cordwood at one end of the room, while as many others lay dead in their blankets on the platforms ... the bodies of the people were found everywhere in the village as well as scattered along in a line toward the graveyard for half a mile inland.

room and in the pulpit. The original translators were, quite naturally, the few men who had worked on whaling ships. Their English, at least according to Vene Gambell, the first teacher, appears to have been limited to a colorful mix of expletives and profanities:

> Their few English words, picked up from whale-men and smugglers, were mostly terrible oaths, and still more revolting expressions. As they crowded forward laughing, they poured forth a torrent of this awful language. Of course, they did not in the least comprehend what it signified to us, and later we learned that all this was only their way of making us welcome. But you can imagine how shocked we were, and with what haste I conducted my wife to the school building (Gambell 1910:5).

The school room became almost immediately a training ground for younger, more malleable, and, to the teacher-missionaries, more acceptable translators.

Silook, by his own account, was nine years old when he began his formal schooling. For him, the school, built in 1891, would have been a fixture in the village landscape. As a child, he could have seen the comings and goings of at least two of the village's three teachers and their companions or wives. Silook's experience of school was special. His teacher was Edgar Campbell, a Presbyterian minister and medical doctor who, along with his wife and children, spent 10 years in the community (Campbell 1904-1911). He singled out Paul as one of his most able students. In the beginning, Paul was a kind of teacher's pet, later a teacher's assistant. As one of several "promising" young men, he was chosen by the doctor to be trained as a reindeer herder, an occupation which did not particularly attract him. Campbell encouraged him in more intellectual work as well: "In 1903 I became transferred for translator [interpreter] in woman's class …" (Silook 1940:1).

Eventually, Silook translated not only for Campbell, but for those who followed Campbell into the community: "Then when another teacher comes, I still interpret …" (Silook 1940:1). As a teenager, he traveled by ship to Nome, his trip subsidized by yet another teacher. In 1923, long after Campbell's departure, he served as village teacher as well as substitute teacher, as translator in the church, and as Sunday school teacher. Schooling fueled his endless curiosity about his world and about the outsiders, *laluramka* (white people),[8] who had become more than sporadic visitors in his community. Throughout his life, in fact, Silook was involved with the school and the various *laluramka* teachers and ministers who served in the village.

Although Silook's formal schooling was limited to the elementary years, he found in literacy a release for his restless energies and for his creative intelligence. In a nonliterate society, he became literate in a second language and used his skills not only in the new social institutions described above, but he took it upon himself to become a self-styled ethnographer for his society. He had, in fact, a profound sense of history. In his early years as Campbell's translator in the Presbyterian Church, he had honed his skills. His English, as J. Louis Giddings described it, was shaped by his familiarity with Biblical language and phraseology:

---

8. From the Chukchi, meaning "bearded clan" people.

At Gambell I engaged new helpers, one of whom, Paul Silook, introduced himself in surprisingly good English almost the moment I arrived in the village [September, 1939] ... The teachers had told me that Paul was a good worker and was the "literary" man of Gambell, and it took me only a few hours to learn why he had this reputation. He had learned English by working for the missionaries who had lived there when he was a child. Learning to read from mission books, he had absorbed biblical terms, which he used both in ordinary speech and in the interpreting he did for me when I recorded ethnographic accounts from the oldest people in the village (Giddings 1967:167).

A quiet, seemingly modest and unassuming man, Silook was remarkable in his persistent willingness to seek acquaintance with outsiders. His first active engagement was as translator and cultural facilitator for Riley Moore. Moore, the young graduate student assistant of Hrdlicka, spent the summer of 1912 in Gambell; Silook was then 20 years old. In typical fashion, Silook left no clear account of how he happened to have been recruited by Moore. He does describe, however, how he obtained a local shaman to perform for Moore. Moore had no command of Yupik and must have depended on Silook to explain and to interpret. The result of Moore's visit is a single publication, "The Social Life of the St. Lawrence Island Eskimo," published in 1923. However, Moore left notes of his visit that are now deposited in the National Anthropological Archives at the National Museum of Natural History (Smithsonian Institution). These notes include descriptions of ceremonies and related activities associated with whaling. Given the similarity of these descriptions to those recorded by Silook himself (see e.g., Geist 1928-34 and Leighton 1982), it seems logical to attribute Moore's data to Silook, either in his role as translator or as consultant, or more likely some combination of the two.

Silook's willing cooperation derived, I would suggest, from several motivations. First, he was an intellectual, a quality noted by school teachers, missionaries, and by anthropologists such as Giddings and Dorothea Leighton. It is not possible to know if he had begun to take up his pen at this juncture, but later he became a prolific writer. Second, he had been strongly influenced by the missionary voice of his teacher, Edgar Campbell, and later, he described his own growing skepticism regarding the efficacy of the various hunting and curing rituals his family expected him to perform. Third, he evidenced a willingness not only to discuss what he had seen and heard, but the very unusual quality of willingness to speculate on the possible intent or meaning of the ritual activities common to his community. Thus, he was quite liberal and intellectually daring in a conservative society. Finally, for Moore to have described what he had seen in great detail, he would have had to employ someone who was not terribly threatened by the spirit forces dominant in the traditional religious system. Silook's increased participation in and devotion to the Presbyterian Church presumably freed him to some extent from traditional concepts.[9] In essence, he became a "transitional" man, intellectually and emotionally in transformation between older religious traditions of his society and the new socio-relig-

---

9. Recent conversations with Silook's daughter suggest that Silook was an extremely religious man who read the Bible with his family in the evenings. It seems likely then, that he drew on his new religious conviction as a source of empowerment which allowed him to speak freely of many of the religious ceremonies and practices associated with traditional whaling in the community without undue fear of the spirits associated with the ceremonies.

ious system introduced by missionaries. Silook's writings offer the following brief reference to Moore:

> About in, I forget what year, Dr. Riley Moore came here, and I was writing stories for him. These were my first writings. He asked me to ask one of the sorcerers to come and do some tricks for him. So, I asked Otoiyuk,[10] the same sorcerer that treated me when I had the nervousness.[11] So he consented (Silook 1940:45).

Moore's descriptions of whaling ceremonies, other related hunting ceremonies, gift exchange ceremonies, and curing ceremonies remain among the most detailed records of specific ritual practices for St. Lawrence Island. While much credit certainly goes to Moore for the careful detail apparent in his recorded observations, it is equally important to acknowledge Moore's substantial indebtedness to Silook. Practices related to whaling described by Paul and recorded by Moore included the following:

(a) "Atoghok" (ateghaq)[12] — a term actually used to refer to one segment of the overall ceremony. According to John Apangalook, Silook's younger brother:

> The term used when people left the house to go the boat was *ateghaq*. So the ceremonial clothing was called *ateghaasit*" (Apangalook, in Apassingok et al. 1985:207).

The ceremonial term derives from its more common use, meaning "to go out or go down." It became known somewhat erroneously among teachers and missionaries as "moon worship." The ceremony, which precedes whaling, was performed by each boat crew in conjunction with a specific phase of the moon at the end of January, in February, March, or April. Part of the formalities of the ceremony included food offerings to ancestor spirits and to powerful spirit beings, but most especially Kiyaghneq. The ceremony initiated whaling for lineages, sublineages, and *ramke*.

(b) "Ock kuh sah took" [*aghqesaghtuq*] took place following the close of whaling season. While Moore recorded that it only took place in those years when a whale had been taken, according to elder Lincoln Blassi:

> When the boating season was just about over we would then go and have another ritual. When the ritual was about ready, the boat captain would let his relatives and the elders of the clan know. If they had killed a whale they would take along the *teghrughaaq* [center creased portion of the tail] (Blassi, in Apassingok et al. 1985:243).

---

10. Otiohok probably refers to Jimmy Otiyohok, his uncle, who was a shaman in his early years, but a passionate Christian in his later years.
11. Here Silook refers to a time when he was afflicted with the kind of confusion which often preceded the acquisition of power through the establishment of a relationship with a spirit force. It is possible that Silook was supposed to inherit a spirit assistant or *tughneghaq* from one of his patrilineal ancestors. If that was the case, he had the potential to become a shaman. In such a circumstance, the *tughneghaq* generally approaches the individual and attempts to establish ties. Such occurrences still plague present residents on the island who are in line, traditionally, to inherit such powers.
12. Wherever possible, I give the contemporary spelling of the ritual or ceremony, based on the current orthography and spelling system developed by St. Lawrence Islanders, with the assistance of Steve Jacobson and Michael Krauss of the Alaska Native Language Center at the University of Alaska Fairbanks and David and Marilene Shinen of the Wycliffe Institute, residents of Gambell since 1959.

Its purpose appears to have been both to celebrate the end of the hunting year and to honor the deceased whale by giving offerings through the transforming properties of fire. At the same time, this ceremony included both a purification rite for the successful hunters and their families to guard against future illness and a spiritual feeding of the ancestors in order to secure harmonious relations for the coming year.

(c) "Kameqtook," according to Moore/Silook, was a more generalized ceremony of thanksgiving and respect performed by boat captains and their crews at the end of whaling season. It, too, included offerings to the whale, to the ancestors, and to the spirits of various diseases. It involved elaborate ceremonial gear. In this ceremony, the ritual portions of any whale taken in the hunt may have been formally placed in the captain's sacred hunting pouch, including the tip of the nose, tips of the flippers and flukes, the tip of the penis, and the crystalline lens of the eye. The bag also often held a bird (or alternatively had a bird attached to the outside of the pouch) and a selection of powerful stones — those having unusual surfaces or shapes and somehow involved in the captain's personal search for power relationships.[13] To date, details of this ceremony have not been further elaborated upon by local island historians. From the details of this ceremony, it is possible that it was designed to open the walrus hunt as well as to close the whale hunt. Both walrus and whale were honored in the elaborate ritual performance. It is not clear how much of the information regarding this ceremony comes from Silook, since Moore remarks at one point:

> My notes are not quite clear on the sacredness of the [whale] gum which I believe is because all of the Eskimos are not agreed upon this matter" (Moore n.d., "Komukhtook [Kom okh — Boots]", unnumbered).

Moore also detailed several more generalized exchange ceremonies and celebrations whose purpose appears to have been to confirm harmonious social relations among the various community factions. These exchanges included both persons and objects.

I have described the general scope of Moore's notes here, because Silook's later and more substantial work builds on this base. This early collaboration between the two men, while not usually recognized, brings into focus the problem of ethnographic authority. Specifically, whose ethnography are we reading? And, what is the nature of the filter we encounter? How did Silook "see" the ceremonies? Also, this first work of Silook's appears to have provided, through Moore, an interpretive model which Silook followed and elaborated upon in subsequent work. Silook and Moore apparently agreed on appropriate vocabulary. Certainly, Silook later employs terms which first appear in Moore. For example, we find the term "worshipper" used again and again to indicate a ritual performer. The word "worship" becomes Silook's standard term for ritual action. "Sorcerer" is common to both, as is the use of the word "devils" to describe spirits of all kinds. Moore identifies small ritual carvings as both "idols" and "images." Silook, however, prefers to use "image." Both the similarities and differences in description speak to the process through which textual representation includes a personalized objectification of

---

13. An example of a captain's pouch which once belonged to Lincoln Blassi is in the University of Alaska Museum. Blassi's pouch, a gift from former Gambell school principal, Darrell Hargraves, was given after Blassi sadly gave up traditional whaling ceremonies in the early 1970s, one of the very last men to hold on to the old ways.

cultural performance based on individual experience and observation. Silook had no sense of himself or others in his community as "pagans" or "idol" worshippers. Moore, on the other hand, would conceivably have held this notion as part of his cultural baggage.

Between 1912 and 1927, Silook turned his energies elsewhere. He had married in the intervening years and his talents during this period were directed to raising and supporting his growing family. The island itself received little attention from the outside, with the usual exception of the school teachers who continued to educate children in the building originally constructed under Sheldon Jackson's direction. With the departure of the last school teacher-missionary, Edgar Campbell, the Presbyterian Church remained vacant with the exception of the year 1924-1925. Silook continued to be regularly employed by school teachers to act as a classroom assistant and as a substitute. On one occasion, he served as substitute for an entire year. Occasionally, as he put it, he wrote stories for the teachers. To date, these remain in private family collections or have simply disappeared from the record.

In 1927, life changed for Silook with the appearance of Otto Geist, a German-American immigrant collector and self-styled archaeologist, who took up temporary residence on the island. Once again, Paul Silook found employment. This time, he and several others worked as Geist's field assistants. It was not long, however, before Geist selected Silook from among the others to use as research assistant, translator, and consultant. Geist, by local account, was a difficult man, an adventurer, according to one anonymous consultant. J. Louis Giddings (1967:154) later described him as "... a self-educated man of alternately explosive and sentimental temperament." He was ambitious, given to bouts of temper, opportunistic, and, in addition, he had an eye for the ladies. According to Penapak, Silook's oldest daughter who still lives in Gambell, Geist originally arranged to live in Silook's household. But, according to Penapak, Silook and Geist quarreled, because of the latter's lack of respect for women and girls in Silook's family and his disregard for other rules of household behavior. Geist eventually went to live with two other families closely related to Silook and members of his *ramket* or clan unit, the Okinellos and the Otiyohoks. Silook and Geist patched up their differences and, between 1928-1934, Silook recorded hundreds of pages of ethnographic description for Geist.

Geist himself participated in whaling ceremonies as the result of his incorporation into a whaling crew headed by Booshu, brother to Otiyohok and therefore another of Silook's close relatives. Geist described some elements of that participation in his notes. However, he relied on Silook to provide much of the detailed, written data on whaling. Geist collected widely; flora and fauna samples, descriptions of local customs, and material objects were assembled, but his true interest was the collection of archaeological remains, especially artifacts which he later transferred to the fledgling Alaska Agriculture College and School of Mines for a museum. The scope of Silook's contributions to Geist's endeavors is apparent from the excerpt illustrated in Figure 2 from the Otto Geist Finding Aid[14] of data collected by the former Alaska Agriculture College, now the University of Alaska in Fairbanks.

---

14. When examining Geist's data it is difficult to tell which documents actually emanated from Silook and which resulted from Geist's own observations. This response to a question in a letter, received by Geist on February 26, 1937, gives some clues:

Dr. Otto W. Geist Papers — St. Lawrence Island Section V 2

BOX 3 Folders #49-89
  Folder
    49. Ethnology — Human Teeth (from diary)
    50. Ethnology — The making of intestine raincoats and snowshirts (from diary)
    51. Ethnology — The making of seal poke for whaling
    52. Ethnology — The making of white leather (from diary)
    53. Ethnology — Mixing of ceremonial paint and mortars (from diary)
    54. Ethnology — Moon worship (from diary)
    55. Ethnology — Nose bleeding: fainting: death (con't) (from diary)
    56. Ethnology — Notes on new homes (from diary)
    57. Ethnology — The sacred eye of the right whale
    58. Ethnology — Sacrifice to the dead
    60. Ethnology — Sacrifice to walrus heads (from diary)
    61. Ethnology — Skin canoe worship and sacrifice (from diary see folder #29 for original)
    62. Ethnology — Tattooing (from diary)
    63. Ethnology — Trapping and hunting seasons (from diary see folder #29 for original)
    64. Ethnology — The wealth of some ... hunters (from diary)
    65. Ethnology — Whale worship; whaling charms; contents of charm bag (from diary)
    66. Ethnology — Worship ceremony for fur seal ...
  BOX 4 Folders #90-#125
    35. Silook, Paul Diary. 1928
    36. Silook, Paul Diary. 1932 (June-Oct. 1921)
    37. Silook, Paul Diary (original). Oct. 22, 1932-May 20, 1933
    38. Silook, Paul Diary. 1934 (with note from Geist)
  BOX 9 Folders #39-#53
    39. Silook, Paul Diary. 1934
    40. Silook, Paul Diary. 1935 (?)
    41. Silook, Paul Names of grown-up people at Gambell
    42. Silook, Paul Notes 1927 (indexed)
    43. Silook, Paul Notes (original). 1929 (with Geist appendix)
    44. Silook, Paul Notes 1931-1932 (to be indexed) 92 pp
    45. Silook, Paul Notes on birds
    46. Silook, Paul Notes on lemmings, notes on Burials ...

Figure 2. Excerpt from the Otto W. Geist Collection Finding Aid (Archives, Alaska and Polar Regions Department, Elmer E. Rasmuson Library, University of Alaska Fairbanks).

What is less apparent is that pages of finding aid entries labeled simply "ethnology" are also Silook's work. Again, that work includes extensive descriptions of whaling practices, complete with drawings showing the positions which boat crews took in the ceremonies following a strike, as well as elaborations on his earlier work with Moore. Now, however, Silook is writing his own descriptions. These convey different kinds of details than the original work and include numerous small diagrams. The excerpt illustrated in Figure 3 from the Geist collection is one example.

Henry Collins, an archaeologist for the Smithsonian Institution, also made his appearance on St. Lawrence Island during this period (1928), and Silook immediately sought employment with Collins as well. In Collins, Silook found a scholar model after his own heart. Eventually, Silook kept Collins's work diaries along with his own more personal diaries. He wrote down numerous traditional stories for Collins, and again described traditional practices associated with whaling. At the end of a busy day of digging on the Punuk Islands,[15] Collins, with Silook, James Aningayou, whose conversion to Christianity and devotion to an evangelistic faith strongly affected Silook, and various others gathered in the work site tents. Often Silook alone joined Collins in his tent to talk at the end of a long day. Silook responded to Collins's questions, telling stories and commenting on island practices, while Collins recorded his words in notebooks, often employing a personal shorthand. Silook also made himself available to Collins' assistants, including archaeologists James Ford and Moreau B. Chambers. When Collins departed in the winter, Silook wrote to him at the Smithsonian, often describing some ritual practice which he had neglected to relate when Collins was on the island (Fig. 4).

The body of data, which belonged to Silook and is now found in Collins' papers, is extensive, but since it has only become available since Collins' death, it has not actually been the source of the various Silook descriptions on which so many have drawn. Rather, it reinforces the consistency of Silook's descriptions over time. Below, for example, are several pages from Collins' "Ethnology" notebooks which he recorded from Silook between 1928-1931.

> Jimmie Otiohok, a member of Imaremkeit tribe made a trip to Indian Point [Chaplino], Siberia, a week before the Northland arrived to get a certain kind of plant needed by his brother Burscha [Booshu was one of Silook's relative and the man who allowed Geist to participate in his ceremonies], for a ceremony, a "whale worship" in commemoration of the whale he (B) had killed some weeks before. The Imaremkits claim to have come

---

Question #31 (Author not given, but probably Mrs. O. J. Murie): Is the little black book Silook's own handwriting?
Answer (from Geist): I am not certain but I rather think that the black book is in Silook's handwriting. You will be well able to tell by his style and form of writing. I have asked and paid Paul Silook for all the notes he gathered for me for he had access to many things later on and which I wish to have done in writing for my files and I know that he is about the only who would be able to do this for me because at one time he was employed as an assistant teacher at the Gambell school and during some of my work, he was employed by me. In other words, whenever I heard of a good story, or persons who could tell good stories even though they could not give them to me in English, I asked Paul Silook to get them for me and put them down right away (Dr. Otto W. Geist Papers, Section GV, Folder — Murie, Margaret, Questions and Letters received and answered about St. Lawrence Island).

15. Three small islands located just southeast of Northeast Cape.

Page -7- KILLING A WHALE

Every boat must keep the first head of the female walrus and the baby for the worship that they are going to held after whaling. They keep this in the cellar also. They held this worship every spring after the hunting season is over.

Then the time comes each boat captain held this worship. He invite his boat crews if he had killed a whale he invite anybody especially older folks. He held this early in the morning. He let his wife put a lamp in the center of the house and lit it. He brought in the hunting weapons, his own paddle, the lance and the pouch with their beaks.

He tied a piece of string to a wooden image of a whale, that they have kept for a long time made by their grand parents. He tied it over the lamp. Then he let his boat crew get the flukes from the cellar, while they were away he put his cap on and took a drum and sing his whaling song.

Then all is brought he let them laid the female walrus head on the right side of the lamp and the flukes any where around the lamp. He then cut pieces of all the meat in a small wooden cup and go out throw these pieces to the air, the pieces about this size:

When he entered he and his family would step on these flukes, also his relatives do the same as they had done.

After this he pretended to throw pieces to the lamp naming some names. They believe that in any worship the spirits came around to eat. When all was thrown, the guests would eat as much as they want. After they had feasted the flukes is divided among the boat crews. First the captain gave some to his relatives, what was left he divided among his crews equally.

Not all the boat captain do the same, some has no wooden whale.

After dividing the captain and his crew would go around the lamp, following the direction as the sun travels. They sing a long song while they are going around.

Then when the song is ended the worship is over.

Figure 3. Killing a whale: excerpt from a description of whaling ceremonies written by Paul Silook (Dr. Otto W. Geist Papers, Section V, Box 3, Folder 44?, p. 7. Archives, Alaska and Polar Regions Department, Elmer E. Rasmuson Library, University of Alaska Fairbanks).

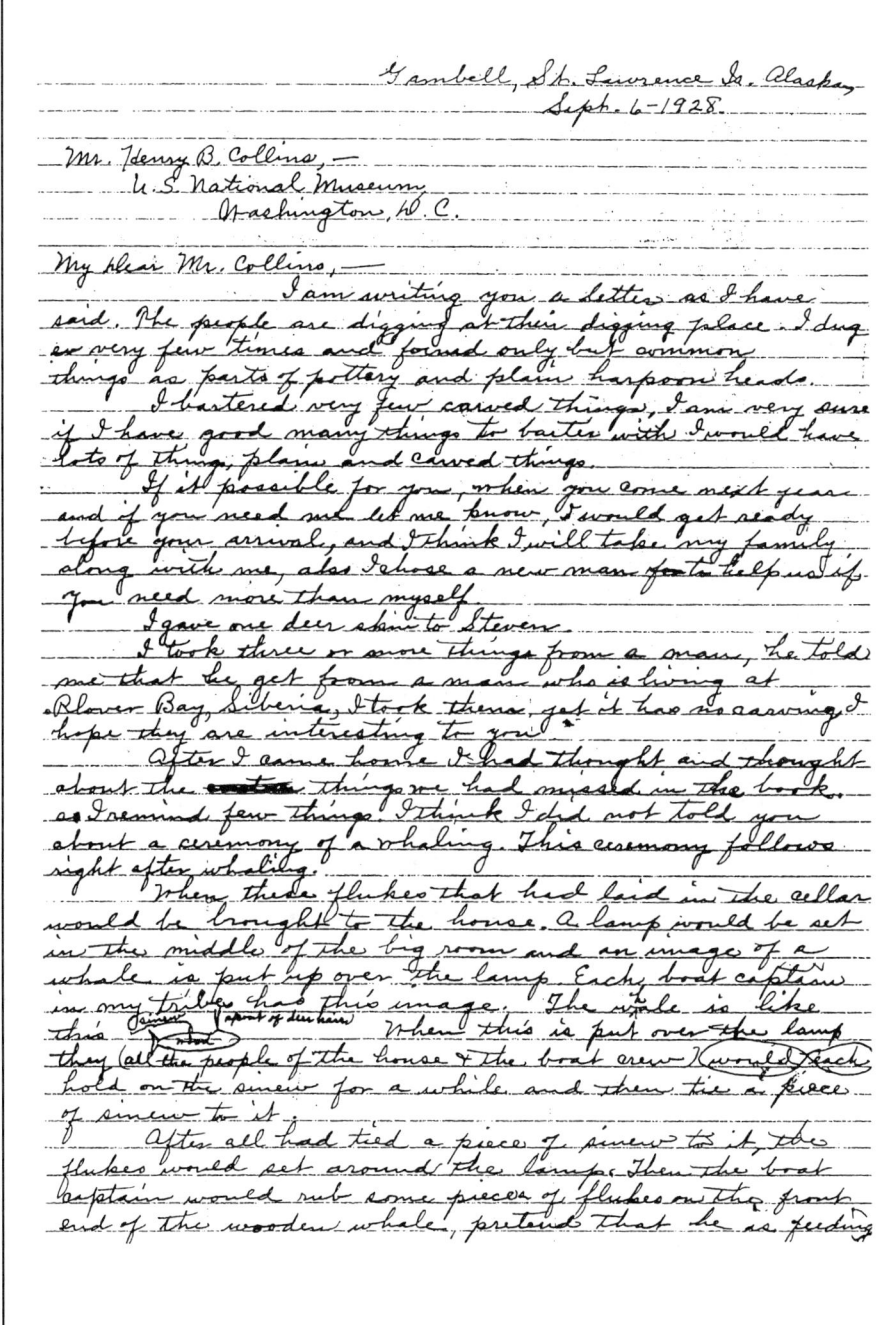

Figure 4. (Front) Letter to Henry Collins from Paul Silook, September 6, 1928: description of a ceremony which follows a successful whale strike (Henry Collins Collection, Box 3, unprocessed — field notes, etc).

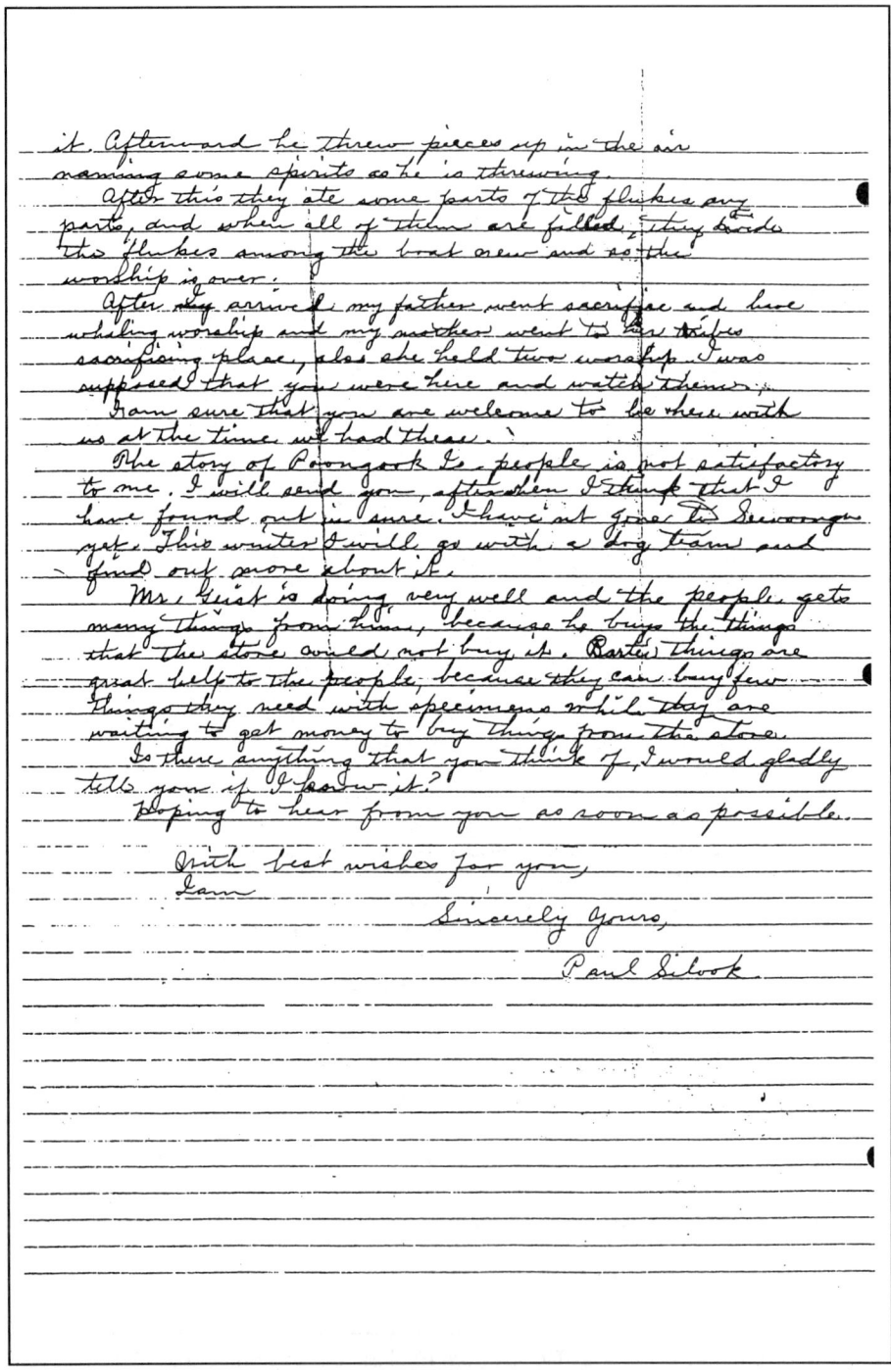

Figure 4. (Back) Letter to Henry Collins from Paul Silook, September 6, 1928: description of a ceremony which follows a successful whale strike (Henry Collins Collection, Box 3, unprocessed — field notes, etc).

from Indian Point and whenever one of them kills a whale he must have for the ceremony that follows this particular plant used by his forefathers. The Imaremkit people say they are descended from a woman who lived at Kukuliak but who was sold to a man at Indian Point. Her grandchildren then returned to St. L. living at Gambell (Sevuokuk; Henry Collins, Unprocessed, Box 3, Notebook: H. B. Collins 1930, Doc. File. Collins 1930.00A: Unnumbered).

* * * * *

U vuk tuk — A ceremony formerly in use among the Imaremkit and Owahlit as a kind of thanksgiving for recovery from sickness or for having been saved from some disaster. Also after killing a whale. In this latter case: the boat capt, when the whale's struck, calls out the name of ceremonies, asking them (sic) to go ahead of the whale and stop it. After the whale is killed the ceremonies called upon are held. Even if the whale escapes the ceremony is held because calling out the name of ceremony is a promise.

In preparation for the Uvuktuk meat is prepared and the principal (worshiper according to Paul's terminology) or man giving the feast throws into the air a tiny piece of meat or even only touching the meat with a hand — throwing it up and rubs a little on the walrus skin (feeding it). Then all sit down on the skin and eat the meat. Believe that if they eat, the Spirit of ceremony will also, but if not then it will not. When throwing the meat in the air he calls out "Uvuktuk tiwa," (man of the ceremony) then calling out also if he wishes the name of some ailment he or some near relative has and the names of ancestors. All these are offered homage in the same sense; the idea is that if the ceremony is fed it will be strong and so able and willing to help out in time of need. Some meat is divided among the old people. After eating & dividing up the meat (about 1/2 of small walrus or a whale, reindeer or seal or white man's food) the principal steps on the walrus skin and is tossed up into the air by the others. Pcs of rope or thong are tied thru the perforations at the edge and it is by these that the skins is held. Then anyone who wishes is tossed up. Some are tossed as high as 30 ft. Men women or children take part, throwing & jumping — Good jumpers come down straight on the feet, some double over in the air like a highdive.

Additions to notes on whaling — Material for sacrifice pertains to the Imaremkit tribe. Others sacrifice different things. Owalit & Nungpagak sacrifice only fermented plant & walrus blubber. Neskok (Aginaktat) sacrifice only willow leaves, walrus blubber & w- meat, raw-

Ceremony A "moon worship" …: Taking sac.[rifice] material to beach — It is taken to the boat in wooden dishes, placed in the boat which was on the rack. It is dragged to the beach. No women in this.

After men return from beach, then women of the boat capts house (wife), daughters, sisters, (daughters-in-law) go to every house in village and give a small pc of the sacrificial food (fermented plants, codfish, tobacco). In return for this they are given something: cartridges, powder or primers, walrus rope, files, saw blades, sugar, tea, etc.

Anyone can go to one of these sacrifices, regardless of tribe, except that the Imaremkit [Silook's *ramket*] do not go to those of other tribes altho others may go to theirs. Say that once they went to a sac of another tribe & everyone died, so no more … (Henry Collins Collection, Unprocessed, Box 3, Notebook: H. B. Collins 1930/Doc.File. Collins 1930.00A: p.4&5)

* * * * *

238 *Carol Zane Jolles*

> The whaling ceremonies "A tagh'hok" are held by each tribe separately, each differing only in minor details from the others. Four ceremonies for a boat captain. The third only if he kills a whale. A. "moon worship" held by certain tribes on certain months, i.e., Feb., March, April; B. if a whale is killed and C. after whaling; D. in November (Henry Collins, Unprocessed, Box 3, Notebook: H. B. Collins 1930, Doc. File. Collins 1930.00A: p.22)

* * * * *

> Senoruuk says: that when the old people used to strike a calf whale far from the village, not having "evinrudes" they would put two walrus harpoons into the whale, one on each side of his head and then, by means of walrus stomach pokes attached to the lines about half way back, frighten him toward the village. Then when they had brought the calf within a mile or two of the village it would be killed. Of course a whale harpoon was also used.
>
> The calf was frightened and thus guided by jerking the walrus lines, causing the walrus pokes to resound on the water. Is guided to the beach by flapping the bladder on whichever side is necessary. If whale turns to right, flap on the left bladder, etc. Illustration of old time whale towing drawn by Silook. ... This method is supposed to make this job very easy.
>
> I would suggest, however, that he has misplaced his wind. Perhaps it should be as in "A." [Ford's handwriting] (Henry Collins Collection, Unprocessed, Collins Field notes etc, Box 4, Unprocessed, James A. Ford 1930 Archaeology Notebook [+ ethnographic notes]: p.6)

Silook developed the habit of keeping diaries for both Geist and Collins, continuing to keep them for a time after the departure of the two men. Those he kept for the two men are among their collected data.[16] His diaries represent not only his ethnographic descriptions, but a pragmatic record of day-to-day events associated with whaling.[17] Note his rather interesting sketches (Figs. 5a,b), worked out with James Ford, illustrating the way in which a struck whale could be harnessed by boat crews to bring itself to shore.

In 1939, Paul again found employment. This time with J. Louis Giddings. Giddings wrote just prior to his death in 1964:

> At Gambell I engaged new helpers, one of whom, Paul Silook, introduced himself in surprisingly good English almost the moment I arrived ... Conversation occasionally became strained between us when he pointed out that I was not measuring and digging precisely as Mr. Collins had done. Paul was so clearly an archeologist at heart, however, that we compromised in our methodology and freely talked, to my great benefit, during each digging day (Giddings 1967:168).

---

16. Silook's diary notebooks are housed in the Otto Geist Collection, Alaska and Polar Regions archives, University of Alaska Fairbanks, and in the Henry B. Collins Collection, National Anthropological Archives, National Museum of Natural History, Smithsonian Institution, Washington, D.C.
17. Excerpts from Silook's diaries in the Collins Collection are located in Appendix A. These detail the calendar cycle of ceremonial whaling events along with the more mundane details of boat repair and weather. Of particular interest is the articulation of whaling activities and successes with children's games and courting games for young men and women on the beach. Also of interest, unrelated to whaling itself, is Silook's daily record of the number of seal oil lamps going in his house, correlated with the number of persons living in the house and the corresponding house temperature.

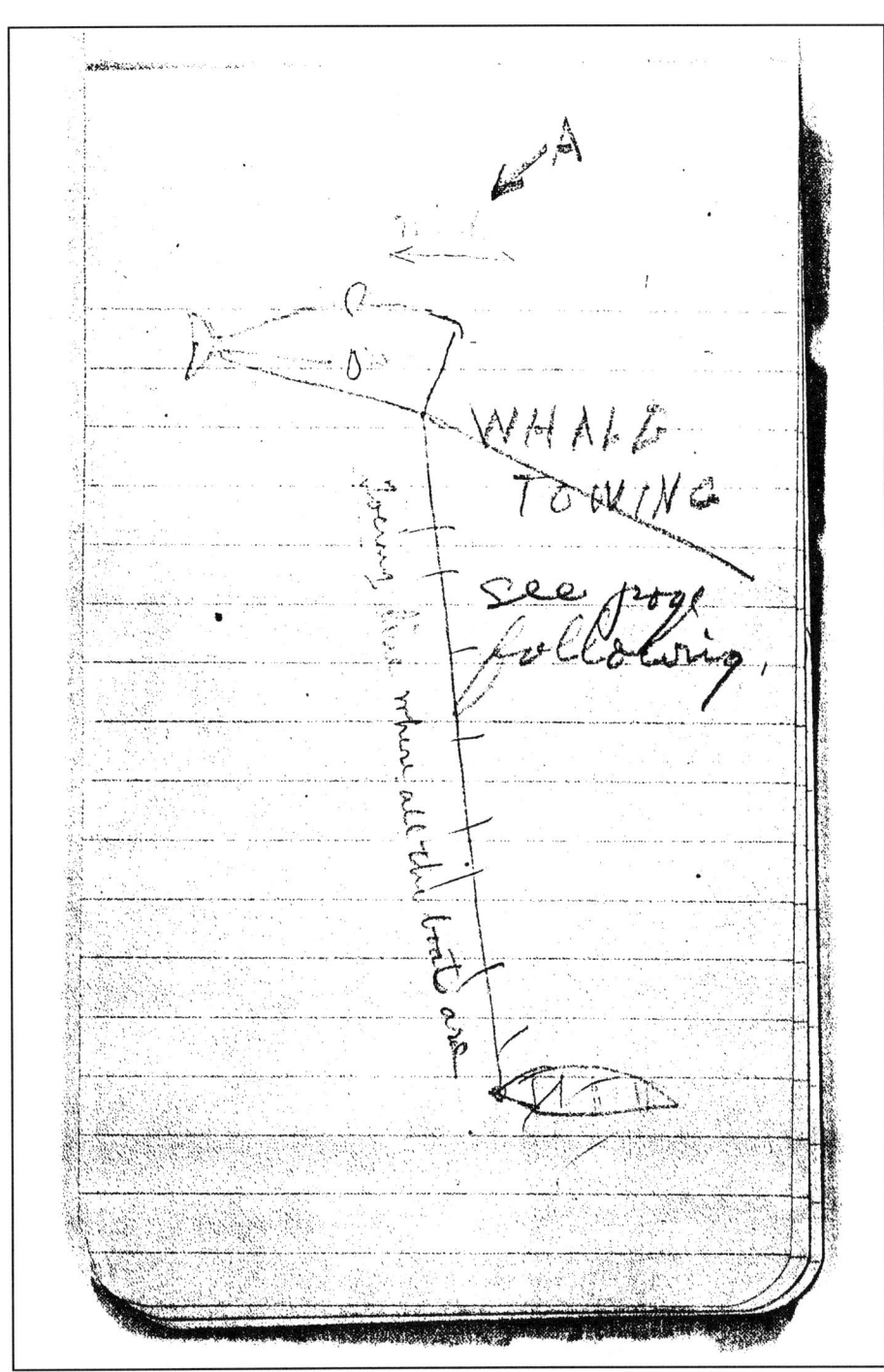

Figure 5a. Harnessing a whale: drawing from James Ford's notebook (Henry Collins Collection, Box 3, Unprocessed — field notes, etc., H. B. Collins-1930, Notebook B).

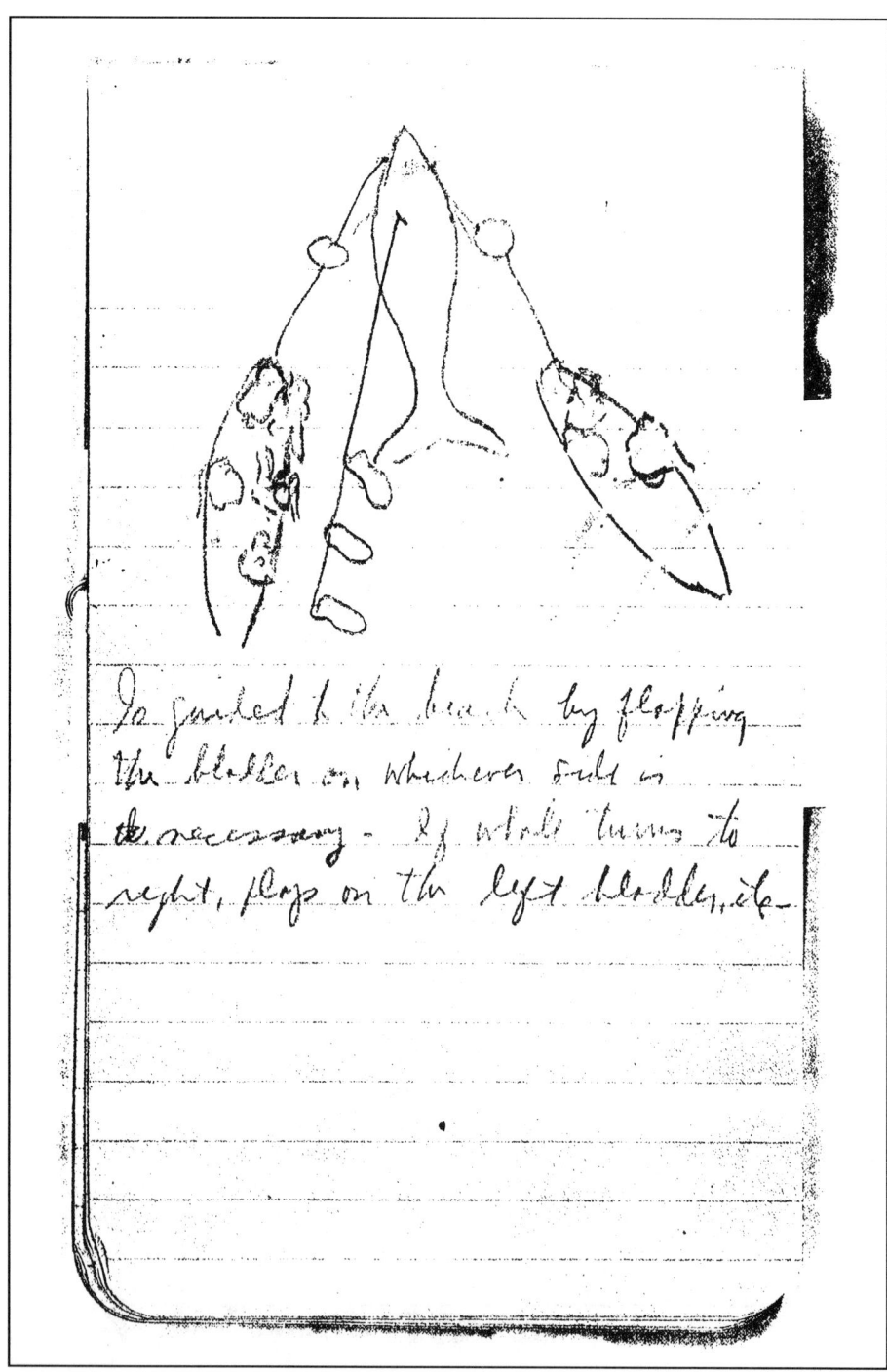

Figure 5b. Using the wind when towing a whale: drawing from James Ford's notebook (Henry Collins Collection, Box 3, Unprocessed — field notes, etc., H. B. Collins-1930, Notebook B).

Giddings, too, collected ethnographic data from Paul in the long summer evenings when the work day was over. However, unlike other researchers, Giddings did not leave behind a collection of written notebooks or journals of the information he accumulated.

Finally, in 1940, Alexander Leighton and Dorothea Leighton (Dorothea Leighton, M.D., Collection 1982) spent the summer in Gambell collecting life history data from local residents. Silook, 48 years old, was one of their subjects (see e.g., Fig. 6). He told his life story to Alex Leighton, filling over 200 double-spaced typewritten pages and once again laboring upon the basic data given 38 years earlier to Riley Moore. His work with Alexander Leighton represents his last formal effort to record data, and his autobiography is certainly the most personal of his written legacy. The diaries and notebooks kept in earlier years are austere and businesslike. His work for Collins is characterized by the degree to which it emulates Collins' own work diaries, even to the handwriting. His personal diaries maintain the same sparse style.

Figure 6. Paul Silook modeling proper hunting gear for Alexander Leighton (Dorothea Leighton Collection, acc. #84-31-35N, Archives, Alaska and Polar Regions Department, Elmer E. Rasmuson Library, University of Alaska Fairbanks).

SILOOK'S LEGACY

All of Silook's work has heretofore been treated as ethnohistorical documentation or has been subsumed by the ethnographic authorship of others. Yet, his work clearly is that of an indigenous ethnographer/historian. Contemporary ethnographers using Silook's data (or data of other indigenous authorities) can hardly avoid the multiple convolutions of textual representations involving levels of subjectivity and objectivity, the nature of the cultural lens, and especially cultural biases. In Silook's case, one is initially overwhelmed by the body of material he enthusiastically recorded for others. Examples of his work, contained in the appendices, give some notion of Silook's contributions to our understanding of whaling practices. His descriptions, found among the papers of Moore, Geist, Collins, Ford, and Leighton, have yet to be fully reviewed and analyzed. His gentle warning, to be aware of the Aymramket bias of his work, has been more or less ignored. Other ethnographers, perhaps caught up in the theoretical debate over the existence of "true" Eskimo clans, have also neglected the issue of *ramket* affiliation. As Silook himself points out, however, we should be concerned that so many have relied on the work of one man whose clan or *ramket* membership defined his knowledge. To quote Silook:

> You know each tribe has its own secrets. They believe that if they reveal them to others, the good luck may be transferred by the giver. So each tribe has a little different ceremony.

While these issues are always with us, they are particularly complex when the person labeled "key informant" by some is an ethnographer in his/her own right. We are brought back to the issue of historicity and the place of the observer who records his/her own society. As Clifford comments:

> Insiders studying their own cultures offer new angles of vision and depths of understanding. Their accounts are empowered and *restricted* in unique ways (Clifford and Marcus 1986:9, emphasis added).

It is critical that we recognize who the contributing authors are to any text on which we have come to depend. We are perhaps in danger of having become a little too embedded in the "daily practice" of anthropology, and texts themselves have become taken-for-granted objects.

There are ethical considerations as well. As already stated, many have used the Silook data, but Silook himself has gone unnoticed. I offer a few reasons for this oversight. As with Giddings, whose acknowledgment of Silook's contributions is as good as it gets, Moore, Geist, Collins, and Ford accepted Silook as a kind of convenient local archaeologist and handy data source. His day-to-day existence was, for them, as it probably would have been for any researcher who came to work in the few summer months, quite remote. Even for Geist, who spent several winters in Gambell, St. Lawrence Island was a resource and so was Silook. The men who worked for him, certainly his companions of the moment, were not ultimately tied to him in his professional life. As is apparent from the few examples of Silook's correspondence, Collins, and, in fact, Geist, sent letters to Paul from time to time and also sent gifts for Silook's growing family. Even so, Silook was but an interesting component of their professional lives.

For Alexander Leighton, on the other hand, Silook simply provided a longer and fuller life history than Leighton's other informants. Leighton never really utilized his data from St. Lawrence Island to any great extent. Rather, he offered the body of data to his student, Charles Hughes, who used it as the basis for his Master's thesis (Hughes 1953) and as the background for his more comprehensive study of St. Lawrence Island between 1954-1955.[18] I think it would be safe to say that Silook was viewed neither as a research partner nor as a full-fledged assistant by the Leightons. Yet, when Leighton offered to record Paul's story, he received the outpourings of a man who had found few satisfactory outlets for his observations of his culture. While Silook's descriptions may have inspired and supported the work of these scholars and informed the work of many more, it does not seem to have served him well. Particularly disturbing is the fact that St. Lawrence Island residents remain singularly unaware of Silook's contributions.

Silook's history leaves us with the interesting question: How much of the work of anthropologists is really the product of self-taught, highly informed, and disciplined local ethnographers, and how should acknowledgment of such contributions be made? Once a more comprehensive assessment of Silook's contributions has been conducted, I believe it will be concluded that his work approaches that of George Hunt, Franz Boas's well-known assistant. Native scholars, like Silook, however remote and local their contributions, deserve new scrutiny and full recognition for their contributions to ethnographic and ethnohistoric research in their communities. Surely they should not have to wait, like Hunt, for belated acknowledgement in the reflected light of a famous mentor.

Acknowledgments. The research for this article was supported primarily by the National Science Foundation's Arctic Social Science Program, Division of Polar Programs (DPP #9122083) and by supplementary grants from the Anthropology Department and Graduate School at the University of Washington. I would like to thank Paul Silook's daughter, Estellle Penapak Oosevaseuk, a respected elder, for her assistance, and the Apatiki family in Gambell for their continuing support and friendship.

REFERENCES

Apassingok, Anders
   1985   Introduction. In: *Lore of St. Lawrence Island: Echoes of Our Eskimo Elders*, Vol. I: *Gambell*, edited by Anders Apassingok, Willis Walunga, and Edward Tennant, pp. xv-xvi. Bering Strait School District, Unalakleet.

Apassingok, Anders, Willis Walunga, Raymond Oozevaseuk, and Edward Tennant (editors)
   1987   *Lore of St. Lawrence Island: Echoes of Our Eskimo Elders*, Vol. II: *Savoonga*. Bering Strait School District, Unalakleet.

Apassingok, Anders, Willis Walunga, and Edward Tennant (editors)
   1985   *Lore of St. Lawrence Island: Echoes of Our Eskimo Elders*, Vol. I: *Gambell*. Bering Strait School District, Unalakleet.

Apassingok, Anders, Willis Walunga, Raymond Oozevaseuk, and Edward Tennant, (editors)
   1989   *Lore of St. Lawrence Island: Echoes of our Eskimo Elders*, Vol. III: *Southwest Cape*. Bering Strait School District, Unalakleet.

---

18. By the time Hughes settled in Gambell to commence fieldwork, Silook was no longer living and he turned to others as his major informants.

Badten, Linda Womkon, Vera Oovi Kaneshiro, and Marie Oovi (compilers)
   1987   *A Dictionary of the St. Lawrence Island/Siberian Yupik Eskimo Language*, 2nd Preliminary Edition, edited by S. A. Jacobson. Alaska Native Language Center, University of Alaska Fairbanks.

Bandi, Hans-Georg
   1969   *Eskimo Prehistory.* University of Alaska Press, College.

Blackman, Margaret B.
   1989   *Sadie Brower Neakok: An Inupiaq Woman.* University of Washington Press, Seattle.

Bogoras, Waldemar
   1904-1909   The Chukchee. *Memoirs of the American Museum of Natural History* 11.
   1913   The Eskimo of Siberia. The Jesup North Pacific Expedition 8(3), edited by Franz Boas. *Memoirs of the American Museum of Natural History* 12.

Campbell, Edgar O. and Louisa K. Campbell
   1904-1911   Unpublished journals: 1-5. MS. on file, Archives, Presbyterian Historical Society, Philadelphia.

Clifford, James
   1986   Introduction: Partial Truths. In: *Writing Culture: The Poetics and Politics of Ethnography*, edited by J. Clifford and G. E. Marcus, pp. 1-26. University of California Press, Berkeley.

Collins, Henry B., Jr.
   1937   Archaeology of St. Lawrence Island, Alaska. *Smithsonian Miscellaneous Collections* 96(1).

Crapanzano, Vincent
   1986   Hermes' Dilemma: The Masking of Subversion in Ethnographic Description. In: *Writing Culture: The Poetics and Politics of Ethnography*, edited by J. Clifford and G. E. Marcus, pp. 51-76. University of California Press, Berkeley.

Cruikshank, Julie
   1979   Athapaskan Women: Lives and Legends. *National Museum of Man, Mercury Series, Canadian Ethnology Service Paper* No. 57.
   1990   *Life Lived Like a Story.* University of Nebraska Press, Lincoln.

Dorothea Leighton, M.D. Collection
   1982   Field Notes (1940) of Dorothea Leighton and Alexander Leighton. On file, Archives, Alaska and Polar Regions Department, University of Alaska Fairbanks.

Fienup-Riordan, Ann
   1986   The Real People: The Concept of Personhood Among the Yup'ik Eskimos of Western Alaska. *Etudes/Inuit/Studies* 10(1-2):261-270.
   1994   *Boundaries and Passages: Rule and Ritual in Yup'ik Eskimo Oral Tradition.* University of Oklahoma Press, Norman.

Fitzhugh, William W. and Aron Crowell
   1988   *Crossroads of Continents: Cultures of Siberia and Alaska.* Smithsonian Institution Press, Washington D.C.

Gambell, Vene
   1910   *The Schoolhouse Farthest West: St. Lawrence Island, Alaska.* Woman's Board of Home Missions of the Presbyterian Church, New York.

Geist, Otto W.
   n.d.   Section GV, Folder, n.d, Murie, Margaret, Questions and Letters received and answered about St. Lawrence Island.

Geist, Otto W. and Froelich G. Rainey
   1936   Archeological Excavations at Kukulik, St. Lawrence Island, Alaska. *University of Alaska Miscellaneous Publications* 2. U.S. Government Printing Office, Washington D.C.

Giddings, J. Louis
   1967   *Ancient Men of the Arctic*. University of Washington Press, Seattle.

Henry Bascom Collins Collection, National Anthropological Archives, National Museum of Natural History, Smithsonian Institution
   1982   Box 3 Unprocessed — [Collins Field notes etc 3]
           1928 — Silook diary
           1929-1930 — Silook diary
           1930-1931 — Silook diary
           H. B. Collins 1930 — Ethnographic notes — Book A (written on back of notebook)
           Unnumbered Box Unprocessed — [Collins Notebooks, Field Notes — Resolute etc Alaska]
           H. B. Collins — Archaeology notebook — 1930 — Gambell, SLI (especially notes following entry Cut 16)
           H. B. Collins 1930 — Ethnographic notes — Book B (written on front of notebook)
           H. B. Collins — Archaeology notebook — 1930 (J.A. Ford's notes) Box 4 Unprocessed — [Collins Field notes, etc]
           Paul Silook — Archaeological Notebook 1930-1931 (for the Smithsonian 1931 Expedition)
           James A. Ford 1930 Archaeology Notebook (+ ethnographic notes)
           Loose-leaf notes from a memo book
           1928 Ethnographic notes
           Collins 1928 diary (continued)
           Paul Silook (H. B. Collins) Folk Lore

Hrdlicka, Ales
   1930   Anthropological Survey of Alaska. *46th Annual Report of the Bureau of American Ethnology*.

Hughes, Charles C.
   1953   *A Preliminary Ethnography of the Eskimo of St. Lawrence Island, Alaska*. Unpublished MA thesis, Department of Anthropology, Cornell University, Ithaca.
   1958   An Eskimo Deviant from the "Eskimo" Type of Social Organization. *American Anthropologist* 60(6):1140-1147.
   1960   *An Eskimo Village in the Modern World*. Cornell University Press, Ithaca.
   1984   St. Lawrence Island Eskimos. In: *Handbook of North American Indians*, Vol. 5, *Arctic*, edited by David Damas, pp. 262-277. Smithsonian Institution, Washington, D.C.

Jolles, Carol Zane, with Kaningok
   1991   *Qayuutat* and *Angyapiget*: Gender Relations and Subsistence Activities in Sivuqaq (Gambell, St. Lawrence Island, Alaska). *Etudes/Inuit/Studies* 15(2):23-53.

Marcus, George E. and Michael M. J. Fischer
   1986   *Anthropology as Cultural Critique: An Experimental Moment in the Human Sciences*. University of Chicago Press, Chicago.

Moore, Riley D.
   n.d.   Notes, National Anthropology Archives, National Museum of Natural History, Smithsonian Institution, Washington, D.C.
   1923   Social Life of the Eskimo of St. Lawrence Island. *American Anthropologist* 25(3):339-375.

Nelson, Edward W.
   1983   *The Eskimos About Bering Strait*. Smithsonian Institution Press, Washington D.C. (1899)

Otto Geist Collection
   n.d.   Otto Geist Collection Finding Aid. On file, Archives, Alaska and Polar Regions Department, Elmer E. Rasmuson Library, University of Alaska Fairbanks.
   1927-34   Dr. Otto Geist Papers. On file, Archives, Alaska and Polar Regions Department, Elmer E. Rasmuson Library, University of Alaska Fairbanks.

Ridington, Robin
   1988   *Trail to Heaven: Knowledge and Narrative in a Northern Native Community*. University of Iowa Press, Iowa City.

Schneider, William
   1986   *The Life I've Been Living*, by Moses Cruikshank. University of Alaska Press, Fairbanks.

Schweitzer, Peter
   1989   Spouse-Exchange in North-Eastern Siberia: On Kinship and Sexual Relations and Their Transformations. In: *Vienna Contributions to Ethnology and Anthropology*, Vol. 5, Kinship, Social Change and Evolution: Proceedings of a symposium held in honor of Walter Dostal, edited by Andre Gingrich, Siegfried Haas, Sylvia Haas, and Gabriele Paleczek, pp. 17-38. Frankfurt.
   1992   Reconsidering Bering Strait Kinship and Social Organization. Paper presented at the 91st Annual Meeting of the American Anthropological Association, San Francisco.

Silook, Paul
   1940   Autobiography of Paul Silook. In: Dorothea Leighton, M.D. Collection, Field Notes (1982) of Dorothea Leighton and Alexander Leighton. On file, Archives, Alaska and Polar Regions Department, Elmer E. Rasmuson Library, University of Alaska Fairbanks.

Walunga, Willis (compiler)
   1987   *St. Lawrence Island Curriculum Resource Manual, Revised and Expanded Edition*. St. Lawrence Island Bilingual Education Center, with the assistance of the United States Department of Education, Gambell, AK.

APPENDIX

[Author's note: The following material contains many omissions, abbreviations, and creative and inconsistent spellings; these are not noted separately.]

Part 1: Selection from the Dorothea Leighton, M.D., Collection 1982 (1940), from "Autobiography of Paul Silook," pp. 18-29:

> ... In summer my mother gathered some greens and filled a seal poke. First she fermented it in a barrel, let it get very sour, then put the greens in a poke, and put this poke up on the rack ... Then in time, when the Siberian people come around here my father gets some deer fat, especially the internal fat of the deer ... When he buys a box of tobacco he saves a pound of that, too. In time of cod fishing, he dried some up and save them for that purpose. So he waits for the time. Also he saves some green extract, for another little ceremony, which takes place in November. So when he start, in November, to have this little ceremony I go with him to the boat. He opened this, the other poke — sometimes my mother fill two pokes — then when he open the poke, he cut little pieces from the contents of the poke, and we go to the boat together. This takes place in the mornings. He has a little wooden cup. He puts the green into this cup. Then he begins to sacrifice by the boat. The way he sacrifices, he just throw little pieces up into the air. First, he name the name of the moon, then afterward while he throw little pieces, he name some of the old people that have died. This means it will bring him good luck during the winter.

Part 2A: Selections from the Henry Bascom Collins Collection: Henry Bascom Collins Collection, Unprocessed, Box 4, Doc. File/Collins 28:1-2: Collins, Field Notes, Etc., Memorandum Book — Ethn — 1928:

> 2. Imaremket — nine houses — six whaleboats. Over 100 people ...
> 2. Imaremket, Paul thinks, is a Chukcha word, because they have none like it. They come from Indian Point, Siberia, and they were the most numerous people ...
> 4. Merooktameet, one house. They are from an old village 3 miles east of Gamble called Merookta. James & Stephen's tribe. James' [*Aningayou*] boat is called this ...
> Bowhead whale. These are no longer hunted with native gear but with whale lance set in whale boat — 50 feet of manila rope from the bomb lance to a larger seal poke (float), then 100 ft. to a second poke, and 100 ft to a third — Originally whales were hunted from the angiak [boat] with the large whale harpoon. Hunted in spring. Usually from one to five a season. Meat is divided, or rather every one comes and cuts off. Flippers and flukes belong to one who strikes it first. Also gets half the whale bone.
> Sacrifice is made by each of the boat owners at different days of the same (new) moon, from (usually) some time from Feb. to April. Fill a poke with fermented plant, a small green plant: 6 to 12 plugs of tobacco; fill the lower part of reindeer stomack "reindeer fat" a dozen or more codfish. All this is prepared for use the night before the time selected for the sacrifice and placed with the guns, harpoons, etc. At sunrise of the following morning they go to the beach and throw small portions of the several things into the water or into the air. Divide the rest of the food, etc., up among the crew of the boat, who have a sort of feast eating up as much of the food as possible.
> When a whale is dead the striker and the captain put on their seal skin peak (somewhat like an eyeshade — Dr. Moore got one or more). Then paddle around the whale in the direction of the sun (E. to W.). The boats proceed to the village, the captor first, second the one who came up first after the striker, and so on.

Tips of the flukes & flippers are cut off & hung from the roof & left for three or four days, during which period man does not do any hunting. Small pieces of the flukes & flippers are then thrown into the lamp. A figure of a whale is then painted on the right side of the bow of the boat (about 3 or 4 inches long). Wife of man wears a snow parka [parka cover of bleached seal intestine decorated with crested auklet feathers] over her clothes, new boots, ties a piece of reindeer hair around her plaited hair. Also wears this outfit during the "moon worship" — Does not wear these shoes or parka on any other occasion.

Part 2B: Henry Bascom Collins Collection, Unprocessed, Box 4 (Collins, Field Notes, Etc.), Doc. File/Collins 28:Ethn:10:

Similar observances to the whale are made on killing a polar bear. Man does not hunt for 3 or 4 days afterward. Cuts off head & keeps it in house, people come in & stories are told. Bear's skull is kept for several years and then taken to place where sacrifices are made to the old dead people, and deposit the skull, making sacrifice of small pieces of meat. Whale is sacrificed to at the same place.

Part 2C: Henry Bascom Collins Collection, Unprocessed, Box 4, (Collins, Field Notes, Etc.), Doc. File/Collins 28:Ethn:10-11:

… At ceremonies man of the house should apply a single streak of graphite from the nose to the cheek. Women two lines on chin. If man is a boat captain (owner of a whale boat) or if the son of a boat captain, he has a little blubber and native plants, cooked, skin of willow roots or leaves on a wooden dish, used only for this purpose. Throw piles of this up in the air & those helping & he eat the rest. Wife brings the sacrifice. She wears the same ceremonial clothes as she wears for the whaling ceremony.

Boat captain & wife never eat the skin or meat of the first whale until after he can go hunting again (4 days later). Boat capt., his wife, the striker (but not his wife), & the capts. of all the other boats & their wives observe this taboo. The capts. in the Imaramket (7 in number) & their wives can eat this meat the day after the whale is killed. Capts. of other tribes must wait the four days, but some wait until the whaling season is over — Rest of crew can eat wherever they please. The apparent privilege enjoyed by the Imaramket Paul explains by saying that it was observed by their ancestors and that the group are the best whalers & always get the most whales.

Two or three weeks after their ceremony, "offering to the whale 'moon worship'", the captains and strikers take turns in watching the whale boats to see that no one tries to put in a piece of human bone or articles belonging to dead people, or a knife that had been used for severing a baby's umbilical cord — the baby dying subsequently. This would frighten the spirit of the whale that they are going to kill. This is not observed so rigidly today.

Part 2D: Henry Bascom Collins Collection, Unprocessed, Field Notes, etc. Box 3 (Notes on Games taken down from Paul Silook), Doc. File/Silook 28.Dia:2:

Few Games for Girls
The only games of the girls is a ball which is played just after whale is killed, but each time the boys would join with them. The young men and young women also played it in the evenings of spring.
The girls can kick a ball when a calf whale is killed …

Ball game can be played anytime after the 1st whale of year is killed. Men throw to the men; girls to girls. Men take this opportunity to kiss the girls. Each sex trys to take the ball from other. Same game as at Marshall on Yukon.

Girls kick the ball up in the air and try to catch it, each of the others trying to get it when it falls to ground. Whoever gets it kicks it up again. They stand around in a circle. Can play it now regardless of calf whale being killed. [End]

Part 2E: Henry Bascom Collins, Unprocessed, Box 3 (Notebook: H. B. Collins 1930), Doc.File. Collins 1930.00A: p.7:

Between the moon worship and the close of whaling the striker's mother must not knock her feet together as is done when entering the sleeping room to knock the snow from the feet. If she should do so, any whale her son might strike would flap the water with its flukes & hinder its final capture ...

Crew of whaling boats do not wear parkas trimmed with wolverine when in the boat. Same applies to hunting seal or walrus on the ice. Same applies to wearing a bird skin parka in which there is more than a single kind of bird skin used. Also do not take cooked meat out hunting for whale, seal or walrus in early spring. When the white breasted auklets first appear in May, cooked food may be taken along. Same applied formerly to use of sails on their umiaks. Sails were made of walrus stomach & intestine. In paddling (whaling) only one mitten is used, the upper hand. Also for rowing. When the first snipe (phalorope?) is seen the upper mitten may be left off also. Is worn up to that time.

Part 2F: Henry Bascom Collins Collection, Unprocessed, Doc. File (Paul Silook Notebooks), Silook 29.Dia:6, 8, 9, 10, 11, 13:

Feb. 17 Monday
N. clear all day. Womkon prepare to have dancing ceremony which they have every year. Andrew & Ernest each caught a fox.
Feb. 18 Tues.
N. clear all day the wind is not so high. Womkon is having a dancing worship. No body go hunt.
Feb. 25 Tues. 2 p.m. 70
[degrees inside of house] 4 l. [lamps] b.[burning] 8 p. [people staying in house]
N. Clear. Oscar killed 3 seals. Womkon & his people came to sing at 2: p.m.; 3: p.m. 88 3 l.b. 26 p. some dances. N. cloudy. They all leave out at 5: p.m.; Observe at 6: p.m. 84.
Feb. 27 Thurs. 82
4 l.b. 11 p. S.E. cloudy and snowing all day.
Womkon went to sing in Koonooku's house & Dicks.
Feb. 28 Fri 78
4 l.b. 11 p. N.E. cloudy & snow. Yavoghseuk prepare to have a moon worship — I killed 1 seal. Lost 1 mukluk & 1 walrus. This day we get our traps from its places.
March 1 Sat. 6 p.m. 72
4 l./b. 11 p. N. high wind cloudy & snowing all day. Yav. held a moon worship.
March 2 Sun. 6 p.m. 74
4 l.b. 10 p. N.E. low wind part clear & part cloudy. Few men killed seals. Some went to get their traps. Andrew caught 1 fox. Lawrence prepare to have moon worship.
March 3 Mon. 6 p.m. 74
4 l.b. 11 p. N.E. high wind cloudy. It was clear in the morning. Lawrence held a moon worship. Lloyd & Henry each caught 1 fox. Arrival of Pungowiyi, Noongwook, Penin,

Annogiyok, Rookok & Mr. Sam Troutman & Marvin [school teacher and son] from Seevoongu.

March 4 Tues 6 p.m. 74

4 l.b. 8 p. N. high wind clear but sleet.

The directors meet at school house. They talked about having a cool storage which will [be] built at Boxer Bay.

March 5 Wed 74

4 l.b. 11 p. N. clear but high wind. The Seevoongu men did not leave. Several men killed seals on the beach. I killed 1 seal.

Every evening the young boys and girls having tag between Jimmie's old house and James'.

March 24 Mon. 6:p./m. 78

4 l.b. 11 p. N.E. high wind clear., no hunting. I sew on my sail & boat cover ready for whaling next month.

March 25 Tues. 6 p.m.

4 l.b. 11 p. N. low wind clear all day. 9 dog team went to Seevoongu in the morning. Booshu killed a walrus. Some Seevoongu people arrived.

March 26 Wed. 6 p.m. 68

4 l.b. 12 p. N.E. high p. cl. p. clear still Seevoongu people here. They sing at Otiyohok's old house. Gambell people arrived from Seevoongu & 2 Seev. people came with them.

March 29 Sat 6 p.m. 80

4 l.b. 12 p. Tooli is with us N.E. cloudy & snowing. No hunting. People preparing for whaling.

March 30 Sun 6 p.m. 78

4 l.b. 11 p. N. Part clear part cloudy. It was clear in the morning snow at 7 a.m. clear at noon. Seevoongu people went back.

March 31 Mon. 6 p.m. 74

4 l.b. 10 p. N. Sleet all day. Andrew prepare to have a moon worship tomorrow morning. Iyaketan prepare also.

April Tues. 6 p.m. 78

4 l.b. 11 p. N. Clear but blowing. Andrew & Iyaketan held their moon worship this morning.

April 5 Sat. 6 p.m. 80

4 l.b. 11 p. Calm clear all day. Everybody prepare for whaling. Seevoongu people are still here. Soonogurok came back from Seevoongu.

April 6 Sun 6 p.m. 76

4 l.b. 9 p. S.E. high wind clear in the morning, so the people of Seevoongu went back. cloudy & snowing in the afternoon.

April 7 Mon. 6 p.m. 82

4 l.b. 8 p. E. cloudy cool snowing in the morning ceased in the afternoon. Shore ice break in acc't of waves. Little open water. Calm in the evening.

April 8 Tues. 6 p.m.

4 l.b. 9 p. W. low wind cloudy & foggy. Not good for hunting. Seevoongu people arrived. Lawrence & Harold also arrived from Tamnik hauling a small whale boat.

April 9 Wed. 6 p.m. 84

4 l.b. 11 p. E. cloudy & blowing. Everybody prepare their whaling weapons. Seevoongu people still here. I sew my sail and finish it. Oscar working on motor.

April 10 Thurs. 6 p.m. 80

4 l.b. 11 p. N. Snowing but clear. No good for hunting. I went to Walunga house and listen to some stories.

April 11 Fri 6 p.m. 70
4 l.b. 11 p. N.E. cloudy & Snowing Sevoongu people went back. Singing at Hokk house. I & Lawrence work on motor.
April 12 Sat 6 p.m. 80
4 l.b. 8 p. N.E. low wind cloudy & foggy but the fog disappear a while. Irrogoo killed 1 walrus. Womkon killed mukluk with baby. Everybody paint their whaleboat.
April 13 Sun 6 p.m. 80
4 l.b. 11 p. S.E. cloudy & snowing all day.
April 14 Mon. 6 p.m. 82
4 l.b. 9 p. S.W. low wind Several boats went out hunting. Hokh. lost a whale. The striker missed it. Only Lawrence killed a walrus. Emily bare a baby boy named Noosookuk, father, Oseuk.
April 18 Fri. 82
4 l.b. 9 p. E. Snowing. This morning we all went hunting in whale boat, after working on boat road …
May 8 Thurs. 78
4 l.b. 10 p. calm wind all day, young ice. We all the boats went out hunting in whaleboat. James, Womkon, Booshu, & Koonooku each killed a mukluk. Me & Shoolook each killed 2 mukluk.
Little boys begin to use slings …
May 13 Tues. 80
4 l.b. 6 p. S.E. very high wind. It was not so windy in the morning, so we went hunt in whaleboat. 5 whale boats came back with a load. Booshu killed a whale they toe it to the shore against the high wind. Work all night.
May 14 Wed. 82
4 l.b. 11 p. E. high wind, cloudy. Every body still cutting up the whale blubber and meat. In the afternoon we moved the whale to my boat road, where is a safe place.
May 15 Thurs. 80
4 l.b. 9 p. N.E. cloudy high wind. Not much ice around. No cutting, the whale is still on the beach. Several men went to Seevoongu this morning. No hunting. Play ball begins.
May 16 Fri. 74
4 l.b. 8 p./ N.E. cloudy high wind big open water, waves on the shore because no ice near by. Many dog team arrived from Seevoongu for mungtuk. Two bunch play ball on the south side and the other bunch on the northern bank. In the evening the Seevoongu people returned to Seevoongu.
May 17 Sat. 78
4 l.b. 5 p. N. cloudy high wind still there is no ice. George climb up the hill to see if there's ice near by, but no ice. Cutting up the whale all day. Play ball in the evening.
May 24 Sat. 80
4 l.b. 11 p. S.E. cloudy & rain no hunting, high wind. Only play ball in the evenings.
May 26 Mon. 80
4 l.b. 8 p. N.E. by E. part cloudy and part clear. No hunting in account of high wind and surf on the beach, but not so big as yesterday. Play ball and some shooting murres on the north beach.
May 30 Fri. 82
2 l.b. 8 p. N.E. cloudy rain in the morning and it ceased at noon, but begins to a fog in the afternoon. Big singing at Womkon's house all day. No more dog team because the trail is very bad for a team. Several person traded off their whale bone to the store. My dog have pups.

June 2 Mon. 82
2 l.b. 8 p. N.E. low wind p. clear p. cloudy. Young folks play ball.
I worked in store taking inventory. In the evening I helped Peter, writing an order from Sears, Roebuck & Co.
June 9 Mon. 80
4 l.b. 10 p. Calm. W. wind in the morning and it diminish in the afternoon. I worked in the store packing goods. Play ball every evening by the young folks.
June 10 Tues. 78
4 l.b. 9 p. N. — wind cloudy. Every body hung their meet to dry. I worked again in the store packing.
June 11 Wed. 78
3 l.b. 10 p. N. part clear part cloudy. Worked in the store. Yoghok bought the house of the store and begin to tear the shed down. Big sing at Womkon lumber house.
June 16 Mon. 82
2 l.b. 9 p. Calm clear. Every one works all day some continue on cleaning up. I cut some blubber of a walrus skin. In the morning my father & Yoghok worship whaling worship. Big singing at Womkon's lumber house.
June 27 Fri. 74
3 l.b. 7 p. just entered. N. cloudy and rain nearly all day. Taking measurement about 14 boys & girls. Afternoon singing at Womkon's house. Collins & Ford went to watch.

Contemporary Alaska Eskimo Bowhead Whaling Villages

Stephen R. Braund and Elisabeth L. Moorehead
Stephen R. Braund and Associates
P.O. Box 1480
Anchorage, AK 99510

Abstract. *This paper provides an overview of the 10 Alaska Eskimo bowhead whaling communities and their approaches to the still-central activity of bowhead whaling, including some detail about each community for comparison. This overview includes both historical and contemporary aspects of aboriginal bowhead whaling. These 10 communities share a common history of contact with commercial whalers, who introduced Eskimos to the technology of Yankee darting and shoulder guns that is still used today in subsistence whaling. Throughout the dramatic economic and political changes occurring since the 1800s, Alaska Eskimos have continued to hunt bowheads. Since 1978, bowhead whaling has occurred under a quota system imposed by the International Whaling Commission and implemented by the Alaska Eskimo Whaling Commission. Aboriginal bowhead whaling has always been one element of a geographically varied and resource diverse subsistence seasonal round. However, of all subsistence pursuits, bowhead whaling is the one on which the communities concentrate the most time, effort, money, group organization, cultural symbolism, and significance.*

INTRODUCTION[1]

The highly labor-intensive activity of bowhead whaling has been central to the lives of Eskimos in northern Alaska for centuries. Several papers in this volume describe elements of whaling as it occurred in prehistoric and historic times, but what is the context for modern subsistence bowhead whaling? This paper provides an overview of the 10 Alaska Eskimo bowhead whaling communities and their approaches to contemporary bowhead whaling, including some detail about each community for comparison.[2]

---

1. Portions of this paper have appeared in other documents (e.g., Braund 1992; Stephen R. Braund & Associates 1991; SRB&A and Institute of Social Research 1993a and 1993b; Braund et al. 1988c; Alaska Consultants, Inc. and SRB&A 1984).
2. Because this paper is a contemporary overview, not all of the older ethnographic literature about these communities and/or bowhead whaling is cited.

The 10 Alaskan communities that are currently recognized by the Alaska Eskimo Whaling Commission (AEWC) and the International Whaling Commission (IWC) as bowhead whaling communities can be organized by region and include: the Saint Lawrence Island villages of Gambell and Savoonga; the Bering Strait villages of Wales and Little Diomede; the Chukchi Sea communities of Kivalina, Point Hope, Wainwright, and Barrow; and the Beaufort Sea villages of Nuiqsut and Kaktovik (Fig. 1).

HISTORICAL BACKGROUND

Archaeological evidence indicates that the Eskimos of northwestern Alaska have hunted bowhead whales since approximately A.D. 800, and the activity developed 1000 years earlier on Saint Lawrence Island and the Siberian coast near Bering Strait (Bockstoce 1977). By the time European explorers arrived in the 19th century, the whaling tradition in several of these villages was very well established. According to Marquette and Bockstoce (1980:5), the Eskimos shared a presumably stable ecosystem with the bowhead whales until 1848, when European whalers discovered the rich bowhead whaling grounds north of Bering Strait. Over the next 60 years, the commercial whaling industry harvested over 19,000 bowhead whales from arctic waters for the whales' oil and baleen (Bockstoce 1978).

In the 1800s and early 1900s, most Inupiat lived in a seminomadic fashion. Their seasonal round entailed moving from coastal winter communities to different camps during the year to take advantage of harvest opportunities offered by those locations (e.g., a good fish camp in the summer, a whaling and marine mammal hunting camp in spring and early summer, a caribou/fish/berry picking camp in the fall). Some of these prime places were used off and on for centuries. Traders, government agents, and missionaries typically initiated permanent settlement by constructing a store, a school, or a church, often in or near a location already used by Natives during some part of the year. Local Native families would then move to that location and establish a permanent home, depending in part upon work or trapping and hunting associated with Euroamericans. These settlements evolved into formal communities, most of which continue to exist today. Although more settled now than before contact, Eskimos continue to use a variety of seasonal camps for hunting particular resources.

Residents of northwestern Alaska first had contact with non-Natives in the early 1800s. However, regular contact with non-Natives did not ensue until later in that century, when commercial whalers sought bowheads in the Chukchi Sea. In the 1880s, when commercial whalers established shore-based whaling stations and employed Inupiat whalers, contact increased considerably. Non-Native contact initiated a difficult period for Alaska Eskimos, due to the introduction of disease and alcohol and through the overharvesting of resources upon which the Inupiat depended, mainly bowhead whales and walrus. Additionally, in the 1880s the caribou population declined severely, and in 1885-1886 so also did most of the other local subsistence resources, resulting in starvation for many residents.

In 1884, commercial whalers established the first of 13 Alaska coastal whaling stations at Point Barrow. The whaling stations had year-round populations of non-Natives: a core crew of Yankee whalers with their whaling equipment. Because they were not dependent on the arrival of the fleet from the south, these shore-based whalers were

Contemporary Alaska Eskimo Bowhead Whaling Villages 255

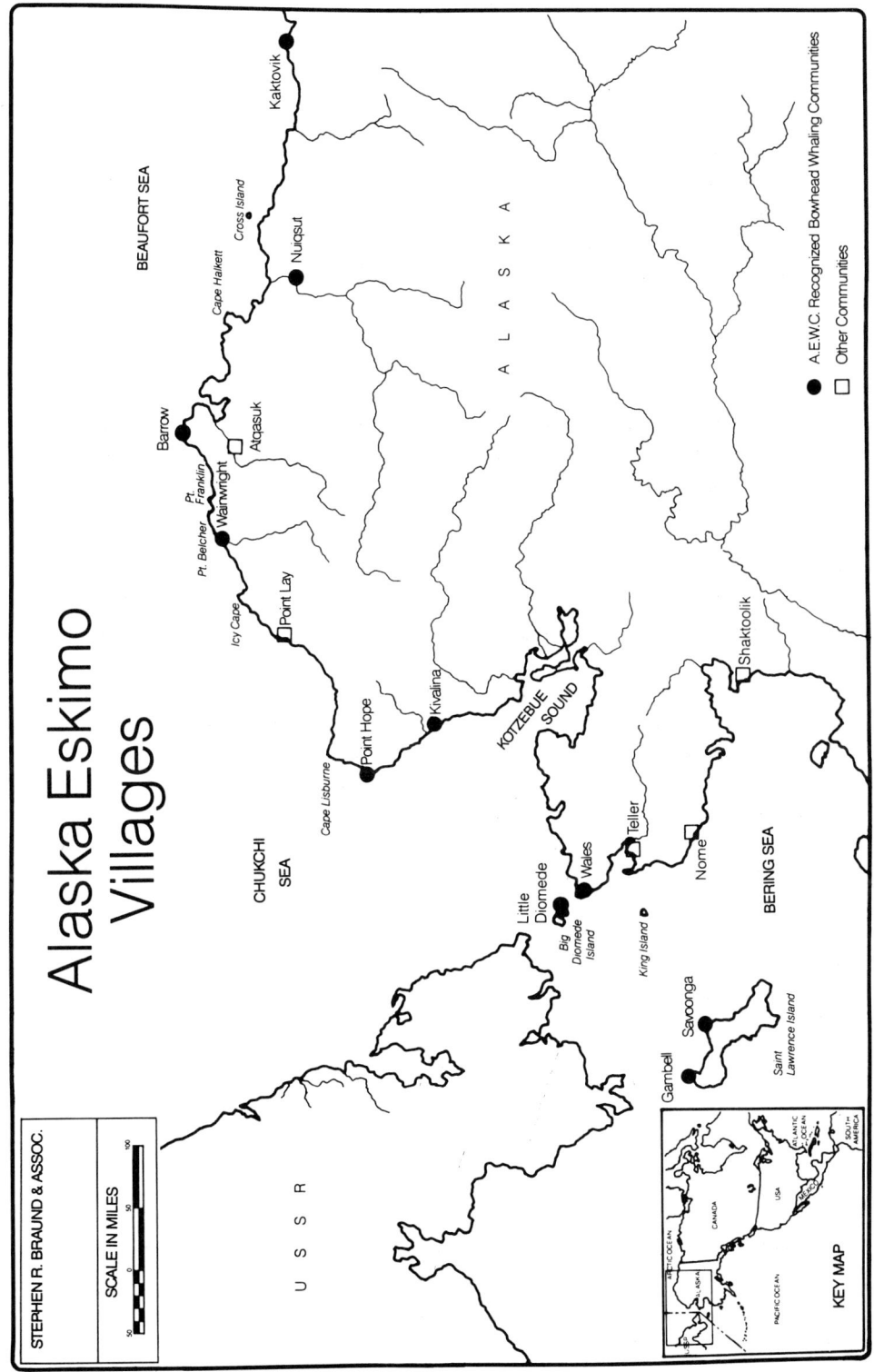

Figure 1. Map of Alaska Eskimo Whaling Villages.

able to organize bowhead hunts in the early spring, before the fleet could arrive, by employing Eskimo crews to hunt in the traditional manner (Bockstoce 1978). The crews would wait at the edge of the landfast ice, as the bowheads migrated north through the open leads to their summer feeding grounds in the eastern Beaufort Sea. Using darting and shoulder guns and traditional skin covered boats or Yankee wooden whale boats, the crews would intercept the bowheads on their northward migration. Due to the high price of baleen, the demand for labor exceeded the local labor pool. As a result, Eskimos from as far away as the Siberian coast, Saint Lawrence Island, and interior Alaska came to the coast to work in the whaling industry (Bockstoce 1986). The economic opportunities associated with the whaling stations resulted in the aggregation of Inupiat at certain locations along the coast.

Throughout the commercial whaling period (1848-1910), the Eskimos continued to use bowheads for subsistence purposes, as the Yankee whalers were primarily interested in the baleen and oil. After the baleen had been stripped from the whale, the carcasses were given to the Eskimos for food (Marquette and Bockstoce 1980). The Eskimos shared the edible portions through ritualized distribution patterns. Thus, a mutually beneficial arrangement evolved. The Eskimos hunted bowheads for pay (they received money, staples, manufactured items, or other trade goods in exchange for their labor), the Yankee whalers received the baleen (by the time the shore stations were established, baleen was the primary product they sought), and the Eskimos had the carcasses for food. By 1909, the commercial whaling industry collapsed due to the replacement of whale oil and baleen with petroleum products and spring steel (Marquette and Bockstoce 1980:6).

It was during this post-1885 period that the Eskimos were introduced to Yankee technology, including the darting gun and shoulder gun, which fired small bombs that exploded within the whale's body and increased the chances of a successful kill and recovery. The darting gun has a detachable harpoon and bomb attached to the end of a wooden shaft. It is thrown at the whale from a very close distance, and serves two purposes: (a) it harpoons the animal by attaching a line connected to one or more floats that are designed to slow the whale and also to indicate where it may surface, and (b) it shoots a small bomb into the whale that is designed to explode quickly, deep inside the whale after the harpoon strikes it. This bomb is intended to kill the whale instantly or gravely injure the whale so that it cannot escape. The shoulder gun, usually made of brass, fires a similar bomb that can be used at short distances.

After 1909, Eskimo whalers continued to hunt bowheads for subsistence purposes. Having been introduced to commercial whaling equipment, Eskimo crews continued to use the darting guns and shoulder guns introduced by the commercial whalers. Eventually, all crews adopted this equipment, which today is standard.[3] Marquette and Bockstoce (1980:9) characterize whaling from 1910 to 1969 as having a relatively low but steady level of activity.

---

3. In response to efforts to increase their struck to landed efficiency and to respond to "humane killing" concerns by International Whaling Commission member nations, the Alaska Eskimo Whaling Commission is in the process of developing and implementing a "penthrite" bomb to replace the old black powder bomb. This new bomb is for use in the 19th century shoulder and darting guns that have undergone minor modifications to accommodate the new bomb.

MANAGEMENT BACKGROUND

In 1931, the Convention for the Regulation of Whaling instituted an international ban on commercial bowhead whaling, but no limit was placed on the Alaska Eskimo subsistence harvest. Since its formation in 1947, the IWC has regulated the commercial hunting of whales. However, the exemption for subsistence hunting, continued from 1931, enabled Alaska Eskimos to continue their subsistence harvest of the western arctic bowhead stock free of IWC regulation until 1977. In recent years, subsistence hunting has been regulated by the IWC to permit Alaska Eskimos to continue their harvest on a quota basis during the IWC's moratorium on commercial whaling.

Due to the excessive commercial harvests of the late 19th and early 20th centuries, the IWC in the mid-1970s determined that all bowhead stocks, including the western arctic population upon which the Alaska Eskimos depend, should be considered seriously depleted (Tillman 1980). Beginning in 1972, the IWC requested that the United States provide data regarding the status of the western arctic bowhead population and the Alaska Eskimo hunt of this whale. As a result, the U.S., through the National Marine Fisheries Service (NMFS) of the Department of Commerce, began to gather biological information on the bowhead and to monitor the Alaska Eskimo subsistence hunt in 1973.

In the mid-1970s, the number of documented whaling crews nearly doubled in three major Alaska whaling communities, and the number of landed bowheads increased considerably from the previous decade, as did the number of whales struck and lost. With bowhead stocks still depleted from commercial whaling and little data available about the size and health of the bowhead population, the IWC banned Alaska Eskimo bowhead whaling in June, 1977, for the 1978 hunting season. Whaling captains from the nine active whaling communities (all but Little Diomede) formed the Alaska Eskimo Whaling Commission (AEWC) and developed a bowhead whale management plan, attended IWC committee meetings, initiated a bowhead census, and worked with the U.S. delegation to the IWC to build a case for rescinding the subsistence whaling moratorium. The U.S. government proposed a limited hunt to the IWC at a special December, 1977, meeting, and the IWC, based on the cultural and subsistence dependence on the bowhead by Alaska Eskimos, granted a quota of 12 whales landed or 18 struck (whichever occurred first) for the 1978 season.

Currently, the subsistence harvest of bowheads in Alaska is managed under a quota system through a cooperative agreement between the AEWC and the National Oceanic and Atmospheric Administration. The quota, set by the IWC, is based on both the subsistence and cultural needs for bowheads by Alaska Eskimos (Braund et al. 1988a, 1988b, 1988c; SRB&A 1991, 1992, 1994) and the size and rate of growth of the Western Arctic bowhead stock. The allocation of strikes and landed whales among the member communities is managed by the AEWC.

Based on cultural and subsistence needs and on bowhead population data, the IWC has increased the quota periodically over the last 17 years. The IWC increased the 1978 quota from 12 whales landed or 18 struck to 14 landed or 20 struck (to accommodate the fall, 1978, season). For 1979, the IWC granted a quota of 18 landed or 27 struck and for 1980, 18 landed or 26 struck. Beginning with the 1981 whaling season, the IWC gave quotas for more than one year. For the 1981, 1982, and 1983 seasons, the IWC quota was 45 total landed bowheads or 65 strikes with no more than 17 landed in any one year. For

1984 and 1985, the quota was 45 total strikes, with no more than 27 landed in one year. For 1985, 1986, and 1987, the revised quota was 26 strikes per year, including a provision for unused strikes to be carried over to the next year to a maximum of 32 strikes per year (with the 1988 quota being changed later to 35 strikes). At the June, 1988, IWC meetings, the IWC established a three-year quota for the years 1989, 1990, and 1991 of 41 landed bowheads per year and 44 strikes per year (with provisions to transfer up to three unused strikes to the following year). At the 1991 IWC meetings, the three-year quota for the years 1992, 1993, and 1994 was 41 landed per year and no more than 54 strikes per year (with the total number of whales struck for the three years not to exceed 141.) At the May, 1994, IWC meetings, the latest quota increase was granted to the 10 whaling communities (including Little Diomede): no more than 204 bowheads landed in the four year period from 1995 through 1998, and no more than 68 strikes in 1995, 67 in 1996, 66 in 1997, and 65 in 1998 (with provisions to carry up to 10 unused strikes from one year to the next).

To become an officially recognized whaling community, that community must request recognition from the AEWC. Little Diomede is the only community to be added since the formation of the AEWC by the original nine communities. Other communities or areas with a history of or interest in whaling, such as Shaktoolik, Shishmaref, and Kotzebue Sound, have not made any effort to join the AEWC. Shaktoolik residents landed a bowhead whale in 1980. Aklavik, a village located in the Mackenzie River Delta in Canada with a tradition of bowhead whaling, staged a bowhead hunt in 1991 and successfully landed a whale (Freeman et al. 1992). Other communities in western and eastern Canada have a history of whaling but the activity is currently dormant.[4] Some Russian communities on the Chukchi Peninsula and Big Diomede Island also hunted bowhead whales, although most ceased bowhead whaling between 1935 and 1950 (Stoker and Krupnik 1993).

CHARACTERISTICS OF ALASKA ESKIMO WHALING COMMUNITIES

Certain characteristics of these communities are common to the group. For example, the majority of each community's population is Native. The Native residents are predominantly Eskimo. The Natives of Saint Lawrence Island are Siberian Yup'ik Eskimos, while the Natives of all the other whaling communities are Inupiat Eskimos.

Another commonality is that all of the communities are located along the coast, with the Saint Lawrence Island villages farthest south along the whales' migration path and Kaktovik at the far end of the Alaska migratory route. The fall and spring bowhead migration routes dictate when each village can whale. Bowheads migrating north in the spring travel near each of the whaling villages, except Nuiqsut and Kaktovik. After passing Point Barrow, the whales continue traveling east without following the shore. Consequently, Nuiqsut and Kaktovik are not positioned to hunt bowheads in the spring. After spending the summer at their feeding grounds near the Mackenzie River Delta and Amundsen Gulf in Canada, the bowheads' fall migration brings them generally west

---

4. Igloolik reportedly took a bowhead whale in September, 1994.

along the Beaufort Sea coast until they pass Point Barrow, where they continue west, far out to sea, toward the Russian coast, then south through Bering Strait.

During this fall movement, some whales come within striking distance of the Beaufort Sea coast villages. Hence, Nuiqsut, Kaktovik, and Barrow are typically the only communities situated to hunt bowheads in the fall, and Barrow is the only community that, because of its advantageous location, regularly hunts bowheads during both the spring and fall migrations. Occasionally, Saint Lawrence Island crews have also hunted bowheads in November and December (Stoker 1984), but their current access to bowheads at this time of year depends as much on a bureaucratic event as on an environmental one: the availability of leftover strikes under the IWC-imposed bowhead whaling quota. Savoonga landed a bowhead in December, 1990, and struck and lost a bowhead in December, 1991. Stoker (1984) also reported that Wales hunted bowheads in November and December, although the peak period was during the spring migration. Wainwright occasionally has hunted whales in the fall under favorable conditions. Historically, Point Hope whalers held spring and fall whale hunts (Rainey 1947). This variation in hunting effort reflects the opportunistic nature of subsistence activities: if resources are present and harvest conditions are favorable, subsistence hunters will take advantage of available wildlife, especially if they have failed to harvest complementary species in adequate numbers. Thus, while acknowledging exceptions, we have categorized the whaling communities by their primary whaling seasons, spring or fall (or both, in the case of Barrow; Table 1).

These communities also have in common their reliance on harvesting local natural resources for much of their food, also referred to as "subsistence." In addition to using bowhead whales, most of these coastal Eskimo communities rely on a variety of other marine mammals such as seals (bearded, ringed, and spotted), walrus, polar bear, and beluga whale. Some communities hunt caribou and most of the villages also hunt waterfowl and catch fish. The subsistence year is considered to begin in the springtime (April), when the long winter dearth of resources ends with the return of most migratory species to their summer breeding and feeding grounds: bowheads, ducks, and geese. Throughout the summer and fall, families harvest fish, caribou, seals, walrus, and waterfowl. Several species can be harvested during the winter, such as ringed seals, polar bears, fish (through the ice), and ptarmigan. However, peak harvests of most species tend to be in the period between April and October (Stephen R. Braund & Associates [SRB&A] and Institute of Social and Economic Research [ISER] 1993a, 1993b).

The tradition of harvesting natural resources has been a consistent and significant element of Eskimo culture from prehistoric times to the present. Social relationships within a community are reinforced through cooperative hunting endeavors (e.g., whaling crews), the transfer of traditional knowledge and skills, and sharing the harvests with family members and other important relationships (Worl 1980; ACI and SRB&A 1984).

The high degree of risk, the high level of community cooperation required, and the high volume of product combine to make bowhead whaling one of the most culturally significant activities in each of these whaling communities. As a cooperative enterprise, whale hunting is organized around two basic "socioeconomic units," the crew and the village. Membership in the whaling crew is based primarily on kinship. Crew members are involved in preparing for the whaling season (breaking trail to spring camps, setting up their crew's camp, hunting to supply the camp), hunting the bowhead, receiving and

Table 1. Selected Characteristics of 10 Alaska Eskimo Bowhead Whaling Communities.

	1990 Native Population[1]	1990 Percent Native[1]	AEWC Registered Captains[2]	Primary Whaling Season[3]
Saint Lawrence Island				
Gambell	505	96	24	S
Savoonga	494	95	23	S
Bering Strait				
Little Diomede	167	93	4	S
Wales	143	89	5	S
Bering Strait/Chukchi Sea				
Kivalina	309	97	9	S
Pt. Hope	587	92	20	S
Wainwright	464	94	13	S
Barrow	2217	64	44	S/F
Beaufort Sea				
Nuiqsut	328	93	10	F
Kaktovik	189	84	9	F
TOTAL	5403		161	

1. Alaska Department of Labor 1991
2. Alaska Eskimo Whaling Commission 1994
3. S = spring; F = fall. Represents primary whaling season only; communities may whale at other times of year also.

distributing bowhead shares, and hosting formal celebrations after the crew has landed a whale. Landing a bowhead is celebrated with formalized patterns of sharing immediately following butchering (e.g., shares go to participating crews which in turn provide shares to individual crew members), as well as at later celebrations. Among the Inupiat, a feast called Nalukataq or blanket toss festival, is a community-wide celebration held in June following a successful whaling season. (The fall whaling villages of Kaktovik and Nuiqsut also hold their Nalukataq the following summer). On Saint Lawrence Island, the Yup'ik Eskimos celebrate the end of whaling season with their Whale Feast. Bowhead whale products are also typically featured at Thanksgiving and Christmas community potlucks, as well as other special occasions throughout the year. Hence, landing a bowhead and distributing the meat and maktak (whale skin and the attached blubber) serve to integrate the community and reinforce mutual interdependence and cooperation periodically throughout the year.

Whaling crews typically consist of a captain, a co-captain, harpooner, and navigator, plus several other general crew members. Crews may be as small as five members or range as high as 18 or more. Whaling crew members usually are drawn from the captain's extended family: sons, brothers, brothers-in-law, nephews, cousins, or sons-in-law. Some crew members may be unrelated to the captain and some may be from other communities. Although tending to be fairly stable over time, crew membership is not rigid. Some members shift crews from year-to-year or every few years.

A captain is an experienced hunter who has inherited or purchased the equipment necessary to field a crew: a boat, outboard motor, harpoon, darting gun, shoulder guns, bombs, block and tackle, ropes and floats, tents, snowmachine(s) and sled(s), CB radios, and numerous other items of equipment for the hunt as well as for camp (Worl 1980). A captain will invest several thousand dollars each year on fuel, ammunition, food, repairs and maintenance, and miscellaneous supplies. Prior to whaling, a captain may organize his crew to hunt ducks, caribou, or other species for eating while at whaling camp, and to help with equipment repairs and preparation. He also may organize his crew to hunt bearded seal or walrus after the whaling season in order to obtain skins for recovering the boat (which is necessary approximately every one to three years). The captain's family, with help from crew members and their families, hosts the feasts immediately following a successful landing, and co-hosts (with other successful captains) the subsequent community-wide celebrations. These celebrations involve the production of vast quantities of food: whale products prepared in a variety of ways, as well as caribou or duck soup, fry bread, tea, and other treats (SRB&A and ISER 1993a, b).

Although predominantly male, some crews have female members, and a few crews are captained by women. Women captains usually are the widows or daughters of whaling captains who operate a crew, but who typically do not themselves go out in the boat hunting whales.

VILLAGE PROFILES

Each of the 10 Alaska Eskimo whaling communities has unique characteristics. The following profiles provide a brief description of each community's history and their whaling practices.

Saint Lawrence Island: Gambell and Savoonga

Saint Lawrence Island is situated in the Bering Sea approximately 40 mi from Russia's Chukotsk Peninsula and 124 mi from the Alaska mainland. Access to the island is by airplane from Nome. Gambell is located on the northwestern corner of the island, the point nearest to Russia. Savoonga is also located on the north side of the island, 39 mi east of Gambell. Saint Lawrence Island has been occupied for at least 2000 years. Ongoing contact with non-Natives was established in the mid-1800s. In the 1870s, a series of difficulties, including disease and starvation, caused a decline in the population of 50% or more to about 500 residents. In 1880, Mr. and Mrs. V. C. Gambell established a church at the village of Sevuokak, which was later named Gambell in honor of the Presbyterian missionaries (ACI and SRB&A 1984).

In 1900, a herd of reindeer was shipped to the island under Sheldon Jackson's reindeer program for the purpose of insuring a steady, local supply of food. While most residents lived at the western end of the island, the reindeer concentrated on the eastern end. Consequently, residents established a reindeer herding camp in 1914 at an old village site. This camp was established as the village of Savoonga in 1917 (ACI and SRB&A 1984). As of 1990, Gambell's population was 505 Natives out of a total of 525 residents. Savoonga had a population of 519, of which 494 were Native (Table 1).

Although residents of Gambell have hunted bowheads continuously from prehistoric to historic to modern times, until 1972 Savoonga residents hunted bowheads with Gambell crews. Savoonga did not have its own captains and crews until that year, nor its own whaling equipment, although its residents actively participated in whaling. In 1994, Gambell had 24 whaling captains registered with the AEWC, and Savoonga had 23 registered captains. Each captain represents a whaling crew.

Saint Lawrence Island crews are the first to encounter the bowheads migrating north in the spring along the west side of the island, and their hunt takes place in March, April, and May. Gambell crews generally hunt from their village in the open leads north of town, while Savoonga crews haul their boats and gear about 50 mi across the island to the Southwest Cape. There they set up their camps.

Whaling crews hunt bowheads from boats made of a wooden bent-rib frame lashed together with sinew and covered with split female walrus skin (Braund 1988). The Saint Lawrence Island skin boat or angyapik is unique among Alaska Eskimo skin boats in that it has steam-bent hardwood ribs, is rigged for sailing, has an exterior runner or false keel attached outside of the skin cover which enables the boat to be skidded easily across ice and snow on its bottom without a sled to support it, and has a motor well inside the boat. The skin is painted to retard water absorption and rotting of the walrus skin while the boat is in use and to prevent excessive drying of the skin while the boat is in storage. To minimize noise, which is thought to scare off the whales, the boats are rigged for sailing, and Saint Lawrence Island whalers are unique in sailing the leads in pursuit of bowheads. The boats also are equipped with a motor located in an interior engine well, but the motor is used only when returning from pursuing a whale or while towing a whale back to the ice edge. With the exception of Little Diomede Islanders who use split walrus skins, the use of steam-bent ribs, split walrus skins, sails, an exterior keel, and an interior motor well are unique to Saint Lawrence Island compared to the boats used by other bowhead whaling communities.

Little Diomede

Little Diomede is a small island in the middle of the Bering Strait, just 3 mi from the island of Big Diomede and over 25 mi from the Alaska mainland. The 1990 population consisted of 167 Natives out of a total of 178 residents. The island is accessible by boat during the ice-free summer months or by plane on the sea ice between January and March, weather permitting, and thus is quite isolated from the mainland.

As a small, rocky island, Little Diomede's habitat offers mainly marine mammals and birds for subsistence use. Residents rely primarily on walrus and seals as their main subsistence resources. They supplement their marine mammal diet by fishing for tomcod, crab, and sculpin under the sea ice in the winter and hunting birds and collecting eggs in the summer months (Ellanna 1983). Residents also have a seasonal pattern of going to the mainland (Nome, Wales, Teller, and other Seward Peninsula communities) in the summer to visit kin, to gather additional subsistence resources (caribou, reindeer, salmon, freshwater fish, berries, and greens), and/or to seek temporary employment.

Although residents of Little Diomede have not landed a bowhead since 1937, they actively hunted bowheads until the late 1960s or early 1970s, and bowhead whaling is a significant part of their subsistence and cultural tradition. Russian explorer Kobelev in 1779 described residents of Little Diomede as living on "whales, sea dogs [seals], and walruses" (quoted in Ray 1975:33). In 1881, a century later, Hooper noted the same resource base: "The natives of both settlements [Big and Little Diomede] are great traders, and each summer cross over to the American side and meet the natives that assemble in numbers at Hotham Inlet, for the purpose of trading. They are very skillful at killing whales, walrus, and seals" (Hooper 1884:15).

This pattern of subsistence harvesting has been categorized by Ray (1964:62) as the "Whaling Pattern" and was practiced in the 18th and 19th centuries not only by Inupiat on Little Diomede Island, but also by other central Bering Strait people living at Wales and on King and Sledge islands. In the Diomede variant of this pattern, whaling was conducted primarily in the spring, with occasional fall whaling (Curtis 1930:113, Ellanna 1983:32), and was accompanied by extensive ceremonies before and after a harvest (Ray 1984:289). Early 20th century observers indicated a strong whaling tradition, similar to that found in other Inupiat communities located along the coast of mainland Alaska (Jones 1927; Eide 1952; Curtis 1930). Later, in the 1940s and 1950s, Heinrich wrote that Little Diomede "usually 'fields' three whale hunting boats each year" (Heinrich 1963:403). According to Heinrich (1963:408-409):

> By about the 15th or 20th of April, when the weather is still bitter cold but the days already quite long, the various boats take up their stations at the south end of the island, along the edge of the ice that is solidly frozen in between Little and Big Diomede. The usual stance is for the bow of the boat to project over the edge of the ice, with the harpooner and the bowman in their places and the umialiq [boat captain] in his place at the stern. The other crew members stand at the sides of the boat, or nearby, ready to shove the boat into the water and jump into their places. If the whale is sighted, the boat is quickly, and as silently as possible, put into the water, and paddled as silently as possible to the spot where the whale is next expected to surface.

Heinrich reported that most whaling crews ceased hunting bowheads around the middle of May, when the first female walrus began to appear. Those crews that continued to hunt bowheads would have to stop when the ice between the islands broke up, usually in the last 10 days of May.

The last bowhead known to have been landed at Little Diomede was in 1937 and the last strike was 1953 (not including a reported strike in 1979). During our 1991 fieldwork (SRB&A 1991), we continually asked Diomede residents why they were so unsuccessful in their whaling; they hunted many years and did not successfully strike or land many whales. Virtually all of the hunters gave the same answer. Strong currents, unpredictable and constantly changing ice conditions, clear water, and, in some years, a lack of appropriate open water made it very difficult to hunt bowhead whales at Little Diomede Island.

In 1991 Little Diomede had five skin boats, two aluminum boats, and one wooden boat in the village. Little Diomede residents continue to use primarily walrus skin-covered boats. These boats are traditional Bering Strait watercraft used for hunting marine mammals and traveling to the mainland. Well-adapted to the local environmental conditions of sea ice and available building materials, these boats require large crews, a continual supply of walrus hides, and the traditional skills of splitting and sewing the walrus skins, covering the boats, boat frame maintenance and repair, and seamanship. The boats are passed from father to son. The five skin boat crews that were active in 1991 ranged in size from seven to 14 men, and the crews were largely kinship-based. The crews of the two aluminum and one wooden boat are generally smaller than the skin boat crews but are also based on kinship.

Little Diomede requested membership in the AEWC and was granted membership in 1988. The U.S. Government presented information on Little Diomede whaling to the IWC in 1991 and in 1992. The bowhead whaling quota increase approved by the IWC in 1994 recognized Little Diomede for the first time as a whaling community entitled to hunt bowheads under the quota. Hence, Little Diomede residents likely will resume whaling in the near future. In 1994, Little Diomede had four whaling captains registered with the AEWC.

Wales

The village of Wales is located on the mainland on Cape Prince of Wales, the westernmost point of Seward Peninsula. Extending as far west as it does, the cape is part of the narrowest section of Bering Strait. At the time of contact, Wales was one of the largest Eskimo villages in Alaska and was a hub community for several smaller villages. During this century, the population has declined as a result of disease, particularly the influenza epidemic of 1918 (Ellanna 1983). As of 1990, the population was 161 of which 143 were Native residents of predominantly Inupiat Eskimo ancestry (Table 1).

Being situated on the mainland, Wales has better access to terrestrial mammals for subsistence harvesting than do the island communities described thus far. However, the area lacks large enough rivers to support freshwater fish as a significant resource, or to be used by hunters to travel into the inland areas (ACI and SRB&A 1984). Thus, community subsistence pursuits are oriented mainly around the marine environment, which provides an abundance of resources: bowheads, walrus, bearded, spotted, and ringed seals, belugas, and polar bears.

In addition to marine mammals, Wales residents rely (to a lesser extent) on reindeer (introduced in 1894), moose, various coastal fish species (salmon, tomcod, sculpin, crab, smelt), freshwater fish, clams, waterfowl, pelagic birds and their eggs, and a variety of greens and berries.

According to Ellanna (1983:456), "[t]he large, sedentary population of Wales during the historic period was supported by, and provided support to, the successful exploitation of large marine mammals." Moreover, people from the smaller, surrounding communities would congregate at the main village (now Wales) to participate in whaling (Ellanna 1983). Traditionally, crews used large wooden frame boats covered with walrus skins to hunt walrus and whales. However, in more recent years, the active Wales fleet consisted mainly of plywood or aluminum boats.

After several decades of limited whaling activity, whaling crews from Wales landed a bowhead in 1969 and 1970. In 1980, at least three crews hunted bowheads and successfully landed one. Two of the crews used plywood boats and the third (a former resident of Little Diomede) used a skin boat (Ellanna 1983). Wales has successfully landed bowheads at various times throughout the 1980s. Wales residents hunt bowhead whales in November and December, although the peak period is during the spring migration (Stoker 1984). In 1994, Wales had five whaling captains registered with the AEWC.

Kivalina

Of the 10 bowhead whaling communities, Kivalina is the most southerly along the Chukchi coast. The founding of the village of Kivalina is typical of the manner by which many of the villages were founded. The seed for the modern community of Kivalina was planted when a school was established in 1905 on a barrier beach opposite the mouth of the Wulik River. Inupiat residents of the area settled there, thus establishing the community of Kivalina (Burch 1985).

In 1990, the population of Kivalina was 317 of which 309 were Native (Table 1). Kivalina is within the recently formed Northwest Arctic Borough. Kotzebue, about 80 mi to the southeast, is the regional hub community for Kivalina and several other villages. The construction of the Red Dog Mine, about 50 mi northeast of Kivalina, has provided employment opportunity for Kivalina residents in recent years.

Kivalina residents utilize a diversity of subsistence resources. In addition to hunting bowheads, they also rely heavily on other marine mammals, such as the various seal species, walrus, and beluga whales. Walrus figure less significantly in the Kivalina subsistence regime compared to the Bering Strait communities, because Kivalina is generally some distance from prime walrus hunting areas. Fish, in contrast, are a significant component in the Kivalina diet (Braund and Burnham 1983). The Wulik River is an abundant and reliable source of arctic char and whitefish. Burch (1985) describes the local arctic char as a delicacy in the northwest region, the most important fish species in the area, and a mainstay of the Kivalina economy. Salmon (in lesser numbers) are also available in local waters, as are other miscellaneous species. Of the terrestrial mammals available, caribou is the most important species. The Western Arctic Herd passes through the area, usually in July and in the fall months. In the first half of this century, caribou rarely appeared near Kivalina. Since 1948, they have been a fairly regular resource, although variable in numbers. Kivalina hunters occasionally harvest moose, grizzly bears, and furbearing mammals. They also hunt a variety of bird species and plants.

In the last century, Kivalina whalers hunted bowheads in the spring from hunting camps along the Chukchi Sea coast, including a point called Nuvua, about 27 mi northwest of the village, where spring leads often formed in the ice (Burch 1985). They stopped hunting bowheads here in the early 1880s when famine struck Kivalina. Kivalina residents resumed bowhead whaling in the late 1880s, when commercial shore-based bowhead whaling stations were set up that employed Native hunters. Kivalina whalers participated in this endeavor at the whaling station near Point Hope. After the demise of commercial whaling, Kivalina whalers continued to hunt bowheads for subsistence purposes by traveling to Point Hope and whaling with crews there. Finally, in 1966, Kivalina whalers began regularly hunting bowheads from their own waters in the spring, landing their first whale in 1968. Kivalina also held its first whaling feast of modern times in 1968. When Burch conducted research in Kivalina in 1966, his impression was that the community was in danger "of becoming a dying village" (Burch 1985:9). In 1982, he found the community virtually thriving. He explained, "[a] number of factors were involved in these trends, but it is possible to date the beginning of the transformation quite specifically to the spring of 1968. That was the year when a Kivalina crew took a bowhead whale for the first time in several decades. The whaling feast that year was a profoundly moving experience for all who attended it; it was a galvanizing event" (Burch 1985:11).

Inupiat from nearby NANA region communities, including Kotzebue, Noatak, Selawik, and Noorvik, increasingly participate in Kivalina whaling. Kivalina is the only community in this region that has a bowhead quota, and nearby villagers monitor Kivalina's success or chances of success in order to participate in either the hunt, the butchering, or the whaling festival. Bowhead maktak and meat are prized and desired throughout the region.

Some of the boats used by Kivalina whalers are made of old skin boat frames that have been covered with plywood. Kivalina whaling crews continue to hunt bowheads each spring from camps set along the open leads in the ice west of their community. In 1994, Kivalina had nine registered whaling captains.

Point Hope

Located at the tip of a peninsula extending into the Chukchi Sea, Point Hope has been occupied by Eskimo peoples since about 500 B.C. The continuous use reflects the availability of diverse and year-round resources, making it a prime location for human settlement. The warm Chukchi Sea currents provide for regularly occurring periods of open water throughout the winter, facilitating marine mammal hunting during that part of the year when few other resources are available (ACI and SRB&A 1984; Pedersen 1979; Shinkwin and NSB 1978.)

In 1887, commercial whalers established a shore whaling station just southeast of Point Hope and brought Inupiat whalers from the south (Seward Peninsula and the Kobuk and Noatak regions) to hunt whales, since Point Hope whalers did not agree to participate in this venture. The Episcopal Church established a mission in Point Hope in 1890 (Burch 1981).

After the collapse of the bowhead whaling industry, selling furs became an important source of income in Point Hope. The U.S. Government became increasingly involved in the welfare of Point Hope, with programs such as reindeer herding and education and later through the military establishment of a Distant Early Warning installation (ACI and SRB&A 1984). Currently, Point Hope is part of the North Slope Borough and the Arctic Slope Regional Corporation, both of which have prospered from North Slope oil revenues. Consequently, Point Hope has a large, modern school facility, bus service, one large store and some smaller ones, a city government, a tribal government (Native Village of Point Hope), and the Tigara Native Corporation. The 1990 population of Point Hope was 639 of which 587 were Native.

Despite continuous change since the 1800s, Point Hope residents have consistently hunted bowheads, and that tradition continues to be one of the most important elements of Point Hope culture. Point Hope's creation story explains that the peninsula is where the "raven person" landed a "whale-creature" and "the whale became land. And that land was Tikigaq [Point Hope]" (Lowenstein 1981:8-9). As explained by Impact Assessment, Inc. (1990:PHO-99), "[w]haling is seen as the undertaking that gives the community of Point Hope its identity. In an important sense, to be a fully competent adult male in Point

Hope, to have "made it" in Point Hope society, is to be a whaling captain. Jobs may come and go, but whaling is seen as a very strong, unbroken link to the past." Community identity as whalers extends even to the school mascots. The local school's boys teams are the Point Hope Harpooners and the girls teams are the Harpoonerettes.

Point Hope whalers begin preparing for the spring whaling season in late March, taking their boats to camps on the ice in late April when the first bowheads have been sighted in the open lead. Crews set up camps along the lead, which is typically south of the point, 1-3 mi from the village in a good year. In other years, the lead may be seven or more miles away (Lowenstein 1981:43). The whaling season lasts from late April until early June, with early to mid-May being the most intensive part of the season (Lowenstein 1981:54).

The Point Hope whaling boat is a wooden frame covered with five to six bearded seal skins and capable of carrying eight people. (In recent years, some captains have covered their wooden frames with fiberglass.) A typical crew might consist of 11 members: eight whalers plus two cooks and a "boyer" (an apprentice boy whaler; Lowenstein 1981:48). In 1994, Point Hope had 20 whaling captains registered with the AEWC. After a successful season, the community holds a whaling feast, Nalukataq, at which whale products are distributed to everyone attending. Friends and relatives from other communities are always in attendance.

With regard to subsistence in general, Point Hope is, according to Stoker (1984:A-28) "in the enviable position of having a resource base sufficiently diverse so that no one species is of primary importance." His research indicated that from 1962 through 1982, caribou harvests constituted the largest proportion of the subsistence harvest, followed by bowhead whale, hair seals, and fish (primarily arctic char, as well as whitefish species, herring, smelt, chum salmon, Dolly Varden, arctic grayling, and tomcod). Bearded seal, beluga whale, birds, walrus, and polar bears were also harvested by Point Hope residents.

Wainwright

The Wainwright area was occupied traditionally by two main groups of Inupiat people, the Kuugmiut (people of the Kuk River) and the Utuqqaqmiut (people of the Utuqqaq River, presently spelled "Utukok"; Milan 1964; Ivie and Schneider 1979). As elsewhere, these early residents of the area traveled considerably to obtain the resources available from season to season. According to Ivie and Schneider (1979), caribou migration patterns, which vary from year to year, were the major influence over where the Utuqqaqmiut spent the winter. In the spring, bowhead whaling brought many Utuqqaqmiut to Icy Cape. Walrus hunting kept them on the coast through the summer, until the time came to travel up the Utukok River for fall fishing. The year was punctuated by several festivals that brought people together from their scattered camps to visit and trade.

The Kuugmiut followed a similar cycle prior to the turn of the century, according to Ivie and Schneider (1979), with the principal difference being that the Kuugmiut generally did not travel far from the coast. They hunted whales at Ataniq (at the base of Point Franklin) and other sites (including Point Belcher) and hunted waterfowl in the late spring and early fall throughout coastal areas. Walrus were hunted in the summer. Families moved to fall fish camps along the Kuk River before freeze-up.

In the second half of the 1800s, commercial whalers enlisted Natives of the Wainwright area in commercial shore-based whaling efforts, as occurred elsewhere along the Chukchi Sea coast. Having over-exploited the bowhead whale and walrus populations during their decades along the arctic coast, commercial whalers left the subsistence resources of the Native populations diminished when the whaling industry collapsed in 1909. The caribou shortage in particular had severe impacts on the Inupiat (Sonnenfeld 1956; Bockstoce 1986), with many inland peoples moving to the coast where food sources were more abundant.

In 1904, a school was built at the present location of Wainwright, and reindeer herding was introduced at the inlet (Jackson 1905). These two occurrences encouraged settlement at Wainwright of the various inland and more coastal Native peoples, and thus the community of Wainwright was established. The reindeer project was intended to provide a means of livelihood for the Natives, and evidently it lasted only into the late 1930s or early 1940s. Meanwhile, Natives also began to sell furs as a means of obtaining cash. In 1918, several residents pooled their earnings from the fur trade to establish the Wainwright Native Store so that a variety of supplies could be available locally. The wage employment sector increased significantly in the 1970s when the North Slope Borough was formed, resulting in capital projects in the village that generated seasonal and permanent jobs (ACI et al. 1984).

In 1990, Wainwright's population was 492 residents, 464 of whom were Native. Wainwright operates as a second class city with an elected city council and a mayor. In addition to local institutions, Wainwright residents are represented in a number of regional institutions, such as the North Slope Borough assembly, the Arctic Slope Regional Corporation, borough, state, and federal fish and wildlife advisory committees, and the Alaska Eskimo Whaling Commission, among others. The community has a high school and an elementary school, a clinic, emergency services, a laundromat/water plant, hotel and restaurant, a community center, three stores, and three churches. A Distant Early Warning (DEW) Line site was built outside of town in the 1950s and employed some local residents until 1989 when that facility closed.

Residents rely on a variety of subsistence resources throughout the year. During a two-year study conducted by our firm from April, 1988, through March, 1990 (SRB&A and ISER 1993b), bowhead whales represented 35% of the total subsistence harvest (by usable weight) and walrus represented 27% of the harvest, followed by caribou (23%). Fish (mainly whitefish, rainbow smelt, and arctic grayling) constituted an average of 5% of the annual harvest, while birds were 2%.

The traditional activity of bowhead whaling continues to be an integral part of subsistence and community life in Wainwright today. Nelson (1981:95) observed:

> Although they have entered an era of profound change, the Wainwright Inupiat still focus their lives around the land and the hunt. And among all hunting pursuits, whaling is paramount. It dominates the ethos and orientation of these people as no other single activity does. It is a prime source of status and prestige, a matrix for social and economic networks within the village and the region as a whole, and a measure of Inupiat identity.

Wainwright is located on a shallow bight, and spring leads in the ice generally do not open up near the town. Whalers work in April to cut trails through the ice to the more strategic locations for placing their camps, typically 18 to 20 mi northeast of the village.

They usually begin whaling in April, from their camps along the lead edge, and stop whaling by early June. Later in June, the community gathers, along with visitors from out-of-town, for Nalukataq.

In the late 1980s, Wainwright had approximately 12 whaling crews. By 1994, Wainwright had 13 whaling captains registered with the AEWC. Few, if any, Wainwright crews use skin boats anymore; most use aluminum skiffs painted white and pursue the whales under power rather than paddling. Although not as negotiable in narrow leads and not as quiet, Wainwright whalers have found skiffs to be more practical and efficient overall (Nelson 1981:94).

Wainwright whalers have had a high efficiency ratio (strikes to landed bowheads) since the IWC quota was implemented. Some observers attribute this to their waiting until late in the spring season to hunt bowheads, when there is less ice present and they are able to hunt the whales in open water with aluminum boats and outboard motors. A struck bowhead is not able to escape as easily in the open water as one in the narrow ice leads earlier in the spring. In addition, the hunters are able to quickly pursue the struck whale under the power of outboard motors.

Barrow

The area around Point Barrow has been inhabited for approximately 5000 years, with continuous habitation occurring for at least 1300 years (Dumond 1977). The first Europeans to encounter Barrow Inupiat were British explorers in search of a Northwest Passage. Ongoing contact ensued in 1854, when commercial whaling ships in pursuit of bowheads began making regular stops at Point Barrow to trade firearms, ammunition, and alcohol for baleen and furs. The presence of the "Yankee whalers" stimulated Native trade but apparently did not substantially alter Inupiat economic activity (Murdoch 1891:54).

In 1884, the Pacific Steam Whaling Company established the first shore station at Barrow. Within six years, three additional independent operations, employing more than 400 people and organized into 50 boat crews (10 non-Native crews and the rest Inupiat), were operating out of Barrow (Bockstoce 1986:236). Due to a shortage of local Inupiat crews to meet the demand, Eskimos from as far away as the Siberian coast, Saint Lawrence Island, and interior Alaska made their way to Barrow to work in the whaling industry (Bockstoce 1986:241). Genealogical investigations indicate that many present-day inhabitants of the Barrow area are descended from Inupiat who relocated there from other areas (Worl 1980:307).

Thereafter, numerous developments maintained and increased Barrow's position as a hub community on the North Slope. In 1897, the U.S. Government sent 362 reindeer to Barrow, not only to feed 275 stranded non-Native whalers (whose ships were caught in the ice for the winter), but also to instill an entrepreneurial spirit in the Inupiat by providing them with domestic reindeer herds to manage. The herd lasted until 1952 (Chance 1990:36).

After commercial whaling ceased in 1908, subsistence bowhead whaling continued to the present. However, bowhead stocks were severely depleted following commercial whaling, which affected the success of subsistence crews. Local Inupiat turned to fur trapping for cash. When the Depression caused fur prices to plummet, Inupiat returned to relying on marine mammals and "living off the land" (Spencer 1959:361). The local economy improved with World War II, when the Navy began exploring for oil on the North Slope and hiring dozens of Barrow Inupiat for several years (Chance 1966:17). Several subsequent government projects were begun in Barrow, such as the Naval Arctic Research Laboratory (NARL) and the Distant Early Warning site (DEW Line), both of which employed Inupiat (Chance 1966:17). Inupiat subsistence patterns were not greatly altered between the 1850s and the 1950s (Spencer 1959:358; Sonnenfeld 1956:417).

Today, Barrow is a thriving town. It is the seat of government for the prosperous and powerful North Slope Borough, which has invested heavily in the infrastructure and various programs for communities within its boundaries and, consequently, is a major employer. Barrow is also the headquarters for the Arctic Slope Regional Corporation, the Alaska Eskimo Whaling Commission, and numerous other enterprises and organizations. The 1990 population was 3469 of which 2217 were Native.

In a Barrow subsistence harvest study conducted by our firm for three years (4/1/87-3/31/90; SRB&A and ISER 1993a), residents harvested an estimated average of 702,660 usable lbs/yr of marine mammals, terrestrial mammals, birds, and fish. The average household harvested 750 lbs/yr, or 233 lbs/capita/yr. The average Inupiat household harvested 1171 lbs/yr. Bowheads, caribou, walrus, and whitefish contributed the most to the Barrow subsistence harvest (in order of importance by weight). In the three years of our study, Barrow residents harvested at least 46 species of fish, birds, and marine and terrestrial mammals, and collected berries, greens, water, and ice. While the people of Barrow were largely integrated into a cash economy by this time, the Barrow area offers an abundant diversity of resources, and traditional subsistence activity remained a fundamental component of the local economy and the local Inupiat culture.

In 1994, Barrow had 44 whaling captains registered with the AEWC. As mentioned previously, Barrow is one of the few advantageously situated communities where whalers are able to hunt bowheads in both the spring and the fall migrations. During the three years of our subsistence study, Barrow whaling crews landed an average of nine bowheads per year, five in the spring and four in the fall (SRB&A and ISER 1993a). The IWC quota was in effect during these years. Spring whaling is conducted much as in Point Hope and other communities, in which whaling crews make camp at the edge of an open lead and hunt from wooden frame boats covered with bearded seal skins. The boats are paddled in pursuit of a whale; once struck, whalers then switch to outboard motors to tow the whale ashore. The lead generally opens on the west side of Point Barrow and runs parallel to shore, sometimes within 3-4 mi from Barrow. Spring whaling takes place in April, May, and June; later in June, after whaling is finished for the season, successful crews host the Nalukataq. Nalukataq is a North Slope Borough employee holiday.

Fall whaling takes place in open ocean by whaling crews using aluminum or fiberglass boats with outboard motors. Crews range considerable distances in pursuit of bowheads (e.g., 50 mi). However, they prefer to take them as close to town as possible to minimize the distance that the whale must be towed, as a long tow can result in spoiled meat. Fewer crews participate in fall whaling than spring whaling, mainly because the fall is the most important season for obtaining caribou and fish for the rest of the year. Whaling crews generally do not set up whaling camps in the fall; rather, whalers leave from Barrow in their boats and come home the same day if they do not get a whale.

Barrow's fall whaling season is influenced by the outcome of the spring hunt. Barrow crews are assigned a certain number of strikes for the year; thus, the number of strikes used in the spring season determines how many are left over in the fall. Additionally, a spring whaling village that did not use all of its allotted strikes may transfer their unused strikes to Barrow (or Nuiqsut or Kaktovik) for its fall hunt. Second, if the spring bowhead harvest was unsatisfactory (apart from the number of strikes used), people will be more motivated to hunt in the fall than they are after a successful spring hunt, when ice cellars are full of bowhead.

Nuiqsut

Nuiqsut is located along a channel of the Colville River about 25 mi from the Arctic Ocean. A traditional Inupiat village site that was occupied at least until 1939 (when the U.S. Census reported 89 residents), Nuiqsut was abandoned by 1950 mainly because residents moved to Barrow for employment, school, and other services (ACI and SRB&A 1984:52). Some of the former residents continued to use their cabins at Nuiqsut seasonally for subsistence harvesting, and one family remained in the area year-round (Research Foundation of State University of New York [RFSUNY] 1984:11).

When the Alaska Native Claims Settlement Act (ANCSA) was passed in 1971, Nuiqsut was identified as a traditional village. In 1972, the Nuiqsut village corporation was formed under the provisions of ANCSA and 207 people enrolled (ACI and SRB&A 1984:52). In April, 1973, resettlement of Nuiqsut began as groups of families left Barrow by snowmachine for Nuiqsut. Approximately 27 families moved to Nuiqsut and lived in tents until permanent housing was constructed. Many of these families had lived in this area decades earlier. Resettlement was reportedly motivated by a desire to live more closely to the land and to step away from the increasing urbanization of Barrow (ACI and SRB&A 1984:52). As of 1990, Nuiqsut had a population of 354 of whom 328 were Native.

As a "new" village within the NSB, Nuiqsut has been a beneficiary of considerable public sector revenue, particularly CIP (Capital Improvement Project) monies. CIP projects included housing, a school, an airstrip, utilities, roads, sewage, a city dump, a public safety building, fire hall, and clinic. Nuiqsut has two churches, a village corporation, and post office as well (RFSUNY 1984).

Traditional subsistence activities in the Nuiqsut area have revolved principally around caribou and fish. Marine mammals, moose, waterfowl, and furbearers have also been important although secondary (RFSUNY 1984:175; ADF&G n.d.). Nuiqsut's location on a channel of the Colville River, some 25 mi upstream from the ocean, has been a prime area for fish harvests and less advantageous for marine mammal harvests. The Colville River serves as the major transportation corridor for inland hunting and fishing throughout the year by boat and by snowmachine.

A few of the men who resettled Nuiqsut in 1973 had been whaling captains in Barrow previously. (Additionally, the people who lived in the Nuiqsut area historically had hunted bowhead whales.) Hence, these men assembled crews and maintained the tradition of bowhead whaling in Nuiqsut. Nuiqsut whaling occurs only in the fall when the whales migrate closer to shore; the spring bowhead migration is too far offshore. Whaling crews camp on and conduct their hunt from Flaxman Island, a barrier island off the Canning River Delta, over 100 mi east of Nuiqsut, during the months of September and October. Nuiqsut whalers mainly use aluminum boats. Occasionally, Nuiqsut and Kaktovik crews have collaborated in hunting and landing bowheads. Also, some crew members have returned to Barrow in the spring to hunt bowheads with Barrow crews (Stoker 1984:A-45). In 1994, Nuiqsut had 10 whaling captains registered with the AEWC.

Kaktovik

Kaktovik is located on Barter Island, one of the numerous barrier islands along the Beaufort Sea coast. A large prehistoric village is located on the island with 30 to 40 old house features and numerous whale bones among the sod house ruins that are suggestive of a whaling tradition (Jacobson and Wentworth 1982:3). The name Barter Island reflects the island's history as a trading center, dating back centuries. People from Barrow, the Canadian coastal communities, and inland settlements would meet here to trade. The island was used seasonally by the nomadic Inupiat until approximately 1923, when an Inupiat family decided to settle there and help the merchant, Tom Gordon, establish a trading post as a market for furs. Other families eventually settled there also. Kaktovik residents herded reindeer in the 1920s and 1930s, ending in 1938, and around the same time the market for furs declined and the trading post closed. From this time through the mid-1940s, Kaktovik people relied almost exclusively on subsistence for their livelihood. From 1945 onward, wage employment grew in the community mainly as a result of post-war projects, such as the DEW Line site constructed on Barter Island, Air Force construction of a runway and hangar, and a U.S. Coast and Geodetic Survey mapping project. A BIA school was established in 1951 (Jacobson and Wentworth 1982:3-5).

In 1990, Kaktovik was a village of 224 residents, of which 189 were Native (Table 1). Kaktovik is a second class city with a mayor and city council. It is also within the North Slope Borough and thus benefits from the Borough's well-funded education system, infrastructure development, social services, cultural programs, and employment. Kaktovik has a school, clinic, fire hall, power plant and water plant/washeteria, a hotel and restaurant, community center, two to three stores, a post office and public safety building, and a Presbyterian church. The large runway and hangar were built to serve the DEW Line, and the runway also serves the community (IAI 1990). As of 1989, the main employers in town were the NSB (67%), the village corporation, local government, the federal government (U.S. Fish and Wildlife Service), construction, and the hotel (IAI 1990).

The community lies within the Arctic National Wildlife Refuge, and residents utilize those lands for much of their subsistence activity. Caribou are the staple subsistence item; bowhead whale and seals (bearded, ringed, and spotted) are also important, as are ducks, geese, and several fish species (Jacobson and Wentworth 1982:35-68). Kaktovik's proximity to the Brooks Range gives hunters access to Dall sheep, a staple item of the Kaktovik subsistence diet. Regular use of Dall sheep by Kaktovik hunters is unique among the 10 whaling communities.

Kaktovik is a fall whaling community because, as in Nuiqsut, the fall whale migration is close enough to the coast, in contrast to the spring migration which is too far offshore. Stormy or foggy fall weather frequently impedes whaling. Hunting usually occurs within 10 mi of land, but crews may range as far as 20 mi from land while traveling up and down the coast. Hunters prefer not to range too far from Kaktovik, however, because of the difficulty of undertaking a long tow if they successfully kill a bowhead far away from the village (Jacobson and Wentworth 1982:50). In 1979 and 1980, Nuiqsut whalers hunted with Kaktovik whalers due to poor conditions in Nuiqsut's whaling grounds (Jacobson and Wentworth 1982:49). Whalers hunt from 14-22 ft boats with outboard motors; the majority of the boats used for whaling are aluminum (ACI and SRB&A 1984:123). In 1994, Kaktovik had nine whaling captains registered with the AEWC.

SUMMARY

The 10 Alaska Eskimo whaling villages share a common history of aboriginal subsistence bowhead whaling in which whaling was just one part of a geographically varied and resource diverse seasonal round. These villages also share a common history of contact with commercial whalers beginning in the mid-1800s. Contact introduced several changes to the Eskimos, including trade goods, disease, alcohol, Yankee whaling equipment, trade-based employment at shore-based whaling stations, and severe depletion of the bowhead and walrus stocks, to name a few. As a result of technological contact, Eskimo whalers throughout the Beaufort, Chukchi, and Bering seas modified their whaling practices by adopting the Yankee darting and shoulder guns, introducing explosives for the first time into Eskimo whaling. Although some aboriginal whalers also adopted the Yankee wooden whaling boat, most continued to use the traditional wood frame, skin covered boat.

Following the collapse of the commercial whaling industry, Eskimo whalers continued hunting bowheads for subsistence purposes. Because bowhead and walrus stocks were depleted at that point, hunting these species became more difficult. Eskimos became involved in other commercial endeavors, such as selling and trading furs. Gradually throughout the first half of the 1900s, establishment of the modern communities took place through the efforts of missionaries, government educators, and traders. Although Eskimos' commercial activities and sedentism had increased since contact, they continued to organize their lives around the traditional subsistence seasonal round, using different camps at different times of year to harvest particular resources.

Since 1978, bowhead whaling has occurred under a quota system imposed by the IWC and implemented by the AEWC. Table 2 presents the number of bowheads landed by each community over the past 17 years. It is important to keep in mind that the years 1978 through 1994 are quota years (i.e., the number of bowheads landed was externally restricted by the IWC imposed quotas). Thus, communities could have landed more whales if the quotas had not restricted their harvests. Furthermore, weather and ice conditions substantially affect the whaler's success. Spring whaling is dominated by the presence, absence, or condition of sea ice. Sea ice conditions, ice formations, ice movements, and the presence and location of open water leads are critical factors in spring whaling. The nearshore leads in which the bowheads travel must be accessible to the hunters. The shorefast ice must be thick and stable enough to support the hunters, their camps, and the weight of the bowhead (Worl 1980). Fall whaling is also affected by environmental conditions. Although hunters hunt in the open water before ice forms, fall storms, high winds, and rough seas may prevent them from successfully pursuing or landing bowheads.

Each village's whaling captains are represented by one commissioner on the AEWC. The AEWC meets annually to divide the allotted strikes among the 10 villages. For example, in 1993, the 54 strikes were allocated as shown in the first column of Table 3 below. How the strikes were used is indicated in subsequent columns.

Table 2. Bowheads Landed Since 1978, by Community.

Year:	1978	79	80	81	82	83	84	85	86	87	88	89	90	91	92	93	94
Savoonga	1	0	1	2	1	1	2	1	0	1	0	1	5	0	4	1	2
Gambell	1	0	2	1	2	1	0	1	3	2	2	0	4	1	4	4	1
Wales	0	0	1	0	0	1	0	1	0	1	0	0	0	1	0	0	1
Kivalina	0	0	0	0	0	0	1	0	0	1	0	0	0	1	1	0	2
Pt. Hope	2	3	0	4	1	1	2	1	2	5	5	0	3	6	2	2	5
Wainwright	2	1	1	3	2	2	2	2	3	4	4	2	5	4	0	5	4
Barrow	4	3	9	4	0	2	4	5	8	7	11	10	11	13	22	23	16
Nuiqsut	0	0	0	0	1	0	0	0	1	1	0	2	0	1	2	3	0
Kaktovik	2	5	1	3	1	1	1	0	3	0	1	3	2	0	3	3	3
TOTAL	12	12	15	17	8	9	12	11	20	22	23	18	30	27	38	41	34

Source: Alaska Eskimo Whaling Commission; SRB&A 1988.

These data indicate that the whalers collectively landed the maximum number of whales allowed under the 1993 quota, 41. Once they reached that limit, they had to stop whaling. Comparing the first and last columns indicates that Barrow and Wainwright each used more strikes than they were originally allocated. Communities that did not use all of their strikes during their usual season (e.g., Gambell, Savoonga, Wales, Kivalina, and Point Hope) transferred some of their unused strikes to communities that still had good whaling prospects but no remaining strikes. (Transferring strikes is a decision made by the community whaling captains association, not by the AEWC.) In this particular year, Wales, Kivalina, and Little Diomede experienced severe ice conditions during the spring bowhead migration and consequently did not use any strikes at all. (Little Diomede did not officially have a strike allocation, but was included in the year-end report on whaling activity.)

Barrow's Native population is over three times larger than any other community's Native population, and Barrow has nearly twice as many whaling crews as any other village. Those demographic characteristics, combined with the unique advantage of having access to bowheads passing Point Barrow in both the spring and the fall, are the main reasons that Barrow is allocated the most strikes and often uses even more than their allocation (by receiving strikes transferred to Barrow from other villages unable to use all of theirs). Otherwise, weather and ice conditions are unquestionably the dominant variables affecting the outcome of whaling in any of the villages.

Table 3. Allocation and Use of 1993 Strikes.

	Initial AEWC Allocation	Struck & Landed	Struck & Lost	Total Strikes Used
Savoonga	5	1	0	1
Gambell	6	4	0	4
Wales	2	0	0	0
Little Diomede		0	0	0
Kivalina	3	0	0	0
Point Hope	8	2	3	5
Wainwright	6	5	3	8
Barrow	18	23	5	28
Nuiqsut	3	3	0	3
Kaktovik	3	3	0	3
TOTAL	54	41	11	52

Source: Alaska Eskimo Whaling Commission.

Highly formalized patterns of hunting, sharing, and consumption characterize the modern bowhead harvest. Sharing occurs not only within the whaling crews and within their community, but also between communities. Some crews are composed of members from non-whaling villages (e.g., Kotzebue residents on Kivalina crews, or Atqasuk residents on Barrow crews). Communities unable to use all of their allotted strikes are likely to transfer their strikes to a community that still has an opportunity to hunt that year. Friends and relatives from other whaling villages and non-whaling villages attend the celebratory feasts at the end of the whaling season. Families send *maktak* to relatives throughout Alaska and even outside Alaska. Thus, participation in bowhead whaling and enjoyment of the harvest extends well beyond the 10 whaling villages. Of all subsistence pursuits, bowhead whaling is the one on which the communities concentrate the most time, effort, money, group organization, cultural symbolism, and significance. Indeed, being a whaling community is a large part of a community's cultural tradition and its modern cultural identity.

REFERENCES

Alaska Consultants, Inc., C. Courtnage, and Stephen R. Braund & Associates
    1984   Barrow Arch Socioeconomic and Sociocultural Description. Social and Economic Studies Program, *Minerals Management Service, Alaska Outer Continental Shelf Region, Technical Report No. 101*.

Alaska Consultants, Inc. and Stephen R. Braund & Associates
 1984   Subsistence Study of Alaska Eskimo Whaling Villages. Prepared for the U.S. Department of the Interior.

Alaska Department of Fish and Game (ADF&G)
 n.d.   ADF&G Division of Subsistence Community Profile Database, Communities of Nuiqsut (1985) and Kaktovik (1986).

Alaska Department of Labor
 1991   *Alaska Population Overview, 1990 Census and Estimates*. Juneau, AK.

Alaska Eskimo Whaling Commission
 1994   List of Boat Captains Registered with the AEWC. Barrow, Alaska.

Bockstoce, J. R.
 1977   An Issue of Survival: Bowhead vs. Tradition. *Audubon* 79(5):142-145.
 1978   History of Commercial Whaling in Arctic Alaska. *Alaska Geographic* 5(4):17-25.
 1986   *Whales, Ice, and Men: The History of Whaling in the Western Arctic*. University of Washington Press in association with the New Bedford Whaling Museum, Seattle.

Braund, S. R.
 1988   *The Skin Boats of Saint Lawrence Island, Alaska*. University of Washington Press, Seattle.
 1992   The Role of Social Science in the International Whaling Commission Bowhead Whale Quota. *Arctic Research of the United States Interagency Arctic Research Policy Committee 6:37-42*.

Braund, Stephen R. and Associates
 1991   Subsistence and Cultural Need for Bowhead Whales by the Village of Little Diomede, Alaska. *International Whaling Commission Report* IWC/44/AS 2. Prepared for the Alaska Eskimo Whaling Commission, Barrow.
 1992   Quantification of Subsistence and Cultural Need for Bowhead Whales by Alaska Eskimos-1992 Update Based on 1990 U.S. Census. Prepared for the Alaska Eskimo Whaling Commission, Barrow.
 1994   Quantification of Subsistence and Cultural Need for Bowhead Whales by Alaska Eskimos-1994 Update Based on 1992 Alaska Department of Labor Data. Prepared for the Alaska Eskimo Whaling Commission, Barrow.

Braund, S.R. & Associates and Institute of Social & Economic Research
 1993a North Slope Subsistence Study-Barrow, 1987, 1988 and 1989. Prepared for U.S. Department of Interior, Minerals Management Service and the North Slope Borough. (Also published as: *U.S. Department of Interior, Minerals Management Service Technical Report No. 149.*)
 1993b North Slope Subsistence Study — Wainwright, 1988 and 1989. Prepared for U.S. Department of Interior, Minerals Management Service and the North Slope Borough. (Also published as: *U.S. Department of Interior, Minerals Management Service Technical Report No. 147.*)

Braund, S. R. and D. C. Burnham
 1983   Kivalina and Noatak Subsistence Use Patterns. Stephen R. Braund & Associates. Prepared for Cominco Alaska.

Braund, S. R., W. M. Marquette, and J. R. Bockstoce
 1988a Data on Shore-Based Bowhead Whaling at Sites in Alaska. Appendix 1 In: 1988 Quantification of Subsistence and Cultural Need for Bowhead Whales by Alaska Eskimos, S. R. Braund, S. W. Stoker, and J. A. Kruse. Stephen R. Braund & Associates,

Anchorage. Prepared for the Bureau of Indian Affairs, Department of the Interior. *International Whaling Commission TC/40/AS2.*

1988b   Data on Shore-Based Bowhead Whaling at Sites in Alaska. Document SC/40/PS 10 submitted to the IWC Scientific Committee.

Braund, S. R., S. W. Stoker, and J. A. Kruse

1988c   Quantification of Subsistence and Cultural Need for Bowhead Whales by Alaska Eskimos. Stephen R. Braund & Associates, Anchorage. Prepared for the Bureau of Indian Affairs, Department of the Interior. *International Whaling Commission* TC/40/AS2.

Burch, E. S., Jr.

1981   *The Traditional Eskimo Hunters of Point Hope, Alaska: 1800-1875.* North Slope Borough, Barrow.

1985   Subsistence Production in Kivalina, Alaska: A Twenty-Year Perspective. Prepared for Alaska Department of Fish & Game, Division of Subsistence.

Chance, N.

1966   *The Eskimos of North Alaska.* Holt, Rinehart, and Winston, New York.

1990   *The Inupiat and Arctic Alaska: An Ethnography of Development.* Holt, Rinehart, and Winston, New York.

Curtis, E. S.

1930   *The North American Indian.* Vol. 20. The University Press, Cambridge, MA.

Dumond, D.

1977   *Eskimos and Aleuts.* Thames and Hudson, London.

Eide, A. H.

1952   *Drums of Diomede: The Transformation of the Alaska Eskimo.* House-Warven, Hollywood, CA.

Ellanna, L. J.

1983   Bering Strait Insular Eskimo: A Diachronic Study of Economy and Population Structure. *Alaska Department of Fish and Game, Division of Subsistence Technical Paper No. 77.*

Freeman, M. M. R., E. E. Wein, and D. E. Keith

1992   Recovering Rights: Bowhead Whales and Inuvialuit Subsistence in the Western Canadian Arctic. *Studies on Whaling* No. 2. Canadian Circumpolar Institute and Fisheries Joint Management Committee, Edmonton.

Heinrich, A. C.

1963   *Eskimo Type Kinship and Eskimo Kinship: An Evaluation and a Provisional Model for Presenting Data Pertaining to IWC Kinship Systems.* Unpublished Ph.D. dissertation, University of Washington, Seattle.

Hooper, C. L.

1884   *Report of the Cruise of the U.S. Revenue Steamer Thomas Corwin, in the Arctic Ocean, 1881.* U.S. Government Printing Office, Washington, D.C.

Impact Assessment, Inc.

1990   Northern Institutional Profile Analysis: Chukchi Sea. *U.S. Department of Interior, Minerals Management Service, Technical Report No. 141.*

Ivie, P. and W. Schneider

1979   Wainwright Synopsis. In: *Native Livelihood and Dependence: A Study of Land Use Values Through Time.* North Slope Borough Contract Staff. Prepared for NPR-A, Work Group 1, U.S. Department of Interior, NPR-A, 105 (c) Land Use Study. Anchorage.

Jackson, S.
- 1905   Fourteenth Report on Introduction of Domestic Reindeer into Alaska with Maps and Illustrations. *58th Congress, 3d Session, Senate Document No. 61, U.S. Government Printing Office, Washington D.C.*

Jacobson, M. J. and C. Wentworth
- 1982   *Kaktovik Subsistence: Land Use Values through Time in the Arctic National Wildlife Refuge Area.* U.S. Fish & Wildlife Service and Northern Alaska Ecological Services. Fairbanks.

Jones, B.
- 1927   *The Argonauts of Siberia: A Diary of a Prospector.* Dorrance and Company, Philadelphia.

Lowenstein, T.
- 1981   Some Aspects of Sea Ice Subsistence Hunting in Point Hope, Alaska. A report for the North Slope Borough's Coastal Zone Management Plan.

Marquette, W. M. and J. R. Bockstoce
- 1980   Historical Shore-Based Catch of Bowhead Whales in the Bering, Chukchi, and Beaufort Seas. *Marine Fisheries Review* 42(9-10):5-19.

Milan, F. A.
- 1964   The Acculturation of the Contemporary Eskimo of Wainwright, Alaska. *Anthropological Papers of the University of Alaska* 11(2):1-81.

Murdoch, J.
- 1891   Ethnological Results of the Point Barrow Expedition. *9th Annual Report of the Bureau of American Ethnology for the Years 1887-1888,* pp. 19-441. Washington, D.C.

Nelson, R. K.
- 1981   Harvest of the Sea: Coastal Subsistence in Modern Wainwright. A Report for the North Slope Borough's Coastal Management Program.

Pedersen, S.
- 1979   Point Hope Synopsis. In: *Native Livelihood and Dependence: A Study of Land Use Values Through Time.* U.S. Department of the Interior, National Petroleum Reserve in Alaska 105(c) Land Use Study. Anchorage.

Rainey, F. G.
- 1947   The Whale Hunters of Tigara. *Anthropological Papers of the American Museum of Natural History* 41(2).

Ray, D. J.
- 1964   Nineteenth Century Settlement and Subsistence Patterns in Bering Strait. *Arctic Anthropology* 2(2):61-94.
- 1975   *The Eskimo of Bering Strait, 1650-1898.* University of Washington Press, Seattle.
- 1984   Bering Strait Eskimo. In: *Handbook of North American Indians,* Vol. 5, *Arctic,* edited by D. Damas, pp. 285-302. Smithsonian Institution, Washington, D.C.

Research Foundation of State University of New York
- 1984   Ethnographic Study and Monitoring Methodology of Contemporary Economic Growth, Socio-Cultural Change and Community Development in Nuiqsut, Alaska. *U.S. Department of Interior, Minerals Management Service, Technical Report No. 96.*

Shinkwin, A. and the North Slope Borough Planning Department
- 1978   A Preservation Plan for Tigara Village. North Slope Borough Commission on History and Culture, Barrow.

Sonnenfeld, J.
- 1956 *Changes in Subsistence Among the Barrow Eskimo.* Unpublished Ph.D. dissertation, Johns Hopkins University.

Spencer, R. F.
- 1959 The North Alaskan Eskimo: A Study in Ecology and Society. *Bureau of American Ethnology Bulletin 171.*

Stoker, S. W.
- 1984 Subsistence Harvest Estimates and Faunal Resource Potential at Whaling Villages in Northwestern Alaska. Appendix A, In: Alaska Consultants, Inc., with Stephen Braund & Associates: Subsistence Study of Alaska Eskimo Whaling Villages. U.S. Department of the Interior, Washington, D.C.

Tillman, M.
- 1980 Introduction: A Scientific Perspective of the Bowhead Whale Problem. *Marine Fisheries Review* 42(9-10):2-5.

Worl, R.
- 1980 The North Slope Inupiat Whaling Complex. In: Alaska Native Culture and History, edited by Y. Kotani and W. B. Workman, pp. 305-320. *Senri Ethnological Studies* No. 4. National Museum of Ethnology, Osaka.

Sex and Size Composition of Bowhead Whales Landed by Alaskan Eskimo Whalers

Howard W. Braham
National Marine Mammal Laboratory, AFSC
National Marine Fisheries Service, NOAA
7600 Sand Point Way, NE
Seattle, WA 98115

Abstract: *From 1973 to 1992, 423 bowhead whales* (Balaena mysticetus) *were landed by Eskimo whalers at 10 villages located along the western Alaskan coast. From this catch, demographic information on body length and sex ratio of the whales was analyzed for trends. The sex ratio of the catch was equal. In addition, 49.0% of the spring catch consisted of small animals, 9.5 m or less (67.1% of the catch was of immature animals), which was the size of whales landed by Eskimos prior to contact with Yankee whalers. These data suggest that modern Eskimo whalers take larger bowhead whales than did the precontact natives. An analysis of the annual catch data from this century shows that the fewest whales were taken just after World War II and the greatest number were landed in the 1990s (31.8 whales/year), a reflection of the increasing quota, and during the early 1970s (30.3 whales/year), when there was heightened community interest in the Eskimo whaling culture.*

INTRODUCTION

Native subsistence whaling for bowhead whales (*Balaena mysticetus*) in the western North America Arctic can be divided into three periods: archeological or early subsistence (2000 B.P. to 1885), commercial shore-based (1885 to 1909-1915), and modern 20th century subsistence (Marquette and Bockstoce 1980; Ackerman 1988). Only limited anecdotal data exist concerning the harvest or use of bowheads by native people before the 20th century, and these come from archeological data or records of whalers and clergy (see Larsen and Rainey 1948; Rudenko 1961; Giddings 1967; Marquette and Bockstoce 1980; McCartney 1984). Data on the demographic composition of bowheads landed by Eskimos prior to the whale's "discovery" by Yankee whalers in the Bering Sea in 1848 (i.e., the precontact period) are largely unknown.

Throughout the recorded history of whaling in the Western Arctic, few details have been given on the composition of the catch, such as size and sex, although inferences have been made on the number of whales taken by commercial and shore-station whalers. The estimates are based on the amount of whale oil and baleen processed (see e.g., Bockstoce and Botkin 1983; Bockstoce 1986), catches, number of whaling crews, and whales struck and lost (see e.g., Marquette and Bockstoce 1980; Braund et al. (1988). Since 1973, however, the total number and the sex and size distribution of bowheads harvested by Alaskan Eskimos has been documented.

This paper presents an analysis of the number, sex ratio, and size composition of bowhead whales landed by Alaskan Eskimo whalers at nine principal (plus one other) coastal villages in western Alaska. The objective of the study was to document and analyze the number, lengths, and sex of bowhead whales landed by village each year in order to assess trends in bowhead whaling over the past 20 years and to make inferences about whaling practices based on the results of the bowhead harvest during the 20th century. Although not covered in this paper, the study also addressed whether the harvest had affected the recovery of the bowhead whale population in the Western Arctic (Braham 1989).[1]

Harvest data from the bowhead hunt were analyzed for the years 1973 to 1992 (some of the 1970-1972 and 1993 data were also available) in order to determine trends, differences in the sex ratio and age class distribution, and differences in sizes among struck and lost whales, resulting in differential age-class mortality in the hunt. These data are also expected to help us evaluate the significance of the size composition of presumed catches in precontact bowhead whale samples (see McCartney, this volume). In addition, the data were tested for trends in the composition of the catch relative to sex and body lengths, village, and time of year (season). Data from other years, dating back to 1915, were also used to provide a historical perspective on changes in the catch this century.

BACKGROUND: THE PROBLEM

During the early 1970s, concern was expressed by scientists, resource managers, law-makers, and the general public that (a) the bowhead population in western Alaska was very small, when compared to the start of commercial whaling in the mid-1800s, and (b) that the Alaskan Eskimo subsistence harvest might be jeopardizing the bowhead's recovery (see e.g., Scarff 1977; Tillman 1980).

Two critical events in 1977 sharply focused the attention of the public on native whaling and the status of the presumed rare bowhead. (These two events would dramatically alter the political and sociocultural fabric of the Alaskan Eskimo, and occurred at

---

1. As part of a team carrying out a systematic study of the general ecology and biology of bowhead and gray whales in Alaska, we also monitored the native hunt. Some of the results of that work are presented here. These data plus preliminary measurements of about 200 bowhead and gray whale skulls near Gambell and Kialegak, St. Lawrence Island, led me to develop some of the ideas in this paper. A complete analysis of the osteological material on St. Lawrence Island needs to be done before they wash away or are lost to science and the native community altogether. A comparative analysis of the sizes and antiquity of these skulls with others in western Alaska and along Chukotka may help us better understand utilization patterns of the local communities, past and present, and the influence the bowhead whale had on the demographic patterns of native settlements across the Arctic.

the time of renewed heightened cultural identity of native peoples throughout the Americas and concurrent with the emergence of an environmentally vocal segment of society seeking protection for whales.) First was the pronouncement by the international marine mammal community that the bowhead whale was in danger of extinction and that the population in the arctic waters off Alaska was particularly vulnerable because it was subject to an Alaskan native take (allowed by an exemption provided in the U.S. Marine Mammal Protection Act for subsistence purposes and by regulations of the International Whaling Commission). The second event was the two-fold increase in the number of whales harpooned by Alaskan Eskimos in the mid-1970s over the previous 6-year period and a nearly six-fold increase per year over the previous 60 years. In just two years (1976-1977), over 200 bowheads had been harpooned (i.e., landed or struck and lost), a total nearly equal to the entire number of whales landed during the preceding two decades.

The result was implementation of a zero catch quota by the International Whaling Commission (IWC 1978), and a mandated U.S. domestic management scheme. The centuries-old subsistence hunt for bowhead whales was, thus, now center-stage in international politics and initially thought to be in jeopardy. However, rather than establishing a complete ban on subsistence whaling, a research and management plan was developed (National Oceanic and Atmospheric Administration 1978), and a quota system was instituted by the U.S. Government, based on catch limits established by the IWC (1978). This quota system was started with catch levels that were significantly lower than the take in the previous several years. They were close to recent historic levels, but not large enough to cause a dramatic decline in the bowhead population (Tillman 1980). In 1982, experts in sociology, cultural anthropology, and wildlife biology met with the Eskimo whalers and international resource managers to determine what impact whaling restrictions might have on the traditional life of native Alaskan whaling communities. They concluded that the most important loss of a bowhead whale hunt would be cultural in nature (see Donovan 1982).

With this background, I set out to assess what impact, if any, the Alaskan Eskimo hunt had on the status and trend in the recovery of the bowhead population. To do this, I investigated the take in relation to trends in sex and size of whales landed, the number struck and lost but then recovered (to determine if larger whales were lost), and the existence of trends relative to years and villages that might suggest a possible impact on bowhead recovery. The initial presumption was that the bowhead population was not large and that it might be declining (cf. Mitchell and Reeves 1980), and, thus, any removals might be adverse to recovery. However, we have since determined that the size of the bowhead population is much larger than first thought (Krogman et al. 1989), that the population is growing in size (Zeh et al. 1991), and that the current harvest is not jeopardizing the recovery of the bowhead in the Western Arctic (Braham 1989).

WHALING DATABASE AND STATISTICAL ANALYSES

Harvest data from 1973 to 1981 were collected by scientists from the National Marine Mammal Laboratory (NMML). The villages of Point Hope and Barrow were routinely monitored, but Gambell, Savoonga, Wales, Kivalina, Wainwright, Nuiqsut, and Kaktovik were visited intermittently or data were submitted to us by the whaling captains from these villages. Beginning in 1982, all harvest records were supplied by the North Slope

Borough (NSB) and Alaskan Eskimo Whaling Commission (AEWC) as part of a cooperative management agreement with the National Marine Fisheries Service. The data files that I used were copies (located at NMML) of the original field records of each whale landed from 1973 to 1982, or copies of reports of the AEWC, 1983-1993. Data from 1970-1972 and 1993 are summarized in several places but were not used in any statistical analyses of trends. Catch data were compiled by Marquette (1977) and Marquette and Bockstoce (1980) and updated by Braund et al. (1988). A summary of the total catch from 1848 to 1987 is also reported in Sonntag and Broadhead (1989). Further analysis of the accuracy of the catch data was performed by Withrow and Angliss (1994) and Suydam et al. (in press). Any errors in my reporting of these data are mine. A few discrepancies were found between the field logs and the published versions, so I relied on the original data sheets or contacted the whalers or biologists at the scene of the harvest to clarify missing or unclear data. The size of most landed whales was measured or estimated in feet and converted here into meters.

To determine whether those whales that were struck (harpooned) and lost were larger (had longer body length) than those struck and landed, the lengths of whales struck and lost and ultimately recovered were compared to whales landed soon after being struck. The common name for whales struck and lost is *dauhval* (literally, Norwegian for dead whale). The local name for recovered lost whales is "stinker," derived from the smell associated with the breakdown of muscle tissue and internal organs as they putrefy from overheating.

To evaluate trends in the data, I analyzed the size and sex composition of whales landed and those struck and lost for changes in sex ratios and length distribution. Statistical tests performed were (a) linear regression weighted by the inverse of the variance, for analyzing temporal trends in mean annual length of whales in the catch, (b) chi-square goodness-of-fit tests for differences in lengths (particularly mature versus immature animals) and sex, or (c) the t-statistic for averages associated with length measurements and years. An analysis of variance (ANOVA) of length by years, village, and sex was also performed. I investigated the following features of the hunt using ANOVA: whales landed by year, date, village, body length, and sex; whales struck and landed versus those lost and later recovered; and, to a lesser extent, the number of crews by time and village (no statistical analyses were performed). Some analyses on changes in catch by village and years are in Braham (1989).

The harvest of bowheads in western Alaska consists of two distinct "fisheries": (a) a spring hunt associated with the shore-fast ice and pack ice leads of open water (i.e., bowhead migration corridors) in the Bering and Chukchi seas, usually occurring in April and May, although a few whales were caught in March, June, and July, and (b) an autumn hunt associated with open water south of the pack ice in the Chukchi and Beaufort seas, usually in September and October, although a few whales have been landed in August and December.

The nine main whaling villages are (from south to north and east) Gambell and Savoonga on St. Lawrence Island, Wales, Kivalina, Point Hope, Wainwright, Barrow, Nuiqsut, and Kaktovik on Barter Island. One whale was landed at Shaktoolik in 1980 for the first time ever for that village. Some whaling also took place at Little Diomede Island (in the Bering Strait) and Point Lay and Icy Cape along the Chukchi Sea. Prior to the mid-1960s, Barrow, Wainwright, and Point Hope were the main whaling villages. The history of "modern" subsistence whaling at native Alaskan villages, and an overview of arctic whaling in general, has been reviewed by Marquette (1977), Durham (1979), Braham et al. (1980), Donovan (1982), Marquette and Braham (1982), McCartney (1984), Bockstoce (1986), Ackerman (1988), Stoker and Krupnik (1993), and others.

DEMOGRAPHIC PATTERNS

Catch statistics for bowhead whales landed by Eskimo whalers between 1973 and 1992, where body length and sex were reported (n=343), are presented in Table 1. The total catch of whales from 1970 to 1993 (n=551), including animals for which information on length or sex is lacking, is reported in Table 2. The total number of whales struck but lost (n=113) is reported by Withrow and Angliss (1994).

The majority of the whales (75.2%) were landed in spring, during the greatest whaling effort on the part of the hunters, and the rest (24.8%) were taken in the autumn hunt. The approximate distribution of the landings were:

Spring Hunt		Autumn Hunt	
March	< 1.0%	July	< 1.0%
April	25.3%	August	< 1.0%
May	49.2%	September	12.3%
June	1.4%	October	10.6%
		December	< 1.0%

Whalers at Barrow and Point Hope landed the largest number of whales (about 66% of the total). A 10.1 m whale was killed in Norton Sound in 1980, using high-powered rifles and gasoline cans as floats, for the first time, presumably, since the time of first contact. The whale was landed and cut up at Shaktoolik with the help of an elderly whaling captain from Barrow, who happened to be in the village visiting relatives.

Most whales were measured and sexed at the time they were landed, although some length measurements had to be approximated, because the whale had not been pulled out onto ice or land. In some cases, the length and sex were provided by the whalers after the whale had been cut up. Generally, we found these estimates to be acceptable for comparing relative length categories, but not for analyses of life history patterns (see e.g., Nerini et al. 1984).

Sex Ratio in the Hunt

An analysis of landed whales of known sex (from Table 1) was conducted to establish the approximate sex ratio of the harvest. The number of females in the catch (n=183; 8.8/year, s=4.5) was not significantly different than the number of males (n=176; 9.1/year, s=5.2), resulting in a sex ratio of 0.96:1.00 males to females (Table 3). This is essentially the same as the 1961-1982 sex ratio of 1.17:1.00, and the 1973-1982 ratio of 1.20:1.00, reported by Nerini et al. (1984). There also was no difference in the sex ratio of the catch when comparing the years before (1973-1977) and after (1978-1992) the quota system was put in place. Because of the large sample size, I presume that this equal sex ratio is a true reflection of the sex ratio of the bowhead population in general. The sex ratios of

Table 1. Bowhead Whales Landed by Alaskan Eskimo Whalers, by Date, Village, Body Length and Sex, 1973-1992. See text for data sources.

Year	Date	Whaling village	Length (m)	Sex
1973	7. May	Barrow	6.7	M
	"	"	8.2	M
	"	"	9.2	F
	12. May	Barrow	8.2	M[1]
	"	"	9.2	F
	"	"	9.2	M[1]
	20. May	Wainwright	8.3	F
	23. May	Barrow	8.6	F
	24. May	"	8.1	M
	"	"	8.8	F
	"	"	-	M
	25. May	"	9.8	F
	27. May	"	8.2	F
	28. May	"	7.6	M
	6. June	"	4.6	M
	"	"	15.3	F
1974	20. April	Point Hope	8.1	M
	21. April	"	9.1	-
	"	"	12.2	-
	29. April	Gambell	12.2	-
	30. April	Barrow	9.8	-
	2. May	"	12.2	-
	2. May	Point Hope	8.7	M
	4. May	Barrow	6.7	M
	11. May	Point Hope	7.8	F
	12. May	Barrow	8.1	M
	16. May	"	11.4	F
	23. May	Point Hope	15.2	M[1]
	25. May	"	15.5	M
	29. May	Barrow	7.2	F[1]
	"	"	13.9	M[1]
	29. Sept	"	9.8	M
	"	"	-	M
	3. Oct	"	10.7	M
	8. Oct	"	8.2	F
1975	24. April	Point Hope	11.0[2]	-
	26. April	"	6.1[2]	-
	5. May	Barrow	8.0	F
	7. May	Gambell	12.8	M
	9. May	Barrow	6.9	M

Table 1 *continued.* Bowhead Whales Landed by Alaskan Eskimo Whalers, by Date, Village, Body Length and Sex, 1973-1992. See text for data sources.

Year	Date	Whaling village	Length (m)	Sex
	10. May	Point Hope	8.5	F
	13. May	Barrow	9.3	F
	14. May	"	8.0^2	-
	15. May	"	8.5	M
	"	Point Hope	11.6	M^2
	16. May	Barrow	16.2	F
	20. May	"	7.8	F
	21. May	"	11.1	F
	23. May	"	7.2	-
	31. May	"	14.0	M^1
1976	. April	Savoonga$_2$	6.1	M
	"	"	7.6	M
	"	"	8.5	M
	"	"	9.1	F^1
	"	"	9.1	M^1
	"	"	10.7	M
	23. April	Point Hope	7.9	-
	1. May	"	10.2	M
	2. May	"	8.5	F
	"	"	11.2	M
	"	"	13.2	F
	"	Barrow	7.5	M
	3. May	Point Hope	8.5	F
	"	"	8.5	F
	"	"	14.7	M
	5. May	Barrow	11.4	M
	6. May	Point Hope	8.3	F^1
	"	Barrow	7.5	F
	"	"	8.0	-
	6. May	Barrow	11.4	-
	7. May	Point Hope	8.9	F
	9. May	"	8.1	F
	"	Barrow	12.4	M^1
	11. May	"	9.8	M^1
	12. May	"	13.7	M^1
	14. May	Point Hope	7.6	M
	"	Barrow	10.7	M^1
	15. May	"	6.9	F
	"	"	11.0	M
	17. May	"	8.5	F

Table 1 continued. Bowhead Whales Landed by Alaskan Eskimo Whalers, by Date, Village, Body Length and Sex, 1973-1992. See text for data sources.

Year	Date	Whaling village	Length (m)	Sex
	19. May	Barrow	11.6	F^1
	28. Aug	"	16.5	M^2
	30. Aug	"	16.3	M^2
	3. Sept	"	16.2	M^2
	"	"	16.5	M^2
	"	"	17.3	F^2
	10. Sept	"	16.0	F
	20. Sept	Barrow	14.1	M
	"	"	14.3	F
	"	"	15.3	M
	"	Kaktovik	13.7	M^2
	27. Sept	"	9.1^2	-
	7. Oct	Barrow	13.2	M
1977	22. April	Gambell	11.0	M^2
	28. April	"	7.9	M^2
	30. April	Point Hope	8.9	F
	2. May	Barrow	11.0	M
	3. May	"	9.1	M
	5. May	"	9.0	M^2
	"	"	9.1	F
	"	"	9.5	F
	"	"	10.6	M
	7. May	"	9.3	M
	8. May	"	7.6	$F^{1,2}$
	9. May	Kivalina	10.4^2	-
	10. May	Barrow	8.4	M
	"	"	8.8	F
	"	"	10.7^1	-
	"	"	10.8	F
	11. May	"	8.0	M
	"	"	8.4	F
	"	"	10.8	M
	12. May	"	8.0	F
	"	"	8.0^2	-
	15. May	"	$10.7^{1,2}$	-
	24. May	Point Hope	14.8	F
	28. May	Barrow	9.7	F
	29. May	Wainwright	8.5^2	-
	"	"	18.9^2	-
	29. Sept	Kaktovik	16.8	M^2

Table 1 *continued.* Bowhead Whales Landed by Alaskan Eskimo Whalers, by Date, Village, Body Length and Sex, 1973-1992. See text for data sources.

Year	Date	Whaling village	Length (m)	Sex
	2. Oct	Kaktovik	6.7	F
	14. Oct	Barrow	8.5	F
1978	16. April	Savoonga	10.9	M
	21. April	Gambell	13.8	M
	1. May	Barrow	8.5	F
	2. May	Point Hope	9.3	F
	"	Barrow	8.4	M
	"	"	8.4	M
	3. May	"	9.8	M
	4. May	Point Hope	9.7	M
	6. May	Wainwright	16.3	F
	19. May	"	15.2	M
	22. Sept	Kaktovik	11.1	M[1]
	26. Sept	"	13.3	M
1979	15. April	Point Hope	9.8	M
	17. April	"	8.5	F
	6. May	"	9.1	M
	16. May	Barrow	8.7	M
	18. May	"	10.3	M
	26. May	"	8.3	M[1]
	30. May	Wainwright	14.6	F
	22. Sept	Kaktovik	12.7	M[1]
	6. Oct	"	10.7	F
	8. Oct	"	10.3	M
	10. Oct	"	10.8	M
	11. Oct	"	10.8	M
1980	14. April	Savoonga	14.7	M
	20. April	"	10.7	F
	4. May	Gambell	15.7	F
	"	Wales	9.2	M
	9. May	Shaktoolik	10.1	M
	23. May	Wainwright	7.8	M
	24. May	Barrow	10.8	F
	25. May	"	8.5	M
	"	"	8.5	M
	"	"	10.4	M
	"	"	10.4	M
	"	"	10.8	M
	26. May	"	10.0	F
	27. May	"	8.7	F

Table 1 *continued.* Bowhead Whales Landed by Alaskan Eskimo Whalers, by Date, Village, Body Length and Sex, 1973-1992. See text for data sources.

Year	Date	Whaling village	Length (m)	Sex
	"	Barrow	13.7	F
1981	14. April	Savoonga	16.8	F
	"	Gambell	15.5	F
	21. April	Savoonga	14.2	F
	4. May	Point Hope	9.4	M
	"	"	9.8	F
	"	"	10.0	M
	"	"	12.0	F
	6. May	Wainwright	16.2	F
	"	Barrow	10.7	F
	18. May	Wainwright	17.7	F
	22. May	Barrow	8.0	F
	25. May	"	8.4	F
	27. May	Wainwright	16.5	F
	30. May	Barrow	8.7	M
	7. Sept	Kaktovik	17.4	F^1
	11. Sept	Kaktovik	14.4	M^1
	22. Sept	"	16.1	F
1982	17. April	Savoonga	13.4	F
	30. April	Gambell	7.9	M
	1. May	"	8.8	F
	5. May	Point Hope	8.3	M
	14. May	Wainwright	17.7	F^2
	29. May	"	16.5	F
	23. Sept	Kaktovik	16.0	M
	24. Sept	Nuiqsut	9.0	M^2
1983	7. April	Savoonga	8.5	M
	17. April	Gambell	15.8	M
	3. May	Point Hope	8.2	M
	5. May	Barrow	8.5	F
	7. May	"	14.2	F
	10. May	"	12.2	-
	14. May	Wainwright	16.2	M
	15. May	"	15.9	-
	15. May	Wales	17.7	F
	5. Sept	Kaktovik	14.8	F
1984	19. April	Savoonga	15.7	F
	23. April	"	14.4	F
	24. April	Point Hope	13.8	M
	25. April	Kivalina	10.1	M

Table 1 *continued.* Bowhead Whales Landed by Alaskan Eskimo Whalers, by Date, Village, Body Length and Sex, 1973-1992. See text for data sources.

Year	Date	Whaling village	Length (m)	Sex
	26. April	Point Hope	8.3	F
	18. May	Wainwright	7.7	M
	19. May	Barrow	8.6	F
	"	"	8.9	M
	20. May	"	7.7	M
	21. May	Wainwright	13.5	M
	"	Barrow	8.6	F
	5. Sept	Kaktovik	9.8	F
1985	18. April	Savoonga	11.6	M
	21. April	Gambell	9.3	F
	2. May	Wales	7.3	M
	9. May	Barrow	9.0	M
	10. May	"	12.4	M
	"	Point Hope	7.2	F
	11. May	Wainwright	16.2	F
	16. May	Barrow	9.5	M
	18. May	Wainwright	15.1	F
	28. May	Barrow	8.2	M
	13. Oct	"	6.7	F
1986	11. April	Gambell	11.0	M
	27. April	Barrow	8.5	M
	"	"	8.6	M
	30. April	Gambell	14.6	F
	"	Barrow	8.8	F
	1. May	"	8.7	M
	3. May	Gambell	14.0	F
	4. May	Barrow	8.1	M
	"	Wainwright	15.9	M
	5. May	Barrow	12.2	M
	6. May	"	10.7	M
	9. May	Wainwright	17.7	F
	24. May	Point Hope	13.9	M
	1. June	"	15.2	M
	14. June	Wainwright	15.1	F
	13. July	Barrow	10.7	M[1]
	10. Sept	Kaktovik	7.6	F
	19. Sept	"	17.1	F[1]
	26. Sept	"	10.4	M
	6. Oct	Nuiqsut	14.0	M
1987	30. March	Gambell	8.0	F

Table 1 *continued.* Bowhead Whales Landed by Alaskan Eskimo Whalers, by Date, Village, Body Length and Sex, 1973-1992. See text for data sources.

Year	Date	Whaling village	Length (m)	Sex
	16. April	Savoonga	14.9	F
	23. April	Wales	8.7	F
	24. April	Gambell	16.8	F
	26. April	Kivalina	18.0	F
	30. April	Point Hope	9.1	F^1
	1. May	"	14.3	M
	"	Barrow	9.3	M
	2. May	"	8.9	F
	4. May	"	11.0	M^1
	5. May	Wainwright	10.2	M
	8. May	"	13.5	M
	14. May	Point Hope	-	-
	24. May	"	10.7	M
	15. May	Wainwright	8.2	F
	17. May	Barrow	16.8	F^1
	21. May	Point Hope	7.8	F
	28. May	"	17.2	F/M^3
	2. June	Wainwright	11.4	F
	14. June	Barrow	15.7	F
	5. Oct	Nuiqsut	15.1	F
	26. Oct	Barrow	15.6	F
	29. Oct	"	8.5	M
1988	16. April	Gambell	15.7^4	F
	24. April	Barrow	8.9	F
	25. April	Barrow	7.8	F
	"	"	8.8	M
	"	"	8.9	M
	"	"	9.0	F
	"	Gambell	15.3	F
	"	Wainwright	7.9	F
	26. April	Wainwright	9.1	M
	27. April	Point Hope	7.7	M
	30. April	"	8.3	M
	"	"	9.6	M
	"	"	14.5	F
	"	"	14.7	M
	2. May	Barrow	8.3	F
	5. May	"	8.2	F
	6. May	"	7.5	F
	"	Wainwright	13.4	M

Table 1 *continued.* Bowhead Whales Landed by Alaskan Eskimo Whalers, by Date, Village, Body Length and Sex, 1973-1992. See text for data sources.

Year	Date	Whaling village	Length (m)	Sex
	18. May	Wainwright	15.1	F
	15. Sept	Barrow	14.6	M
	17. Sept	"	15.1	M
	"	"	15.6	F
	24. Sept	Kaktovik	14.9^5	F
1989	23. April	Barrow	9.0	F
	25. April	Savoonga	19.5^6	F?
	15. May	Wainwright	16.1	M
	"	Barrow	14.7	F^7
	27. May	Wainwright	17.3	F
	28. May	Barrow	16.9	F
	9. Sept	Kaktovik	14.6	M
	11. Sept	"	15.1	M
	11. Sept	Nuiqsut	14.0	M
	27. Sept	Kaktovik	11.8	M
	1. Oct	Nuiqsut	14.6	M
	2. Oct	Barrow	14.1	M
	2. Oct	"	14.0	M
	2. Oct	"	13.2	M
	2. Oct	"	14.6	F
	10. Oct	"	11.8	M
	25. Oct	"	8.2	M
	28. Oct	"	8.1	F
1990	7. April	Gambell	9.1	F
	7. April	Savoonga	10.7	F
	9. April	"	7.9	F
	10. April	Gambell	10.4	F
	10. April	Savoonga	12.2	F
	13. April	"	10.5	F
	18. April	Gambell	11.9	F^8
	21. April	Point Hope	9.7	-
	27. April	"	12.2	M
	30. April	"	14.9	M^1
	6. May	Wainwright	7.8	M
	7. May	"	12.5	M
	7. May	Gambell	15.2	F^9
	8. May	Wainwright	8.5	M^1
	9. May	Wainwright	8.4	-
	9. May	Barrow	8.4	M?
	10. May	"	11.2	M

Table 1 *continued.* Bowhead Whales Landed by Alaskan Eskimo Whalers, by Date, Village, Body Length and Sex, 1973-1992. See text for data sources.

Year	Date	Whaling village	Length (m)	Sex
	12. May	Barrow	11.8	M
	13. May	Wainwright	14.6	M
	20. May	Barrow	14.9	F^{10}
	24. May	"	15.9	F
	24. May	"	15.2	M
	11. Sep	Kaktovik	10.5	F^1
	1. Oct	Barrow	8.4	F
	2. Oct	"	12.9	M
	11. Oct	"	12.9	F
	11. Oct	"	13.9	F
	11. Oct	Kaktovik	9.3	F
	14. Oct	Barrow	14.0	F
	12. Dec	Savoonga	9.0	F
1991	17. April	Gambell	9.5	F
	17. April	Point Hope	7.6	F
	20. April	"	15.8	F
	21. April	"	8.5	F
	21. April	"	11.1	M
	22. April	"	9.8	F
	26. April	Wales	8.9	F
	26. April	Point Hope	7.7	F
	28. April	Barrow	7.7	F
	28. April	"	8.6	F
	29. April	Wainwright	14.0	M
	29. April	Barrow	8.8	F
	30. April	"	8.2	F
	1. May	Wainwright	8.5	F
	1. May	Barrow	12.6	F
	1. May	Wainwright	11.4	M
	4. May	"	8.1	F
	10. May	Kivalina	7.9	M
	12. May	Barrow	11.0	M
	12. May	"	14.4	M
	16. May	"	10.3	F
	27. Sep	"	7.9	F
	28. Sep	"	11.2	F
	2. Oct	Nuiqsut	14.4	F
	2. Oct	Barrow	12.0	M
	4. Oct	"	7.0	F
	8. Oct	Kaktovik	10.7	M

Table 1 *continued.* Bowhead Whales Landed by Alaskan Eskimo Whalers, by Date, Village, Body Length and Sex, 1973-1992. See text for data sources.

Year	Date	Whaling village	Length (m)	Sex
1992	12. April	Savoonga	8.2	M
	18. April	"	10.1	F
	21. April	"	11.3	-[1]
	28. April	Kivalina	8.4	F
	30. April	Point Hope	8.5	F
	1. May	"	8.2	M
	1. May	Savoonga	16.2	M
	28. May	Barrow	8.5	F
	29. May	"	15.2+[11]	F
	31. Aug	"	14.6	F
	1. Sept	"	16.2	F
	2. Sept	"	3.7	M
	2. Sept	"	14.6	M
	2. Sept	"	14.2	F
	3. Sept	"	15.7	F
	4. Sept	"	15.0	F
	12. Sept	"	14.6	F
	17. Sept	"	15.0	M
	19. Sept	"	12.0	F
	23. Sept	"	15.0	F
	25. Sept	"	14.5	M
	26. Sept	"	11.7	M
	26. Sept	"	11.0	M
	8. Oct	"	7.5	F
	9. Oct	"	8.8	M
	9. Oct	"	10.5	F
	12. Oct	"	8.5	F
	13. Oct	"	9.6	M
	13. Oct	"	9.8	F

[1] "Stinker" or <u>dauhval</u> (recovered dead whale).
[2] As reported by the whalers.
[3] Contained both male and female sex organs.
[4] Carried a 4.3 m foetus.
[5] Carried a 1.5 m foetus.
[6] Possibly an overestimate; fluke width was 4.4 m suggesting the whale may have been in the 13-14 m range.
[7] Carried a 4.0 m foetus.
[8] VHF radio transmitter retrieved from blubber.
[9] Carried a 3.7 m foetus.
[10] Carried a 3.9 m foetus.
[11] The whale was flensed in the water; length estimated based on 5.2 m fluke width.

Table 2. Summary of the Total Number of Bowhead Whales Landed by Alaskan Eskimos, 1970-1993. Includes additional landings not reported in Table 1 where no information was reported on the sex or length of the whale; 1970-1972 and 1993 data are recent acquisitions and not part of the overall analysis of this paper.

Year	Landed	Year	Landed
1970	26	1982	8
1971	22	1983	10
1972	39	1984	12
1973	40	1985	11
1974	20	1986	20
1975[1]	15	1987	23
1976	48	1988	23
1977	32	1989	18
1978[2]	12	1990	30
1979	12	1991	27
1980	16	1992	29
1981	17	1993	41

[1] Poor ice conditions significantly reduced successful whaling.
[2] Beginning of domestic quota system.

stinkers (Table 4) and non-stinkers in the catch were not significantly different, further supporting the hypothesis of an equal sex ratio in the hunt.

Upon further inspection of the data, I noticed a change in the sex ratio in the catch since 1987, when more females were landed than males:

	Females	Males	s	Whales landed
1973-1986	97	119	216	15.7/yr
1987-1992	86	57	143	23.3/yr
	183	176	359	

$x^2 = 7.99$, df=1 and P=0.007 using Yate's correction for 2x2 tables.

The mean number of whales landed per year between 1987 and 1992, however, is not significantly different (males=9.5/year, s=2.1; females 14.3/year, s=4.2). There is also a possibility that more females than males were landed after the 1978 quota system was put into effect (P=0.098). But, with such wide variation in the number of males and females landed among years, it may be simply random chance. Certainly, the hunters could select larger animals, such as pregnant females or large males that migrate later in spring than do smaller whales (Nerini et al. 1984). I have not carried out a time series analysis, for example, to test whether there has been a temporal shift in the spring catches in recent years, however, an ANOVA of length for all of the data by sex and village (P=0.832) and sex versus year (P=0.377) was not significant.

Size Selection By Hunters

Historically, Eskimo whalers have landed smaller bowheads, presumably those in the first few age classes (cf. Nerini et al. 1984; McCartney, this volume). Between 1973 and 1992, the average annual length of landed whales varied from 8.7 m (1973; n=14) to 13.8 m (1989; n=6). Although there was great variation in the average length of whales among years (ANOVA, P<0.001), the trend over the past 20 years has not been significant for either males or females landed (Fig. 1). A frequency distribution of landed whales shows

Table 3. Sex Composition of the Catch of Bowhead Whales by Alaskan Eskimos, 1973-1992.

Year	Females	Males	Sex Unknown
1973	8	7	25
1974	4	9	7
1975	6	5	4
1976	15	24	9
1977	12	11	9
1978	3	9	0
1979	3	9	0
1980	6	9	1
1981	14	3	0
1982	4	4	0
1983	4	4	2
1984	6	6	0
1985	5	6	0
1986	7	13	0
1987	14	6	3
1988	13	10	0
1989	7	11	0
1990	16	11	2
1991	19	8	0
1992	17	11	1
Totals	183	176	63

Table 4. Bowhead Whales Struck but Lost and Later Recovered as "Stinkers."

	Date Landed	*Lengths (m)*	*Sex*	*Village*
1973	May 16	8.2	Male	Barrow
	May 16	9.2	Male	Barrow
1974	May 25	15.2	Male	Point Hope
	May 29	7.2	Female	Barrow
	May 29	13.9	Male	Barrow
1975	May 31	14.0	Male	Barrow
1976	April 21	9.1	Female	Savoonga
	April 21	9.1	Male	Savoonga
	May 6	8.3	Female	Point Hope
	May 9	12.4	Male	Barrow
	May 11	9.8	Male	Barrow
	May 12	13.7	Male	Barrow
	May 14	10.7	Male	Barrow
	May 19	11.6	Female	Barrow
1977	May 8	7.6	Female	Barrow
	May 10	10.7	(Unknown)	Barrow
	May 15	10.7	(Unknown)	Barrow
1978	September 22	11.1	Male	Kaktovik
1979	May 26	8.3	Male	Barrow
	September 22	12.7	Male	Kaktovik
	October 16	15.2	Male	Barrow
1981	September 11	17.4	Female	Kaktovik
	September 11	14.4	Male	Kaktovik
1986	July 13	10.7	Male	Barrow
	September 19	17.1	Female	Kaktovik
1987	April 30	9.1	Female	Point Hope
	May 4	11.0	Male	Barrow
	May 21	16.8	Female	Barrow
1990	April 30	14.9	Male	Point Hope
	May 8	8.5	Male	Wainwright
	September 11	10.5	Female	Kaktovik
1991	April 30	8.5	Female	Point Hope
	May 16	10.3	Female	Barrow
1992	April 21	11.2	(Unknown)	Savoonga
1993	April 29	8.8	Female	Wainwright
	May 4	10.1	Male	Barrow
	May 27	16.7	Male	Wainwright

Mean body length: Males = 11.9 m Females = 11.0
  s = 2.6   s = 3.7
  n = 21   n = 13

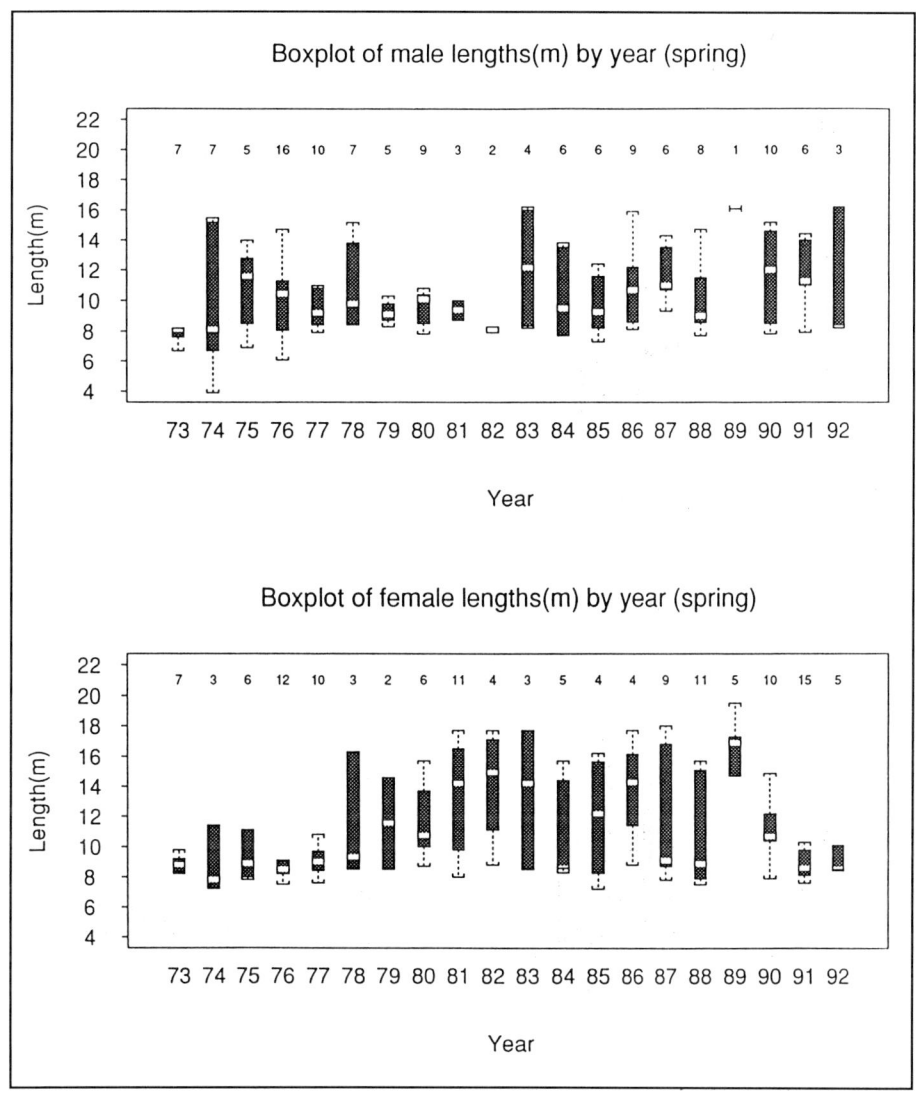

Figure 1. Mean, standard error, and range of lengths of bowhead whales landed at Alaskan Eskimo villages, 1973-1992.

that 49.0% of the spring harvest has been made up of animals <9.5 m in length, presumed 2 year olds, whereas the fall hunt was 21.1% (Fig. 2). These data are similar to those of McCartney and Savelle (1993:Figs. 7-8) for bowheads harvested in the Thule period, although they report fewer larger animals than taken in the Alaskan catch (see their Fig. 12). But, the composition of the 1973-1992 Alaskan Eskimo catch was significantly different (many more small whales in the catch) than that of stranded animals in the eastern Canadian Arctic that predated aboriginal bowhead usage (McCartney and Savelle 1993:Figs. 9 & 11).

300 Howard W. Braham

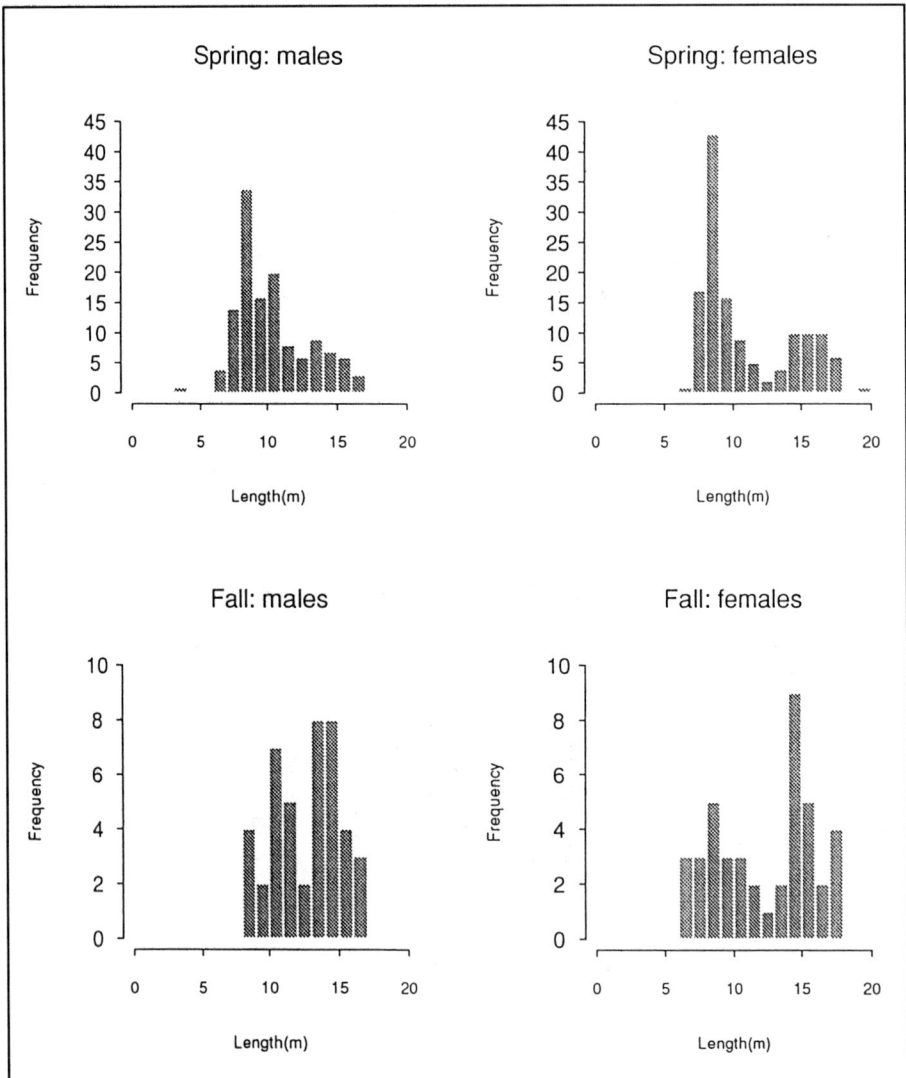

Figure 2. Frequency distribution of the size categories of bowhead whales landed at Alaskan Eskimo villages, 1973-1992.

Because Wainwright has historically taken larger whales in spring than other villages (cf., Reilly and Nerini 1988; this paper), an analysis of the data excluding Wainwright shows that bowheads landed in spring are significantly smaller than those landed in the fall (Fig. 3), except for females taken in October. Bowheads begin their fall migration in September, traveling west across the Beaufort Sea to the Chukchi Sea (Braham et al. 1980, 1984), passing Kaktovik, Nuiqsut, and Pt. Barrow (in that order). The data in Figure 3 suggest that adult animals of both sexes migrate earlier in the fall than do juvenile animals, as animals of both sexes caught in September were significantly larger than those caught

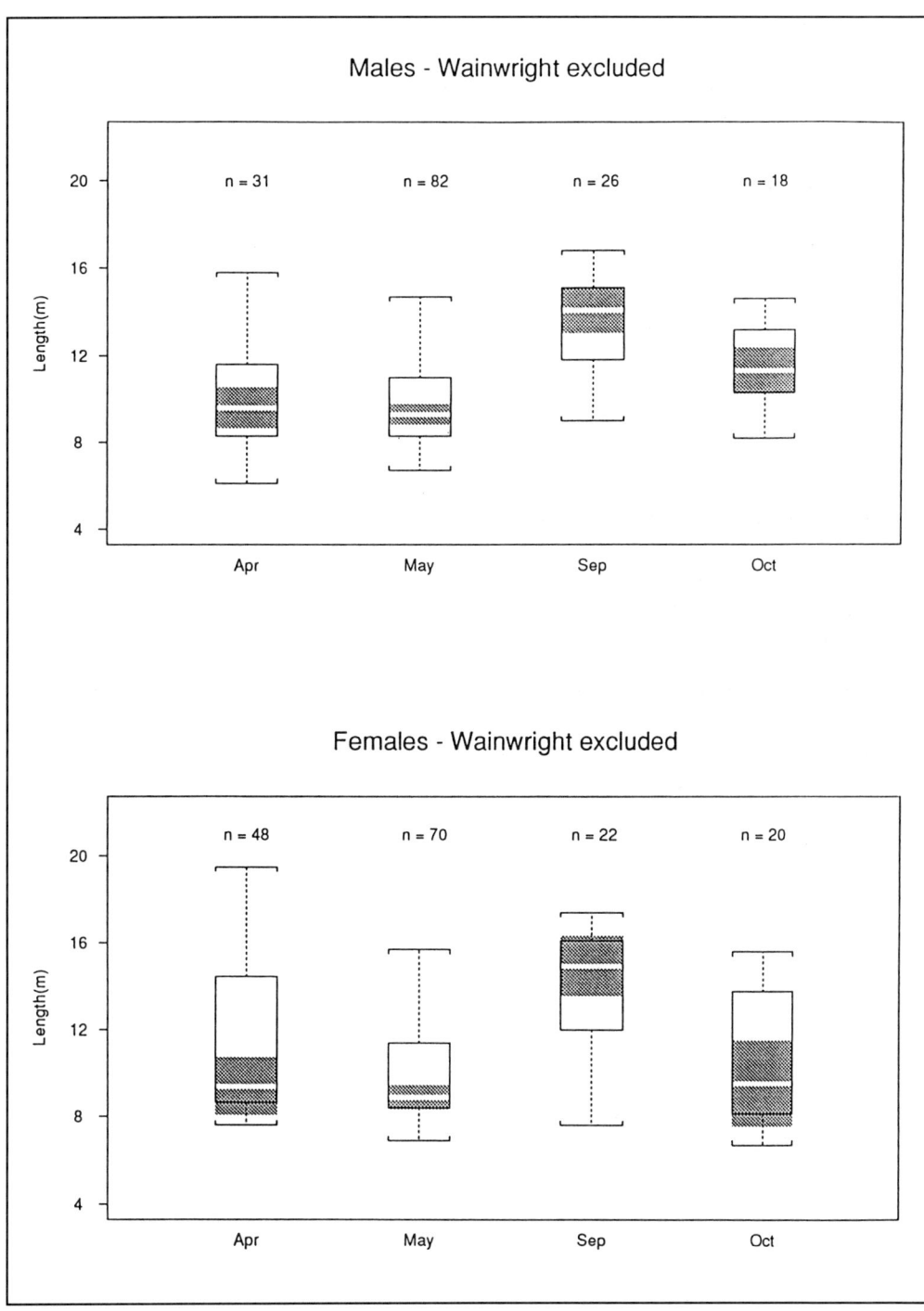

Figure 3. Mean, standard error and range of lengths of bowhead whales by major whaling months of the year (excluding data from Wainwright) to illustrate differences in the lengths of whales landed at Alaskan Eskimo villages, 1973-1992.

in October (P<0.05). This suggests that the whalers might be differentially selecting larger whales during the early period of the fall hunt, which supports the earlier hypothesis that larger animals migrate first (see Marquette 1977; Braham et al. 1980, 1984), or that the whalers could simply be selecting whales at random from the size-classes migrating past the villages.

While addressing the issue of size selection in the hunt, I looked for an independent way of assessing whether large animals were being struck and lost (using spring data) at a higher frequency than smaller animals. The concern was that the struck and lost rate had been increasing during the mid-1970s and, thus, the total kill might affect recruitment in the future. Since 1973, 37 whales of known length were recovered as stinkers after having been struck and lost (Table 4). The fact that there is no significant difference between the sizes of whales struck and lost and subsequently landed as stinkers and those of other landed whales suggests that losing a whale is independent of its size, and instead related to other factors such as ice conditions, poor bomb placement, or failure of the darting or shoulder gun.

Finally, an analysis of the catch of adults (presumably sexually mature animals if ≥ 13 m; IWC 1985) and juvenile bowheads in the hunt was used to (a) compare historical catches, to see if the current hunt differs from past strategies used by native hunters, and (b) evaluate whether greater numbers of adult animals are being taken, even though the IWC had requested that the Eskimos take only small animals (IWC 1981). Table 5 summarizes chi-square 2x2 contingency tables for the spring and fall hunts by sex and

Table 5. Comparison of the Distribution of the Eskimo Catch of Bowhead Whales, 1973-1992, Based on Presumed Mature (≥13 m) and Immature (<13m) Females and Males. The number in parenthesis is the proportion of the total take. The following chi-square statistics use Yate's continuity correction for a 2x2 contingency table.

		Females	Males
Spring	<13 m	92 (.371)	98 (.395)
	≥13 m	41 (.165)	17 (.069)
		133 (.536)	115 (.464)
	$\chi^2 = 7.99$, P=0.005; sex ratio equal		
Fall	<13 m	20 (.235)	20 (.235)
	≥13 m	22 (.259)	23 (.271)
		42 (.494)	43 (.506)
	$\chi^2 = 0.01$, P=0.91; sex ratio equal		

body length. The data indicate that more large females or fewer large males may have been taken than would be expected by chance (P<0.10). At this point I cannot discern whether this suggests that large females are being differentially selected or if it is an artifact of the hunt at different villages. There was no difference in the fall hunt.

Distribution Of The Catch By Village

A summary of the number of whales taken in the hunt at the nine principal whaling villages in Alaska is presented in Table 6. Most whales were landed at Barrow (n=214), where both a spring (n=152) and fall (n=62) hunt are conducted. Barrow, Wainwright, and Point Hope made up 75.4% of the catch. Plots of the distribution of whales in the catch by villages are depicted in Figure 4.

Table 6. Bowhead Whales Landed at Eskimo Whaling Villages in Alaska, 1973-1992 (from data in Table 2 and Withrow and Angliss 1994).[1]

Location	Village	Total landings	Year whaling data started[2]
Bering Sea	Gambell	27	1915[3]
	Savoonga	27	1972
	Shaktoolik[1]	1	N/A
	Wales	5	1915
Chukchi Sea	Kivalina	5	1958
	Point Hope	65	1915
	Wainwright	40	1916
	Barrow	214	1915
Beaufort Sea	Nuiqsut	10	1973
	Kaktovik	29	1964
	Total	423	

[1] Ice blocked the bowhead migration through the Bering Strait in the spring of 1980 resulting in some whales moving into Norton Sound. A 10.1 m bowhead was landed near Shaktoolik for the first time in reported history (see Johnson et al. 1981). Shaktoolik is not a bowhead whaling village.

[2] "Initial" year (near continuous) that whaling began during the modern era of subsistence whaling (1915-present).

[3] Savoonga and Gambell whalers joined forces up to 1972 when whalers from Savoonga moved their whaling operation to Powooiliak at southwest cape on St. Lawrence Island.

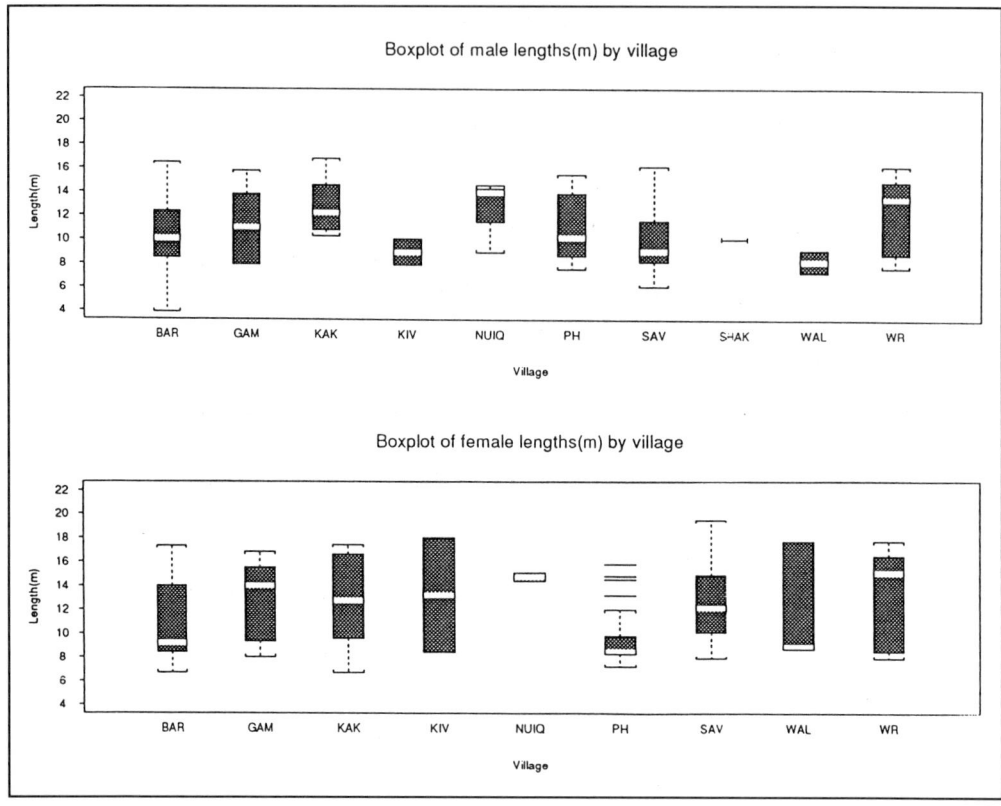

Figure 4. Mean, standard error, and range of lengths of bowhead whales landed at the principal whaling villages in western Alaska, 1973-1992.

Analysis of variance for whale size versus village suggests that larger animals are taken at one or more villages ($P=<0.001$; $df=78$). When the Wainwright data are excluded from the analysis, there is no difference in size by village (spring data only; see e.g., Reilly and Nerini 1988). Whalers at Wainwright use *umiaks* and aluminum boats with outboard motors to pursue larger whales in the often expansive open nearshore lead. This is in contrast to whalers from the villages of Wales, Kivalina, Point Hope, and Barrow, who seldom use aluminum boats and pursue whales that are likely to be closer to the shore-fast ice edge. Further, Wainwright hunters sometimes strike whales on the left side of the whale's body during the pursuit inward towards the shorefast ice. This is uncommon at the other villages.

DISCUSSION

Throughout the history of whaling in the Western Arctic, various changes occurred in the methods used by native American and Russian hunters to land bowhead whales, depending upon the socioeconomic and technological influences at the time. When Yankee whalers sailed into the Chukchi Sea in the mid-1800s, they brought with them darting and shoulder guns that were incorporated into the village subsistence hunt by the late 1800s (Murdoch 1885). These weapons are used today as "traditional" whaling implements, even though they have been in use for scarcely 100 years. Changes in the growth, development, and redistribution of the native population, or changes in the traditional values of the culture, have also resulted in shifts in subsistence practices and successes, including how and where resources are acquired (see e.g., Sapronov 1985; Stenback 1987; Freeman 1988; Krupnik 1988, 1993). As the native villages entered into a predominantly cash economy in the 1960s and early 1970s, for example, there was less dependence on subsistence products.

As subsistence whaling moved into the 20th century, changes were seen not only in the total number of whales landed or lost after having been struck, but also in the size composition of the catch. Such changes could have further influenced the biological productivity of the whale population depending upon whether the population was large and robust or small and declining. Other factors can also play an important role in the distribution, abundance, and availability of whales to the native communities. Some of these include predators (such as killer whales, *Orcinus orca*), prey, local and regional climatic and weather patterns, and the movements of water and ice (Vibe 1967; Dekin 1972; Mitchell and Reeves 1982). Political influences can also have a profound effect on the native's ability to hunt (see e.g., Scarff 1977; Lynge 1992).

Archeological evidence suggests that predominantly smaller bowheads are found in midden remains or near the surface of abandoned villages in Alaska, Chukotka (Russia), and Canada (see e.g., Nerini et al. 1984; Krupnik 1987; McCartney and Savelle 1993; McCartney, this volume). Smaller animals were hunted from shore or ice near coastal settlements, whereas larger bowheads were probably too difficult to exploit routinely. Modern native whaling practices follow this pattern, except that larger whales can be landed at villages when several crews (or the whole village) cooperate to help pull the whale out of the water and onto the ice. The fact that larger whales are taken in greater numbers in the autumn hunt is a result of the use of motorized cabin cruisers that can tow large animals from far offshore more easily than the traditional *umiak* used among the spring ice floes. The exception is the spring hunt from Wainwright, where many whalers now use aluminum boats with outboard motors, rather than *umiaks*, to cover the large expanses of water that are frequently found in the nearshore leads near the village. This may also explain why larger whales have been taken at Wainwright and why this hunt occurs somewhat later in spring than those conducted by the closest villages of Point Hope and Barrow.

Trends or shifts in the harvest data in all likelihood reflect changes in sociocultural or ecological conditions. Influences such as socioeconomic attitudes, cultural resurgence in whaling for bowheads (but not for other whales), political pressures and unaccustomed social demands, or developments in whaling practices and equipment may all have been important. The two most talked about contributions to the increase in the number of bowheads harpooned in the 1970s were a heightened cultural awareness of the importance of the hunt to the community (a return to former days when they did not have to rely on Western food) and the greater availability of money to purchase supplies and equipment and hire crews to engage in hunting. Sources of this newfound wealth were compensation for land rights claims and expansion of oil exploration settlements. It might also have been a move away from hunting the dwindling caribou herds, or a general sense that there were more bowhead whales available to hunt. These issues are addressed in the special IWC book on subsistence and whaling, edited by Donovan (1982). For example, although a large number of coastal villagers do not directly engage in hunting caribou, caribou are important in the total mix of subsistence species used by the nine whaling communities except Wales, where bowheads are seldom taken (Table 3.1B in Donovan 1982). However, in the first several years of my research, during stays at Gambell, Savoonga, Point Hope, Wainwright, and Barrow (and visits with whalers from Wales, Kivalina, Nuiqsut, and Kaktovik), not once did a whaling captain tell me that there were "more" bowheads than prior to my visits (personal communications with whalers from all nine whaling villages between 1975 and 1983). Most said there were about as many as usual during their whaling years. Only two, that I recall, said there were fewer bowheads than before, and these answers may have been merely a reflection of recent years of poor ice conditions. The recent analyses of Krogman et al. (1989) and IWC (1992) suggest that the bowhead population is not only larger than first estimated in 1978 (see e.g., IWC 1981), but that it has been increasing at a rate of about 3% per year (Zeh et al. 1991). This could account for the "rejuvenation" of whaling in the 1970s, but such a slow increase would only be noticed by villagers who had been whaling for many years. Because there can be some variation in the timing and distribution of the bowhead population migrating past any one village, I doubt that an increasing population of bowheads is a good explanation for the increased effort put into the hunt. I think it was simply a resurgence or reidentification of the native bowhead whaling culture.

Historically, bowheads landed in the spring hunt have been small, usually immature animals (see Nerini et al. 1984), because it is safer for the Eskimos to pull smaller whales out onto the ice than it is larger whales. However, in the late 1970s, larger whales were being landed, and the struck and loss rate was increasing. Up until 1977, the harvest had not been closely monitored, and there was concern expressed by the IWC that adult bowheads were being struck and lost at a greater rate than immature animals. If so, then adult animals would not be proportionately represented in the total number of whales killed in the hunt. Advice had been given by the Commission since 1978 "… that removals of any kind should be … of sexually immature animals … in order to maximize reproduction in the short term …" (IWC 1981). The IWC scientists concluded that a higher rate of loss of adult whales would not be a desirable management strategy for recovery of the population; this was in parallel concern to whether the harvest, in total, was greater than the annual production of calves. The data suggest that larger animals were not being taken in the early 1980s but were after about 1985. The analyses here confirm this and further suggest that adult females were landed in greater numbers than expected based on historical native whaling practices.

The analysis of the bowhead catch during the modern subsistence whaling era of this century (Table 7) demonstrates that Eskimo hunting practices experienced four major passages (Table 8). During the post-commercial whaling era (1915-1969), about 12 whales per year were landed with few differences over the five-decade period. This was followed by an intensive whaling period (1970-1977) when more than 30 whales per year were landed and a large number were struck and lost (it is estimated that more than 100 whales were killed in 1977). Only 12 whales were landed per year during the remaining two periods (1978-1985 and 1986-1993), which occurred during the initial heavy whaling restrictions imposed by the IWC quota system. These restrictions remained in place until

Table 7. Bowhead Whales Landed Per Year by Decade, 1915-1992 (1970-1992 data represent mean landings based on monitoring of villages). Beginning in approximately 1915, Yankee whalers no longer operated land stations for commercial profit outside of Alaska, and for the most part, whales were landed for local subsistence consumption. Data summarized from Tillman (1980); Marquette et al. (1982); Braund et al. (1988: Appendix 1), and Table 2. The 1915-1969 estimates are approximated (records are frequently incomplete) and thus possibly underestimated; 1970-1993 catches are rounded to the nearest whole number.[1] (Crew sizes for 1915-1949 are estimated based on the highest total for any one year during each decade.)

Decade	Number of villages with reported takes[2]	Average number of whales landed per year	Total landings	Approximate annual number of whaling crews[3]	Whales landed per village per year
1915-1919	8	10	50	49	1.3
1920-1929	8	16	157	47	2.0
1930-1939	9	12	122	52	1.3
1940-1949	9	12	120	36	1.3
1950-1959	7	10	101	41	1.4
1960-1969	8	14	144	62	1.8
1970-1979	10	27	266	85	2.7
1980-1989	10	16	158	95	1.6
1990-1992	9	32	127	>100	3.6

[1] The actual catch statistics for 1970s-1990s are as follows:
 1970-1979 = 26.7 whales per year (s=12.7)
  70-77 = 30.3yr. (s=11.4) pre-quota
  78-79 = 12.0/yr. (s= 0.0) first 2 years of quota
 1980-1989 = 15.8/yr. (s=5.4)
 1990-1993 = 31.8/yr. (s=6.3)

[2] Only Barrow, Point Hope and Wainwright have had active whaling crews for virtually the entire period.

[3] Not all crews actively assemble each year, because of weather, economics, or other reasons. The literature is incomplete on crew size and, like the recorded number of whales landed in the early years, is underestimated. I suggest that the estimates for villages such as Barrow, Wainwright, and Point Hope (where most of the whales have been landed this century) are more accurate because of their established trading activities and permanent settlement by clergy who often kept records of village activities. Data collected for the periods prior to 1950 are much less reliable than for later years.

Table 8. 20th Century Landings of Bowhead Whales by Alaska Eskimos During Four Periods of Modern "Aboriginal" Whaling.[1]

	No. Landed	Per Year	S	N Years	
1915-1969	694	12.6	2.8	55	Post-commercial whaling subsistence era
1970-1977	242	30.3	11.4	8	Period of "Eskimo" cultural resurgence
1978-1985	98	12.3	3.0	8	Early years of IWC regulated quota[2]
1986-1993	211	26.4	7.3	8	Increased quotas based on revised bowhead status and trends estimates

[1] Villages that no longer participate in bowhead whaling but were active earlier (year when the last whale was landed) totaling 37 bowheads landed: Little Diomede Island (1955), Icy Cape (1940), and Point Lay (1940). A total of 37 bowheads was landed at these three villages.

2 In 1978, the International Whaling Commission withdrew the aboriginal subsistence exemption for bowhead whales. The National Marine Fisheries Service (NMFS) in cooperation with the Alaska Eskimo Whaling Commission (AEWC) developed a domestic quota system based on agreed IWC strike limits (i.e., "quotas").

greater research was carried out on subsistence needs and the status and trends of the bowhead whale population. Since 1986, the quota has increased significantly, and is still increasing, as a result of data indicating that the bowhead population is recovering. Consequently, more than 26 whales per year have been landed since 1986.

Throughout much of the first half of the 20th century, the number of whaling crews in the villages remained surprisingly stable, although they were not always active each year. The number of crews declined in the 1940s and 1950s, but then doubled in size by the end of the 1970s. This did not, however, result in a significant increase in the number of whales harvested until the mid-1970s. The average number of whales landed per village by decade (Table 7) shows that the relative amount of food obtained by the Eskimos would not have changed a great deal if only small whales (<9 m) were landed. For villages, such as Barrow, where the native population has been growing, this could be important. However, the current quota system administered by the Alaska Eskimo Whaling Commission (located in Barrow) takes into account local need, and, thus, village quotas are dispersed accordingly and shared (transferred) regularly when not used. A relatively equitable distribution of the catch has, therefore, been developed. Such cooperation was probably not common in precontact periods except between adjacent villages (e.g.,

Figure 5. Photograph of a large bowhead landed at Point Hope in 1975; the entire village participated in hauling this whale onto the ice using the block and tackle shown in the foreground (photo by Willman Marquette).

Gambell and Savoonga; Point Hope and Kivalina commonly share landed whales and, more commonly before the 1980s, often hunted together), or when visiting common trading groups (see e.g., Larsen and Rainey 1948; Rudenko 1961; Krupnik 1988; personal observations). If larger whales continue to be landed, when compared to the sizes of bowhead landed prior to the mid-1980s, then a relatively larger proportion of meat and *muktak* are being obtained than when the quota allocation scheme was first introduced. Further inquiry about the relative contribution of whale products to the current whaling communities might be a useful exercise.

CONCLUSIONS

(a) The sex ratio of the catch from 1973 to 1992 was equal (183 females, 176 males), and presumably this reflects the actual sex ratio in the population.

(b) The lengths of animals struck and lost and ultimately recovered as stinkers were not significantly different than the remainder of the catch. This suggests that larger adult whales are not more likely to be lost, once struck, than immature whales.

(c) The proportion of adult females in the spring harvest has increased from 4.7% in 1973-1977 to 20.4% in 1986-1992. Whales less than 13 m were in greater proportion in the spring catch than in the summer and autumn catch.

(d) The largest number of whales landed in the Eskimo hunt are immature animals (presumably those <13 m). Prior to establishment of the IWC quota system in 1978, the catch was approximately 80% immature animals. Since 1978, the catch has been approximately 60% immature animals.

(e) There was no significant trend in the sex ratio of the catch by time or village, although chi-square statistics suggest that a greater number of larger adult females were harvested since 1986 than during the previous 15 years.

(f) Less than 25% of the catch has been of animals presumably less than 2 yr of age (<9.5 m), suggesting that modern native whalers select larger whales than did those of precontact periods (prior to about 1849).

Acknowledgments. During my early years in Alaska (1975-1983), several individuals helped me immensely: Bud Fay, Conrad Oozeva, Cliff Fiscus, Thomas Brower, Mathew Iya, and John Burns. I want to thank especially Bruce Krogman and Mary Nerini for stimulating suggestions and Bob Everitt and Dave Rugh for keeping me out of trouble. I also thank Jeffrey Breiwick for his statistical analyses, Bill Marquette for comments from Eskimo whalers which supplemented or confirmed the data found in the published and unpublished literature, Dale Rice and Breiwick with tightening the final manuscript, and Kathy Cunningham for text processing. As always, Marcia Muto skillfully edited the paper to help make it intelligible (where it's not, I'm to blame). Lastly, I dedicate this paper to the memory of Bud Fay, who critically challenged many of my early ideas about bowhead whaling. Our long discussions, often with Eskimo colleagues, helped me better understand the natural history of seals, whales, and walrus and their use by arctic natives.

REFERENCES

Ackerman, R. E.
    1988   Settlements and Sea Mammal Hunting in the Bering-Chukchi Sea Region. *Arctic Anthropology* 25(1):52-79.

Bockstoce, J. R.
    1986   *Whales, Ice and Men: A History of Whaling in the Western Arctic*. University of Washington Press, Seattle.

Bockstoce, J. R. and D. B. Botkin
    1983   The Historical Status of the Western Arctic Bowhead Whale (*Balaena mysticetus*) Population by the Pelagic Whaling Industry, 1848-1914. *Report of the International Whaling Commission, Special Issue* 5:107-142.

Braham, H. W.
    1989   Eskimos, Yankees and Bowheads. *Oceanus* 32(1):54-62.

Braham, H. W., W. M. Marquette, T. W. Bray, and J. S. Leatherwood (editors)
    1980   The Bowhead Whale: Whaling and Biological Research. *Marine Fisheries Review* 42(9-10):1-96.

Braham, H. W., B. D. Krogman, and G. M. Carroll
    1984   Bowhead and White Whale Migration, Distribution, and Abundance in the Bering, Chukchi, and Beaufort Seas, 1975-1978. *NOAA Technical Report*, NMFS SSRF-778.

Braund, S. R., W. M. Marquette, and J. R. Bockstoce
   1988   Data on Shore-Based Bowhead Whaling and Sites in Alaska. Appendix I in Doc. SC/40/PS10, presented to the 40th meeting of the Scientific Committee of the International Whaling Commission, San Diego, California, May, 1988.

Dekin, A. A., Jr.
   1972   Climatic Changes and Cultural Change: A Correlation Study from Eastern Arctic Prehistory. *Polar Notes* 12:11-31.

Donovan, G. P. (editor)
   1982   Aboriginal/Subsistence Whaling. *International Whaling Commission, Special Issue 4.* Cambridge, England.

Durham, F. E.
   1979   The Catch of Bowhead Whales (*Balaena mysticetus*) by Eskimos, with Emphasis on the Western Arctic. *Natural History Museum of Los Angeles County, Contributions in Science* 314.

Freeman, M. M. R.
   1988   Tradition and Change: Problems and Persistence in the Inuit Diet. In: *Coping with Uncertainty in Food Supply*, edited by I. Garine and G. A. Harrison, pp. 150-169. Clarendon Press, Oxford.

Giddings, J. L.
   1967   *Ancient Men of the Arctic.* Alfred A. Knopf, New York.

International Whaling Commission
   1978   Chairman's Report of the Twenty-Ninth Meeting. *Report of the International Whaling Commission* 28:18-129.
   1981   Report of the Sub-Committee on Other Protected Species. *Report of the International Whaling Commission* 31:133-139.
   1985   Report of the Scientific Committee. *Report of the International Whaling Commission* 35:31-152.
   1992   Report of the Bowhead Whale Assessment Meeting, Annex E. *Report of the International Whaling Commission* 42:137-155.

Johnson, J. H., H. W. Braham, B. D. Krogman, W. M. Marquette, R. M. Sonntag, and D. J. Rugh
   1981   Bowhead Whale Research: June 1979 to June 1980. *Report of the International Whaling Commission* 31:461-475.

Krogman, B., D. Rugh, R. Sonntag, J. Zeh, and D. Ko
   1989   Ice-Based Census of Bowhead Whales Migrating Past Point Barrow, Alaska, 1978-1983. *Marine Mammal Science* 5(2):116-138.

Krupnik, I. I.
   1987   The Bowhead vs. the Gray Whale in Chukotka Aboriginal Whaling. *Arctic* 40(1):16-32.
   1988   Economic Patterns in Northeastern Siberia. In: *Crossroads of Continents: Cultures of Siberia and Alaska,* edited by W. W. Fitzhugh and A. Crowell, *pp. 183-191*. Smithsonian Institution Press, Washington, D.C.
   1993   *Arctic Adaptations: Native Whalers and Reindeer Herders of Northern Eurasia.* Dartmouth College, University Press of New England, Hanover, NH.

Larsen, H. and F. Rainey
   1948   Ipiutak and the Arctic Whale Hunting Culture. *Anthropological Papers of the American Museum of Natural History* 42.

Lynge, F.
   1992   *Arctic Wars, Animal Rights, Endangered People*. Dartmouth College, University Press of New England, Hanover, NH.

Marquette, W. M.
   1977   The 1976 Catch of Bowhead Whale (*Balaena mysticetus*) by Alaskan Eskimos, with a Review of the Fishery, 1973-1976, and a Biological Summary of the Species. NWAFC Proc. Rep., 80 p. National Marine Mammal Laboratory, NMFS, NOAA.

Marquette, W. M. and J. R. Bockstoce
   1980   Historical Shore-Based Catch of Bowhead Whales in the Bering, Chukchi, and Beaufort Seas. *Marine Fisheries Review* 42(9/10):5-19.

Marquette, W. W. and H. W. Braham.
   1982   Gray Whale Distribution and Catch of Alaskan Eskimos: A Replacement for the Bowhead Whale? *Arctic* 35(3):386-394.

Marquette, W. M., H. W. Braham, M. K. Nerini, and R. V. Miller
   1982   Bowhead Whale Studies, Autumn 1980-Spring 1981: Harvest, Biology, and Distribution. *Report of the International Whaling Commission* 32: 357-370.

McCartney, A. P.
   1984   History of Native Whaling in the Arctic and Subarctic. In: *Arctic Whaling: Proceedings of the International Symposium on Arctic Whaling*, edited by H. K. s'Jacob, K. Snoeijing, and R. Vaughan, pp. 79-111. Arctic Centre, University of Groningen.

McCartney, A. P. and J. M. Savelle
   1993   Bowhead Whale Bones and Thule Eskimo Subsistence Settlement Patterns in the Central Canadian Arctic. *Polar Record* 29(168):1-12.

Mitchell, E. D. and R. R. Reeves
   1980   The Alaskan Bowhead Problem: A Commentary. *Arctic* 33(4):686-723.
   1982   Factors Affecting Abundance of Bowhead Whales *Balaena mysticetus* in the Eastern Arctic of North America, 1915-1980. *Biological Conservation* 22:59-78.

Murdoch, J.
   1885   Natural History. In: *Report of the International Polar Expedition to Point Barrow, Alaska*, pp. 89-200. U.S. Government Printing Office, Washington, D.C.

National Oceanic and Atmospheric Administration
   1978   *Bowhead Whales: A Special Report to the International Whaling Commission*. U.S. Department of Commerce, NOAA, June 1978, Washington, D.C.

Nerini, M. K., H. W. Braham, W. M. Marquette, and D. J. Rugh
   1984   Life History of the Bowhead Whale, *Balaena mysticetus* (Mammalia: Cetacea). *Journal of Zoology* (London) 204: 443-468.

Reilly, S. and M. Nerini
   1988   The Relationship Between Lengths of Captured Bowhead Whales, Year of Capture and Village. *Report of the International Whaling Commission* 38:115.

Rudenko, S. I.
   1961   The Ancient Culture of the Bering Sea and the Eskimo Problem. *Arctic Institute of North America, Anthropology of the North, Translations from Russian Sources* No. 1. University of Toronto Press, Toronto.

Sapronov, V.
   1985   Whaling and Nutritional Needs of the Aboriginal Population of the Chukotka Peninsula. *Arctic Policy Review* 3(3):5-8.

Scarff, J.
    1977   The International Management of Whales, Dolphins, and Porpoises: An Interdisciplinary Assessment. *Ecology Law Quarterly* 6(2):323-427, 574-638.

Sonntag, R. and G. Broadhead
    1989   Documentation for Revised Bowhead Whale Catch Data (1848-1987). *Report of the International Whaling Commission* (Annex G, Appendix 5) 39: 114-115.

Stenback, M.
    1987   Forty Years of Cultural Change Among the Inuit in Alaska, Canada and Greenland: Some Reflections. *Arctic* 40(4):300-309.

Stoker, S. W. and I. I. Krupnik
    1993   Subsistence Whaling. In: The Bowhead Whale, edited by J. J. Burns, J. J. Montague, and C. J. Cowles, pp. 579-629. *Society for Marine Mammalogy, Special Publication* No. 2..

Suydam, R., R. Angliss, J. C. George, S. Braund, and D. DeMaster.
    n.d.   Revised Data on the Subsistence Harvest of Bowhead Whales (*Balaena mysticetus*) by Alaska Eskimos, 1973-1993. *Report of the International Whaling Commission*. In press.

Tillman, M. F.
    1980   Introduction: A Scientific Perspective of the Bowhead Whale Problem. *Marine Fisheries Review* 42(9/10):2-5.

Vibe, C.
    1967   Arctic Animals in Relation to Climatic Fluctuations. *Meddelelser om Grønland* 170, No. 5.

Withrow, D. E. and R. P. Angliss.
    1994   Timing of Bowhead Whales Landed or Struck and Lost by Subsistence Hunting in Alaska between 1973 and 1993. Doc. SC/46/AS6 presented to the 46th Meeting of the Scientific Committee of the International Whaling Commission, Puerto Vallarta, Mexico, May, 1994.

Zeh, J. E., J. C. George, A. E. Raffery, and G. M. Carroll
    1991   Rate of Increase, 1977-1988, of Bowhead Whales, *Balaena mysticetus*, Estimated from Ice-Based Census Data. *Marine Mammal Science* 7:105-122.

Speaking of Whaling: A Transcript of the Alaska Eskimo Whaling Commission Panel Presentation on Native Whaling

Edited, with comments, by:

Carol Zane Jolles
Department of Anthropology, DH-05
University of Washington
Seattle, WA 98195

Abstract. *The 1993 meeting of the Alaska Anthropological Association brought together a unique forum of Native whaling practitioners and northern researchers. The transcript of that forum documents Native whaling captain concerns about the continuation of their traditional livelihood as well as providing information about the socioreligious aspects of whaling. Of particular interest is the detailed description of the duties of a Barrow whaling captain's wife. The transcript is supplemented with commentary and recommendations for further research on whaling in western Alaska. Additionally, the presentation by the Alaska Eskimo Whaling Commission members and its executive director and the questions from the audience of scholars showcase the differing discourses employed by researchers and practitioners when dealing with Native resource issues.*

INTRODUCTION

In *Man the Hunter* (Lee and DeVore 1968), one of the more fascinating sections is the transcript of conversations among those who attended that now-famous 1965 symposium on the role of hunting in human evolutionary history. In that same tradition, I have transcribed, edited, and commented upon the panel presentations by the board members of the Alaska Eskimo Whaling Commission (AEWC) on issues which concern them as Native whale hunters in western Alaska. The comments by board members and the executive director of the AEWC are offered with only minor editing, in order to preserve as much of the flavor as possible of the presentations while accommodating the demands of written text. Whaling is often on the minds of archaeologists and those concerned with subsistence technologies and strategies in western Alaska (Fig. 1). While it might seem that whaling has been considered almost too thoroughly, many issues still demand our attention. The panel discussion is particularly important because it critically examines current whaling practices, it gives voice to concerns of contemporary hunters, and it serves as a forum for men and women at the very center of western Alaska's whaling tradition. They serve in the capacity of professional contributors, who are vitally engaged in cultural and economic practice and in the global politics of marine hunting and fishing negotiations.

When Burton Rexford of Barrow, chairman of the Alaska Eskimo Whaling Commission (AEWC), and Leonard Apangalook of Gambell (Sivuqaq), his vice-chairman, were approached with invitations to participate in the conference, they graciously accepted along with their commissioner delegates from Alaska's nine whaling villages. When the session was convened, each commissioner, along with Maggie Ahmaogak of Barrow, executive director of the AEWC, prepared statements concerning contemporary whaling in their respective communities and answered questions put to them from the symposium audience. The edited transcript of the panel presentations and the question and answer session which accompanied the presentations follows below.

The transcript begins with a brief introduction by Allen McCartney, session host of the AEWC panel discussion. He, in turn, introduces Maggie Ahmaogak, who then

Figure 1. St. Lawrence Island whalers making a strike (1922). Photograph by Otto Geist, Historical Photograph Collection, St. Lawrence Island, acc. #64-98-1204N, Archives, Alaska and Polar Regions Department, University of Alaska Fairbanks.

introduces Burton Rexford and the members of the Commission. Several references are made to absent members of the Commission. The Alaska Anthropological Association meeting was scheduled at the height of the spring whaling season and, as a result, some commissioners had returned to their villages. Those who remained, as the transcript shows, clearly longed to be at home and hunting.

THE TRANSCRIPT[1]

ALLEN MCCARTNEY: This symposium is a continuation of the Native Whaling Symposium. A very important part of the symposium is the participation of the Alaska Native Whaling Commission representatives, including all of the officers and the executive director. It is certainly our pleasure to welcome them to this forum. Maggie Ahmaogak, executive director of the Commission in Barrow, will introduce all of the participants and invite them to make some opening remarks. There is a video tape that the commissioners would like to show, which I think will give us all an orientation to whaling; that tape can then lead us to more discussion and presentations by our panel members. Eventually, of course, we would like to open this discussion to the floor in order to have questions posed, however specific or general, having to do with whaling. This is a wonderful opportunity, a unique opportunity, to have this kind of expertise made available to anthropologists at this meeting, who can then engage in conversations with the whaling specialists. We very much appreciate all of the individual sacrifices that you are making by not being someplace else today. Let me introduce you to Maggie Ahmaogak, who is going to introduce each of the members who will have an opportunity to make some remarks.

MAGGIE AHMAOGAK: Good morning. First of all, I would like to thank Roger and Allen for inviting the Alaska Eskimo Whaling Commission to the symposium and I'd like to make some introductions of my commissioners. First of all, our chairman, Burton Rexford, is from Barrow; our vice-chairman, Leonard Apangalook, is from Gambell; next, we have Frank Long, Jr., from Nuiqsut; John Hopson from Wainwright; Raymond Hawley from Kivalina; Joseph Kaleak from Kaktovik, which is Barter Island; and, Elijah Rock from Point Hope. In addition, we have Herb Anungusak who has joined us as a participant from Wales. We had two commissioners, from Savoonga and Wales, who left us because they're right in the middle of their whaling season. I would like to introduce chairman Burton Rexford, who will give a brief history of the AEWC.

BURTON REXFORD: Good morning, and as you can see, we respect our women. She is the first speaker. I'll begin with an overview of the Alaska Eskimo Whaling Commission, which started back in 1977. A lot of you know Eben Hopson. He was the man that created the Alaska Eskimo Whaling Commission because of the moratorium on whaling; he wanted a broader view of the whaling issues, as I understand it. He incorporated Savoonga, Gambell, Wales, Point Hope, Wainwright, Barrow, Kivalina, Nuiqsut, and Kaktovik into the Commission. Nuiqsut, at the time, was a newly formed community. I

---

1. Introductions and conversations relating to the practical business of running the symposium have been severely edited.

guess his view was to have a wide range of Alaskan whaling communities represented, in order to justify his cause of continuing whaling. Eben began this effort in 1977, and, in 1978, we formed the Alaska Eskimo Whaling Commission. All of these villages participated in a meeting in Barrow that year. We came to the conclusion that we should have this commission. So, since that time, 1978, we have had this commission and co-management responsibility which Senator Stevens had introduced to NOAA (National Oceanographic and Atmospheric Administration). We do have a co-management, role and we have been very successful in marine mammal management. Not knowing the scientific side of the issue, we had great difficulty during the first few years with the quota that was imposed for the first time in our lives. At that time, we were used to hunting without a quota. And, it was very difficult for us to experience that. We went through hardship the first six years, because we didn't really know how to deal with the quota system. In other words, there were some of us that wanted to go over the quota. The first quota we received from the IWC (International Whaling Commission) was 12 landed [bowhead] whales and 18 strikes.

In the first three years, some of the communities faced hardship for the first time, because no whales were landed under the new system imposed upon us. Barrow, for one, had a very hard year. The quota for Barrow at that time was four, but not one whale was landed. This happened because of the quota system. And, with so small a quota, there was a lot of competitiveness imposed upon us as the hunters, and that's one of the reasons why the "struck-and-lost" was experienced in Barrow. When you are in that situation as a hunter, it is very alarming to go back to the community without a catch. It takes many, many years for a captain to be recognized in the community. The respect which our communities give bowhead whaling is very important to us, as captains.

We are out there providing for our children, our grandchildren, the wives, the elders. We are even, as captains, directed by the elders to land a certain type of bowhead whale, which is considered a very great delicacy by the elders. But with the quota system, it's difficult to deal with that, the issue of the elders' request. The first thing you want to do as a whaler is to make a strike when it is possible. In all the communities, we went through this hardship during the first six years of the Alaska Eskimo Whaling Commission. The Commission has a very long history of cooperating with IWC and NOAA, as well as with scientists from around the world. Every three years, we negotiate a quota for the nine communities. The last negotiation meeting was in Iceland in 1991. It wasn't easy in 1991. We had difficulty in getting our quota. 1988 in New Zealand, from my viewpoint, was not that difficult, because we had non-whaling hunters from Australia who made an amendment on the floor for us. 1991, however, was a difficult year. The first day of the meeting we experienced that we were not going to be honored with flexibility — with what I call flexibility. In 1988, we managed to get the quota increased to 41 landed bowhead whales and 47 strikes, plus a carry-over provision. In 1991, we wanted more flexibility with the strikes. Our first proposal was, I believe, 57 struck and 41 landed. We got hit and knocked down to 54; we were satisfied with this amount of 54 strikes in order to land 41 whales.

The AEWC is a leader in our communities and a well respected organization. We respect the women, our elders; and, as a captain, we can't do any more than offer respect to our people. The disciplining was very difficult for us in making the transition from the non-quota years to quota years. I, for one, as one of the captains, well, I started in 1977 as a captain and it was unreal for me to go through that process. I guess I'm one of the slower people to become disciplined. I wanted to go over the quota and I tried and, luckily, I didn't make a strike and I didn't get over the quota [laughter]. In general, all the whaling communities are about identical in their culture. We have slightly different systems of how we treat the bowhead whale in each community, and we respect each community as to how they treat the bowhead whale, spiritually and culturally. I'm sure that some of the commissioners here will tell you about modern day whaling as well as the traditional way of whaling. The whaling for us on the North Slope as it was passed down in the old days is considered to have been passed down as part of our history. According to our traditional history, one of the Native hunters picked up along the beach a driftwood that's good in your grip, what I would call a wooden shaft. He had the idea that he would go after the big animal, this one particular hunter, and made the driftwood long enough and sharpened the point of the driftwood in order to strike the whale from the abdomen and go in under the rib cage, driving the shaft all the way to the heart of the whale.

This is a very inhumane killing of the whale as we understand it today. But at that time, they didn't have any modern day equipment, as we do now. The Commission is involved in the development of weaponry equipment. We are working on a penthrite [bomb] because of the humane issue which we are charged with as members of the international community. We are working with Norwegian people on what's called a penthrite. And it's still in the development stage. We have been developing this for several years. We're still making adjustments to it. As you know, adapting to new weaponry for us is not something new. Right now, the weight, the balance of the new weapon is considered off target, to be what I call off-target, in making a strike. Because of its weight, it has a tendency to lose velocity as a hand-thrown harpoon. So, therefore, you are not accurately landing your harpoon where you want it. And, these are the reasons why we have many discussions over this new weaponry.

Some of our equipment and the weapons are over 100 years old and still functioning. I, for one, have one shoulder gun which is 115 years old. In fact, we're doing some repair right now in the shop because of its wear and tear.

If you have any questions, I'm more than willing to answer on the overview. Yes.

QUESTION: Is it really necessary to have a quota right now?

REXFORD: Well, last night the statewide Native leaders of different commissions had a meeting with U. S. Senator Stevens. I made a statement that we should keep the bowhead whale in the Endangered Species Act. Yes, we do need to manage the renewable resources, yes. Yes, it's necessary to have a quota, even without IWC, yes.

At this time I will turn the discussion over to the vice-chairman. He will be covering the Native way of hunting. And, he can talk about the use of rawhide. We don't use that any more in Barrow and, in his community, they are still using the traditional way of hunting. Vice-chairman, Leonard (Apangalook).

LEONARD APANGALOOK: Thank you, Burton. I come from Gambell, the southern-most community of the nine whaling villages in Alaska. I'm sure most of you know that Gambell is on St. Lawrence Island out in Bering Strait. Yesterday, I think Carol Jolles and a few other people showed some maps showing where that was. I come from that community. I think Burton covered most of the information relating to the overall organization and where we came from and where we are now. But, I think I can speak mostly at my community level, since our villages are somewhat diversified because of distance and we're so far apart.

Gambell is one of the more economically deprived communities in Alaska. Primarily we are so isolated out there. We're 200 and some odd miles from the mainland of Alaska and 38 miles from Siberia. But we have been whaling just like all of the other communities from time immemorial, and we claim that we are one of the older whaling communities. There was recently mentioned the Bering land bridge, and we claim that we pushed those whales which were at one time land animals into the Bering Sea when the land washed out. And, that's where we are.

I do think there is a lot of documentation of whaling activities on our island. I think I heard some time back that there were some artifacts, including whaling tools and implements, weapons, and other objects, that were excavated out there and carbon-dated to be extremely old. I don't know the exact number of years B.C. So, we do consider ourselves one of the older communities. And, we still use a lot of traditional equipment that Burton mentioned earlier: rawhide ropes, and, of course, we do use what we call the modern whaling bombs, although the design originated some time before the 1900s. And, for the sake of traditional use, I think that is mandated under government regulations that we use these.

Some time back, I remarked to our legal counsel in Washington, D.C. that our country has the technology to develop an effective weapon that can help us. But, their fear is that we will lose traditional status. Having come from a community where no industry has come in and we don't have any local resources developed, we're still somewhat primitive in our method of pursuing the whale. And I wonder sometimes when we will be allowed to progress to where we can more effectively pursue the whale with modern weapons. Burton mentioned the development of the penthrite bomb. We recently suffered a setback on that, but probably in a couple more years that will be improved and in use.

Some years back, I remarked to an assembly that we have hired the best scientists to study the bowhead whale. That was in our quest to increase our quota, and we did get a lot of good information. But I remarked at that time that we need to study us. I'm very pleased to see anthropologists studying the Native people too, because we need studying. I feel that we are victims of transition. We're trying to make this change from our traditional lifestyle to maybe, one day, a total cash economy. I don't know how far away that is. We are in between and suffering the consequences. Like, in my particular community, just this past year, a survey was made where unemployment was over 70%. If it weren't for our subsistence life style activity, it would be a pretty tough situation. We are more or less combining two life styles or cultures to make things work and to survive.

I do want to convey my appreciation to various governmental agencies that Burton mentioned earlier that have made it possible for us to continue our whaling. A lot of times, we feel that the government is imposing some regulations prematurely for some of our communities. Although we are in the Russian side over there at Gambell, we appreciate the fact that we didn't end up another 10 miles further, lest we wouldn't have had an opportunity to work out some of our problems. We do appreciate our system here. That's all I have. Thanks a lot.

QUESTION: You mentioned the Russian side. With Russia changing so much, whaling going on over there is totally different than the way whaling is managed here. Two questions: first of all, what kinds of contacts have you had with folks on the Chukotka Peninsula regarding whaling? and, secondly, has the Whaling Commission entertained any notion of embracing whalers from Chukotka?

APANGALOOK: I think our primary contact with the Russian government in terms of whaling is through IWC. There are a lot of gray whales harvested and they call that traditional whaling, but actually it's commercial whaling, because they use the harvest primarily for feeding the foxes in their fox farms. There are quite a few of those. I forgot your second part of the question.

QUESTION: I talked about including traditional whalers from Chukotka in your discussions.

APANGALOOK: I think Burton prefers to answer that question. Thank you.

REXFORD: The Russian fishermen are members of the IWC, and they have a letter to the secretariat of IWC requesting three bowhead whales for the Natives over there. That's an on-going issue. I'm sure it will arrive next month in Japan. And, we of the North Slope Borough, through our scientists, have made contact with the Native whalers over there in the last two years and we're trying to work with them. We do feel that we don't have all of the count of the bowhead whale. Some of the migration goes through the Russian waters. So, in that respect, we are working with the Russian Eskimos.

The next commissioner is Frank Long. And, I think he has a different perspective on whaling. He will explain how he is impacted by the development, the industry, offshore. Frank Long.

FRANK LONG: Good morning. My name is Frank Long, Jr. I'm from the village of Nuiqsut and I live 70 miles west of Prudhoe. I come from a reestablished village. It was reestablished in 1973. During that time, land settlement was taking place on the North Slope. This was already a community long ago, but the Federal government decided that the Natives needed to go to school and to live in a certain area. So, when the school was erected on the Slope, those people gathered in Barrow. That left the village abandoned with nobody there except for one family.

But anyway, I'm a whaler, I'm a whaling captain. I've been a whaling captain since 1987, but whaling has been going on since before my time in the same area. We don't whale right near our home village like they do in Barrow. We have to go out to the Barrier Islands. It takes at least a day and a half to two days for the first trip out, because it's almost 9 miles away from where we live, in the Prudhoe Bay delta. We base at Cross Island in the middle of the oil industry, and there are activities going on all over the area. We can see what they are doing on a daily basis while we're out there subsisting for whale.

The whalers and the oil industry try to work together. We sign agreements before whaling is started. We have meetings. We negotiate with the industry on what kind of deals we want, and we don't get any financial benefit from all that in any way from the industry. But all of the funds that we use for our whaling, our equipment, and for staying out there, all come from being a hard working, 8-to-5 person.

It's not easy to be a captain in the middle of the oil industry, especially when both sides want to have their own way. I've got the whaling activity and they've got their oil industry, so there's got to be some kind of compromise in order for us both to be successful. I was proud of this last year when we did that. The industry was successful in finding what they were looking for, and we got our three whale strikes. Although we lost one whale, we landed two. So, we were fortunate. I think, at times, we were not very fortunate in previous years, when the activities were so enormous. As Burton stated earlier, we almost went home sad-hearted and uncomfortable from not catching or landing a whale.

As I said before, I've actually been out whaling as a captain since 1987, but I've been whaling since I was 10 years old. And, that's a long time for me. I didn't start in Barrow. My first time out was at Point Hope. We used to live down there. I grew up in the Kotzebue area and migrated like a whale up north. I got as far as Prudhoe Bay. I appreciate this meeting, because it gives me an opportunity to see what's going on in areas other than where I come from. We have people that are very touchy on the issues that concern whaling and the oil industry, but like I said, we can make some compromises and agreements which allow us to keep ourselves intact and continue our life style. We still have a problem in trying to solve most of the issues that need to be straightened out.

I am from a modern day whaling community, because we reestablished in 1973. During that time, in 1974, when we first moved to the village of Nuiqsut, we landed our first whale. But there are whale skeletons still there which have never been touched. They were left on the island by old traditional whalers.

As a whaler in the middle of the oil industry, I see what's taking place, that the pattern of the migrating whale is somewhat diverse and changed since offshore drilling has begun. We, as whalers, have tried to renegotiate those kinds of activities with the oil industry. We deal with the oil companies that want to do some exploratory drilling or "seismic" work in there. I would say they're like a shoe that you don't want to put on the wrong way. So, I'll go as far as that. I appreciate saying a few words to you all and wish everybody good luck.

REXFORD: Well, I had the same experience in Barrow in 1979, 1980, and 1981, of geophysical seismic work in the ocean, and it is a "no-no" to a hunter during the whale migration. I know from experience. There were three of us captains that went out whaling in the fall. In those three years, we didn't see one bowhead whale, and we saw no gray whales, no beluga, and no bearded seal. We traveled as far as 75 miles away from our home on the ocean waters in those three years. And, I know the hardship that Frank Long and his fellow whaling captains go through.

We do address the issue of seismic work with the industry and we will address the issue again on the twelfth of this month in Barrow. We'll start off in Barrow, and then we'll go to Barter Island. We'll address the same issue with the industry. Then we'll swing back to Frank Long's community here, and that will be the final day in addressing this issue.

I have participated in some of the workshops which are now a yearly thing in Seattle. I will represent the Commission. Among all of the scientists that participate there, I am the only non-scientist in the meeting, so I have learned from these meetings not to rebut the issue of disturbance, because the oil industry reacts to the current information before they come to our communities.

The next commissioner is from Wainwright. It has a very similar situation to the one in Nuiqsut. He will be talking about that.

JOHNNY HOPSON: [Some of Mr. Hopson's opening remarks were omitted from the original tape transcript.] Seismic work started in the middle of winter, I think, in our area and they had it going until spring thaw. That was when they couldn't use vehicles out in the ocean any more. That spring nobody saw any whales; Wainwright didn't catch a whale and even the fishing was poor. Villages within 75 miles inland had very poor fishing all summer. Even the ships that stay out in the ocean make a lot of difference to the migratory route of the bowhead. We had a research ship last fall working somewhere between Barrow and Wainwright that was supposed to be 20 miles off shore, but a couple of times it was seen inland. I think it was just going back and forth. We saw no whales last fall; nobody got any. Barrow kept telling us there were whales that were really running. Sometimes we'd go as close as 10 miles outside of Barrow, and then we'd go straight out with whaling boats, but we never saw any whales. (Long pause)

REXFORD: Johnny is right. Two years ago, three years ago, something like that, we had a drilling ship to the east of us about 32 miles from Barrow. In the fall whaling, we had to go further north than we normally do because of the drilling ship just sitting there idling, running the generator. Mr. Hopson is right about the diversion of the migration of the whales.

The next commissioner is Raymond Hawley from Kivalina.

RAYMOND HAWLEY: Good morning. Thank you, Burton. Most of what I might say has been pretty much covered by these commissioners, so I'm going to tell you a little about Kivalina, because I have lived there all my life as a whaling captain. And, sure enough, it has been a hardship since the time the quota began. It has been a very great hardship. I'm glad that the Eskimo Whaling Commission was formed in 1977. In the beginning, that quota was very small. That really hurt our community and the other eight communities as well. You know, the meetings were held annually until 1991. What has made it a lot easier for the communities, you know, has been working together. Sure enough, each community has its own experience as whale hunters. Once you catch a whale, there's a happiness, a lot of activities, that only happen if you catch a whale. There are feasts and there are celebrations afterwards, after the whaling season is over. That's our life. It is still like that, up until now. We will never lose that. It will continue.

As I said earlier, what makes it a lot easier is that each of the nine communities sends representatives to get together in Barrow, and they all have their input, you know, since the quota was set. That is what I would like to say. Thank you.

REXFORD: Thank you, Commissioner Hawley. As he mentioned, cultural and spiritual ties to the bowhead whale are very strong for us. The next commissioner is from Barter Island. His hunting has been impacted also, just as Nuiqsut's hunting has. But he's sort of upstream of the impact of the oil industry now. So, he's much happier today than he was in previous years. Joe Qaleq.

JOE QALEQ: Good morning. I'm from Kaktovik. Yes, I agree with what has been said by the commissioners here from these villages, that almost everything has been covered. Frank Long, Jr., has already covered almost everything concerning what is going on between Kaktovik and Nuiqsut. I've been going out whaling since I was about 12-13 years old, while I was living in Barrow. I'm from Point Barrow, but I have lived in Kaktovik since 1972. When I moved out from Barrow, I became a whaling captain at Kaktovik, beginning in 1972. I wasn't there for the first whale which was caught by a family in August, 1964, in Kaktovik. That was before the quota started. Before the quota started, we had only about three or four whaling captains who started with Kaktovik. There are nine or 10 whaling captains right now. I don't have too many things to say myself. Frank Long almost covered everything that has to do with the problems in whaling. Thank you.

REXFORD: Thank you, Commissioner Qaleq. I think we're open to questions.

[EDITOR'S NOTE: In most cases it was not possible to produce a verbatim transcription of questions from the floor, since questioners did not direct questions from an open microphone. However, the answers to the questions are clear and distinct, and in many cases infer rather clearly the nature of the question posed.]

QUESTION [paraphrase]: Are there attempts being made to document some of what was going on in the past, especially from the elders? What is being done?

REXFORD: I, for one, was afraid of losing history for my grandchildren; so, in the Commission we are working on a documentary grant through the National Science Foundation to do exactly that, to answer your question exactly, to have a Native writer along doing the documentation. Yes, we do need more documentation. It's written down, yes.

QUESTION [paraphrase]: I went to the whaling captains' meeting in 1985 in Barrow and the scientists came to the captains then and asked them for information about large whales. They didn't know if, in the past, there had been large whales that were taken. And, I believe that was the first time the scientists had come to the captains and asked them for that kind of information. I was wondering if there has been more of that since then?

REXFORD: Yes, we are more and more involved in addressing the issue of bowhead whales. And, in fact, in the international community, we are now speaking on behalf of the AEWC in addressing the International Whaling Commission. Particularly in the matter of humane killing. In my address a year ago in Scotland, I told the international community that humane killing of the bowhead whale is nothing new to our hunters. It has been addressed since time immemorial. The hunter does not want to see the animal suffer and therefore the initial strike has got to be accurate. Just like I mentioned with the new weaponry, it has a tendency to lose velocity, because of its weight, because it is hand-thrown.

QUESTION: Just recently, I translated a letter from Sakhalin Island that came to the Alaska Fish and Game Department. One Russian fisherman caught a salmon with a tag from the Juneau, Alaska, lottery, worth $500. What it now means is that Alaskan villages and Russian villages in the Far East share the same resources. As you know, on the Kamchatka Peninsula and also on the Chukchi Peninsula, there are some nuclear plants and nuclear waste sites. So, we share the same North Pacific resources. Is there any cooperation in Alaskan villages with the Russian villages? Is there any progress in Alaskan villages that manage North Pacific resources with other North Pacific countries? Is there anything you can do to influence Russian governments to clean their environment and not to pollute arctic rivers?

REXFORD: A lot of us were in a meeting known as RuralCap. That particular issue that you brought out came up there. That was about a year and a half ago. The pollution that's flowing out to the ocean on the Russian side was brought out and we're doing everything we can to address that. I believe that the Bering Strait Region and NANA Region are communicating more than we are in Barrow. We do address that issue a little. There is also what we call the ICC (Inuit Circumpolar Conference) that's addressing the same issue. I have been invited to the Northern Forum and this will be a part of my interest, economic development. Because of the unrest over there, the meeting was canceled until a later date.

ROSITA WORL: I wanted to comment and ask a question about Frank Long's issue that he brought up. My understanding was that when the oil companies were given their leases, there were some stipulations that specified when they could conduct their activities so that it wouldn't affect the migration of the whale. Recently, I heard that there was an article about Greenpeace where it was asserted that those stipulations had been lifted. And it sounds like the industry now is negotiating directly with you. So, have those stipulations been lifted? Is that something the Borough had agreed to? Because I know the Borough had been taking a very strong position about making sure that those stipulations were in place. It sounds like there have been some changes. Maybe you could talk about that.

REXFORD: Frank, do you want to address that issue?

LONG: There are stipulations that were put in by the North Slope Borough. The Borough still has them in effect in their Zone Management plan which we use in negotiating with the industry, to plan our whaling season. We are fall whalers. We whale in open water and icy water. We try and make these stipulations stick with the industry, but once the document is signed and they are out there, it is very hard to see whether these guidelines and stipulations that we placed in the agreement are being followed.

REXFORD: To follow up on that, Rosita, I mentioned earlier that we are having a letter of authorization created for the community meetings with the industry. On the 12th of

April we will have a meeting in Barrow. From the past history of working and cooperating with the industry, we have learned that when a stipulation like the agreement between Kaktovik and Nuiqsut becomes effective, it is almost never followed exactly by the oil industry. We learned that. Because we learned that, as members of the Commission, we have now in the office of the AEWC a statement of the experience of these whaling communities when dealing with the oil industry off shore with respect to bowhead whaling. One year they had a seismic activity ranging from about 40 miles west of Kaktovik, the area which is called Camden Bay. The industry did work there in August. This is during the migration. They also worked in Harrison Bay, to the west side of Cross Island. Now, for about 120 miles, I believe, from Harrison Bay to Barrow, that particular year of the heavy concentration on seismic work, we found that we had to go further in order to get any bowhead. Also, at the same time, we went too far. We were not able to tow the whales into the community fast enough and the meat got ruined. So, there is a great deal of impact where seismic work is concerned.

MCCARTNEY: For an archaeologist, could you elaborate on some of the size issues and some of the preferences for whales, in terms of your earlier days or even modern times? Is there a preference for a certain size whale? I'd like you to contrast small whales with larger whales. Also, when whales are not caught in the community in a particular year, what alternative foods, Native foods, are then used? What kinds of fall-back is used when whales don't come in?

REXFORD: For me, as a family provider, when there's no whales landed in Barrow, the next source of food from sea mammals would be walrus and seals. Also, caribou are taken.
    For the size of the whale, it is important to satisfy the elders. They want a smaller size whale, which is good and tender and soft. We try to accommodate them the best way we can. Yes, they would rather have a 27 footer or something like that. In the larger whale, the fiber is tougher than the smaller whale. Well, as you notice, I don't have any teeth so I don't want that whale!

[EDITOR'S NOTE: At this point the panel members and audience adjourned to view a whaling video produced by the whaling communities. Following the viewing the symposium reconvened.]

REXFORD: Well, folks, we have with us a commissioner from Point Hope, and Point Hope has a long history of traditional whaling, so we'll give the Point Hope commissioner an opportunity to speak. Elijah Rock.

ELIJAH ROCK: Good morning, everyone. It is an honor to be here participating in this symposium. Point Hope has a history, a long history of whaling, even before I was born. I have participated in whaling at Point Hope pretty much for 45 years as a whole. I have been recognized as a whaling captain for 22 years. From the time you are able to walk, your parents take you out on the ice and train you. It's a training period from the time you learn to walk. You are out there with your parents and they teach you about whaling and to respect things, not to destroy anything. Because this is our resource, our food, they teach us not to take all of them. Don't deplete the resources is what they teach us.

Before me, my father was a whaling captain, and, also, before him, his father was a whaling captain. It goes on generation by generation, and our responsibility is maintained to share with not only the community itself but other outlying communities such as Kotzebue, Norvik, Noatak, Wainwright, Barrow, and all the other villages surrounding our area. And, I have to be here to honor our chairman. If I didn't, I'd rather be at Point Hope whaling right now, because the rest of my crew are out there now. That pretty much covers it.

REXFORD: Thank you, Commissioner Rock. By the way, a lot of you people have known Howard Rock. Howard was Elijah's uncle.

Our captains have mentioned their history of whaling; so, I am the only one that hasn't mentioned mine. It has been over 50 years that I have been whaling. I was first introduced to whaling by my father. I was 10 years old. It interrupted my studies in school, so I tried not to go through the same process with my kids, today.

So for now, we'll recognize the woman's role in whaling in our communities. The hard work starts when it comes to the women folks at home. Maggie Ahmaogak.

MAGGIE AHMAOGAK: Thank you, Mr. Chairman. For those of you that didn't catch my introduction, earlier, I'm Maggie Ahmaogak, and I have a husband who's a whaling captain also. So, I'm a whaling captain's wife. I will try to walk you through the procedures and processes that have been carried on for quite some time, since the whaling for subsistence has started with the Eskimos. As a whaling captain's wife, you learn the process from the elderly women. They hand the knowledge down. The hard work starts when your whaling captain successfully lands a whale. That day you are getting ready and the final butchering is done out on the ice. The captain's crew members bring in the flag and put it on top of your house. Then the women start their work. We get Eskimo donuts made [fry bread]; we're getting ready to feed over 2000 people in one day in one house. And, after that is completed, when the quota or the whaling season is finished, you start what they call *apugauti*, which means beaching of your boat from a successful captain who has caught a whale.

Then, you're preparing again a little celebration for your crew and the family members when they bring in the boat with the flag on it. After that part is over, you are into the real spring, going into the summer. This happens about June. They have the blanket toss celebration, and it's called *nalukataq*. We once again prepare to feed the whole community, and it's not just Barrow itself or just one community. All the villages will be coming in to participate in the distribution of the *muktuk* [whale skin] and meat to all the community.

Once again, that blanket toss process starts early in the morning. They get ready to make all of the windbreakers, so that if there is any chance that there will be a wind that changes direction during the day, they have the big windbreakers that are already made. The blanket itself is prepared down there in the middle of the windbreak, where the people will be sitting all day long during the feast. The women do all of the cutting. The men bring up the meat and the *muktuk* from the cellar to get all thawed out. The women will be doing all of the cutting and the cooking. Again, for distribution of that meat and *muktuk* and other parts of the whale, the heart, all of the major organs of the whale are all boiled and distributed too, as the lunch portion of the celebration.

Then, during the afternoon, at about three o'clock, they go and serve what they call *mikigaq*. It's fermented *muktuk* and meat that has been fermented for about a week, and it's part of the special time of the day when they distribute this. A lot of Inupiaq people where we come from can't wait for that time, because that's their delicatessen.

Then, in the evening about six o'clock, they distribute the *muktuk* and meat to everyone, waiting for those that are working all day to give them a chance to make it too for the distribution. Then, in the evening, after everything is distributed to the community, the blanket toss starts. The blanket toss itself is made out of the boat. The *umiaq* skin has been taken off the boat; they make it into a blanket. The men prepare it and they start the blanket toss and it's their way of celebrating the successful catch. After the blanket toss, between eleven and midnight, they have a final Eskimo dance. It's a way of ending the celebration for the day by having the Eskimo dance.

After the *naluqatak* or the blanket toss is over in June, the seasons will start again. In July, the men will go out seal hunting. Once your crew members or the whaling captains catch their seals, the women do that part of their work again, the process of cutting the *ugruk* skins or the *ugruk* [bearded seal] itself. Its skin is prepared in a special way by the women, using a fermentation process which takes place during the whole winter. After the process of cutting, that skin is prepared by the women, and they put portions of the blubber that is taken off the seal itself into strips and then rolled together and put in a burlap sack for fermentation processing during the winter.

Then, during that summer season, you're also hunting other different game, like the caribou, to be able to store food for the winter. You normally go inland to hunt for that and most of the threads or the sinew is taken from the caribou itself and is prepared by the women also. It's soaked, that sinew from the caribou is soaked in water until the meat is able to drop by itself, and then the back straps or the leg sinew from the caribou is dried. Then, the women take it back home and start the braiding. They braid sinew to make the threads for the *umiaq*. When the women go through a process of collecting, again, when it's time to thaw out those *ugruk* skins, there's a special way of cutting them. The women hand down the process of the cutting to the younger generation. We've been more involved recently, directly working with the elderly women to learn the process of the sewing. An average *umiaq* will have six to seven skins that have been properly processed and cut ready for sewing, using the caribou sinew braided thread.

During the winter again, you have Thanksgiving and Christmas which also we prepare along with the men. This includes the cutting of a third or other portions of the whale that have been stored in ice cellars. It's taken out of the ice cellars and cut up and made ready for distribution to the community again during the winter. Then you go through the Thanksgiving and Christmas process for distribution through the church to the community.

Then your springtime comes around again and your process starts again for the women to start preparations, especially the clothing. We prepare the men's clothing so that they're warm. They have fur parkas, their snow shirts, and all of their food that has to be prepared for the captains to go out for their annual process again for the whale hunting.

Again, the process is, like I said, it's like a procedural or systematic thing; it goes year-round and once you go through that, it starts all over again. It's the women that always are anxiously waiting for their captain to catch that whale. Your process will start again.

If there's any portions I missed, maybe Mr. Chairman, you could elaborate a little bit.

REXFORD: As Maggie mentioned, I want to put some points of clarification into the process of cutting the *ugruk* hide using what I call the rule of thumb. It's a traditional way: two men stretch a skin out and you use your thumb as a ruler; so, you cut under your thumb all the way to the arm pit from the belly button of the *ugruk*. It's a very easy process.

As you notice, women do a lot of work in whaling communities. The men will be sitting in the windbreaker which she mentioned just waiting to be served.

Yes, there's not much waste in the whale. For some reason, according to the history itself, we have never tried to eat the liver of the whale. I don't know why. So, like she mentioned, the heart, the kidney, the stomach, the intestine, we do have some evidence there is some contamination visible in the kidney.

If you have questions for Maggie?

QUESTION: I was just wondering about the boat skins. Do you replace the skins in the same way each year?

REXFORD: Well, the skin, itself, if it is properly taken care of, is good for three consecutive years. If for some reason the women fail to treat it properly, we'll have some spots where it has rotted and then we would have to do the sewing ourselves in the ocean. St. Lawrence Island has a bit different way of preparing their boats than us. They use walrus hide.

WORL: Maggie, I have two questions. First, I know that it used to be that not more than eight or 10 women were sewing. Are there more women sewing now than there were a few years ago? Are the younger women learning that?

AHMAOGAK: That's what I was elaborating earlier. The younger women are now being requested to take part, so there are a lot more younger women actually doing the sewing along with the elders.

WORL: Then the second question, maybe you could comment just from your own personal experience. I know that you were one of the earlier women to spend a lot of time out on the ice. I just wanted to know what your reactions were? How did people behave towards you when you first started to go out and spend time on the ice during the hunt.

AHMAOGAK: Actually, when I was very young, I started out with my father's crew. The men were insulting when a woman went down.[2] My brother would always take me

---

2. The convention of referring to the ice as "down" is part of a larger system of referring to elevations in the local landscape. "Down" generally refers to slight differences in relative elevation, usually with the person's home or village center as the reference point. Similarly, one speaks of going "in" land when going away from the sea ice or open water or "out" when referring to going to the ice or open sea.

in land, and I always ended up back down. So, they came to be comfortable with me. They always said that it wasn't a woman's place to go down. As time went by, one day my father had all of his sons in land and I was the only one down, inside the tent. I had to get on the boat myself when my father actually shot a whale. So, they never took me in land again. (Laughter)

REXFORD: Yes, Rosita, to follow up on a woman's participation out in the ocean. It's more visible now, today, than it was in the past.

QUESTION: Of the women who do stay at home, are there things that they follow or that they think about while the men are out? While the crews are out?

AHMAOGAK: Well, actually from my experience, now that I have children that are in school and they're younger, we have children to attend to, plus we now have the VHS radios. We're in constant contact with them, with the captains from our own crews. You never know when they'll contact you, or if they will need something that we need to go to the store and pick up. They'll send a crew member up, so we're always at the home base ready with trying to answer their requests or calls that come in, so that things are ready for the crew member who comes up to take the things back down. It's important to have somebody at your home station.

REXFORD: Rosita, to your question about younger women participating in sewing. I think a good example is Maggie's sisters. You did the sewing one year. How many sisters do you have?

AHMAOGAK: I have 10 sisters, and I have no problem getting them over. There are six of them out in one line and they have six other elderly women on the other side coaching them along.

ELIJAH ROCK: Rosita, in Point Hope it's a tradition that a whaling captain's wife does the sewing and also she teaches their children.

QUESTION: Maggie, have you ever seen an instance of botulism?

AHMAOGAK: No, I have never seen that kind of a case because it's the captain's wife that normally takes care of the fermentation process inside the house. You have to stir the meat *constantly* at least three times a day. If it's not taken care of properly, it can go to waste. If it goes to waste, there has been a lack of responsibility to taking proper care during the fermentation process. It only takes five to seven days for it to process, and then, after that process, it keeps for a while. Then, you put it in a cool place.

REXFORD: The whaling captain's wife, if the fermentation is not right, gets a lot of criticism.

AHMAOGAK: That's right. Plus, your containers have to be very clean. Most of the women prefer stainless steel.

WORL: I have another question, Maggie. I made the same mistake, I think, that other anthropologists made when they have gone to the communities in the past, perhaps with the exception of Margaret. I studied with the men instead of with the women. Just recently I started to study what the women were doing. I know in Barrow that the women, after a whale has been butchered, are allowed to go forward and then cut off a piece of meat (*pitaniaq*). Yes, and then, I know that the women, especially the whaling captain's wife, will also go and contribute to the community to those people in need, such as those who don't have hunters in their families. I am wondering if that is something that the wives in other villages also do?

FRANK LONG: I'd like to respond to that. I'm from the village of Nuiqsut, as I indicated earlier. We are fall whalers, we don't do any whaling in the spring. The women are the most effective, needed partner for the whaling captain or whaling community. They are the brains, the computer, and the whole works all put in one, while you are out there during the hunt. They have the biggest responsibility, perhaps more than the main captain of the crew. They are the generals, as far as I can see, of the whaling captain and crew. They support you of your needs, of what you need, such as bombs when you run out, outboards when they break down. They are there to either purchase or find you a spare outboard when it is needed. In Nuiqsut, we have to fly our grocery supplies through Deadhorse in order to get them out to Cross Island. That's the women's project, and they do a lot of work. By the time I'm done with it, I have a little bit of pity on my wife.

QUESTION: The questioner comments on learning about the role of women in Point Hope. At the time of whaling [a woman known to the questioner] found when she was in the laundromat that she was walking very slowly, and this kind of experience is very traditional. The wife moves slowly because then the whale will be going slowly through the water and not harm her husband. So, in that respect, she is bringing the whale.

LONG: Yeah, I would agree with you, because as a whaling captain we're out with the crew. As I indicated earlier, the woman, the whaling captain's wife, is like a general. Her responsibilities are so great that the captain doesn't go out to seek the whale. To my understanding, the captain's wife, who is the supporter and provider at home, is the main "catcher" of the crew. She "brings in" the whale, like you indicated. She makes it easier for the captain to harvest a whale, but the woman has to be in a proper state of mind, because she is actually the "bearer" of the crew and is called a "crew captain." The woman who stays at home with the children and the family along with the whole community are what I call "callers of whales."

JOHN HOPSON: Also, at Point Hope, women are designated whaling captains. I'm just a hunter. So, that tradition is always recognized. Women have a big responsibility in our whaling. Not only at whaling time. It's a continual process year to year, month by month. You have to share with your community if you are going to be successful in any type of hunt that you're after. Women play a big role in our life. Without them, I wouldn't be here.

REXFORD: The Point Hope commissioner brought out a good point that we also practice in Barrow. That is true. A woman is designated captain. We also practice that in Barrow whaling.

[The audience viewed a video on whaling on the North Slope and then returned to the questions and answer session.]

QUESTION: I was interested in the use of CB radios and what effect it has had on whaling and what changes it has caused, along with the comment in the film on keeping in touch with home base using Cbs.

REXFORD: It's a very effective way of communicating with the people in the community and also at the same time when you are out there, you want to be able to alert the community if you are in a distressed position. In the case which you've just seen, where the ice piled up, you can see the difficult situations which occur out there. We have adapted to this communication system and we're advancing into a broader range. I know that Nuiqsut has something to say about the communications.

LONG: We do depend nowadays on modern technology such as CB radios, VHF radios, GPS units; we have a radio station that is based at Prudhoe. We communicate, but the range of them is so short that once you go beyond the 70 degree line on your map out in the ocean, the VHF is out of range. That occurs at 40 miles. But we go beyond the 71 mile line on the map. When we do that we use the GPS unit that guides us, takes us where we want to go, how long it will take and practically why we're going there. They are very smart units; I like them very much. You can close your eyes and drive as long as you don't have anything to run into. But they are very beneficial to us in today's age. As we are telling ourselves, myself anyway as a whaling commissioner, for our village, we are going beyond the ranges that we usually normally do nowadays, and I enjoy the modern technical equipment that we use.

QUESTION: I have a question about the spirit of the whale. After the whale is killed, what does the spirit of the whale do? Does it go to the community for a while? Just what happens?

REXFORD: After the whale is struck and landed, the first thing we do with the struck and landed whale, if we have an elder in the vicinity, he will do the prayer for the whale; that's the first thing. And, the process goes on where, if we're towing it, it takes us at times maybe 10 hours or less, depending on the distance from our community, and everybody gets on the CB radio during those hours and keeps the spiritual thing going.
   Go ahead, Rock. He wants to add to it.

ELIJAH ROCK: I'd like to add on to that what our chairman has said about the spirit of the whale. At Point Hope, every part of the whale is used for the community and the only part that is put back into the ocean is the head, skull, that part. And when we do that, that tells us that the spirit is good and in the next spring, another whale will be coming back to the community to be used as food. And, when we throw that head into the water, we have a joyful moment. Yeah, and when thinking back, when you throw that part of the whale into the ocean, there's also life under the ocean that needs that food and they eat off of that.

REXFORD: We also in Barrow, especially as we start to land whales, you know. We're the largest whaling community. We start landing two, three a day, four a day, and there's only one in the spring, this is in the spring. And, if there's only one good spot for a butchering area and we'll be using that particular area. Yeah, we also dump it into the ocean.

There's an old saying that even a dead whale, the whale bone, like he said, won't be dry, won't be dead of thirst.

Commissioner Apangalook?

LEONARD APANGALOOK: I'm afraid that in our community we have changed considerably from some of the traditional practices that used to be at one time, perhaps because Gambell is so isolated from the other communities. Our people prefer to save the whale bones, instead of putting them in the water. If conditions will allow, we will save the whale head. That's something that contrasts with some of the practices in other communities.

REXFORD: Thank you, Leonard.

[EDITOR'S NOTE: Conversations among members of the audience and commissioners continued for several minutes, but no further clearly distinguishable questions were formally addressed.]

AHMAOGAK: I'd like to thank all of the commissioners that have come during this busy time when whaling is just starting. I know they went out of their way to come here, and I'd like to thank each and everyone and thank also yourself and Roger for all the help of getting us here.

[Applause ending the session.]

COMMENTARY

The session transcript is interesting in several respects. It brings together on the same platform those whose livelihood, whose practice, is centered on hunting, with those more often engaged in constructing theoretical hunting models or ethnohistorical and archaeological reconstructions of the hunt. Thus, theory and practice are wedded in the same intimate time/space forum. Also, this remarkable dialogue reveals telling distinctions among practitioners and theoreticians through the types of information which the AEWC commissioners felt it most critical to present and through the related but clearly differently focussed questions which persons from the "floor" directed to Ms. Ahmaogak and the commissioners.

The AEWC presentations should be taken, I believe, as strong indicators of Native concern. One might define these generally as a preoccupation with the politics of contemporary cultural survival in mixed/market subsistence circumstances: pollution, unnatural disruptions of the bowhead whale migration routes and breeding grounds, management of renewable resources, and the maintenance of meaningful community-oriented, subsistence-based hunting traditions. Improved technologies, particularly creation of sophisticated weaponry to promote humane, but nevertheless efficient, harvesting of bowheads, continues to concern all commissioners.

Each commissioner spoke of the quotas which regulate Native whaling. The constraints which these have imposed on Native practice surface again and again in recollection, a not so subtle reminder that through no fault of the whaling communities, the ability to provide food for families has been severely curtailed. Not surprisingly, commissioners desire quotas which will allow villages to put food on the table, to practice traditionally conceived Native celebrations, and to meet the newer demands of celebrations incorporated from non-Native traditions such as Thanksgiving and Christmas. Quotas for whales set by the International Whaling Commission (IWC) should, according to the AEWC commissioners, meet local demands of Native hunting communities without placing undue stress on bowhead populations. The concern of all of the commissioners with quotas is evidence of their awareness of the tensions inherent in such needs, particularly when taken together with the less conservative practices of other whaling nations and of the dominating presence of the oil industry in their hunting grounds.

A close reading of the transcript suggests, also, that the present communal, celebratory function of whaling is not always fully appreciated by contemporary scholars. While anthropologists ask questions about "alternative resources" in the absence of whales in the villages or express interest in the preferred sizes of whales, in recent years it is not at all well demonstrated that whaling itself is valued simply because it meets primary subsistence needs. Rather, it appears to be at least as important as an element of a deeply embedded and valuable socioreligious identity. As such, whaling meets needs often identified as religious, spiritual, and/or psychological as well as physical. This is poignantly and telling illustrated in Maggie Ahmaogak's personal description of women's responsibilities and work, which she has performed as a whaling captain's wife.

Both the commissioners and Ms. Ahmaogak appear to identify themselves strongly through the work of whaling. For example, each commissioner began his presentation with a brief personal history of his involvement as a hunter of whales. Each man spoke of his training by respected elders, usually a father or an uncle. And each seemed to feel that his engagement as a whaler shaped him as a man. Thus, an occupation which no longer produces a "cash" income, and which presently puts food on the table at best on a seasonal basis, absorbs tremendous amounts of time and energy (at least two months or more per year) and a devotion to community which must be subsidized through steady streams of outside cash for its continuance. Burton Rexford sees this devotion as a bond of respect which ties together elders, hunters, and their families in an intricate web cemented through long-standing tradition. For example, when asked about the size of whales preferred, Mr. Rexford answered that it is extremely important for a hunter to be able to provide meat which the *elders* have requested. And he added that it is very hard to come home with nothing to give them. Elders, he noted with a smile, don't always have strong teeth [he pointed to his own], and many prefer younger whales whose meat is less fibrous. Of course, this particular traditional preference is not everywhere universal. With characteristic exception, St. Lawrence Islanders sometimes express a preference for larger whales because the skin layer which attaches to the highly preferred *mungtak* (*muktuk*) is much thinner. The meat of the bowhead in the whaling communities of Sivuqaq (Gambell) and Savoonga is secondary to the valuable skin and *mungtak* layer, which once served as a central ceremonial food in older socioreligious traditions.

Eaten fresh, cooked, or aged, from highly preferred younger animals, or equally desirable older animals, the value of the whale goes far beyond that of mere food. Whether it is the meat or the *muktuk/mungtak* which whaling villages seek, it seems important to regard these foods as that quintessentially identified as *niqipiaq* or what people often refer to locally as "real food".[3]

CONCLUSION

In closing, I would like to offer several thoughts. First, the pragmatic concerns of those who live and identify themselves through hunting whales deserve careful and consistent attention by those who work in northern communities but who do not necessarily make the north home. This was illustrated in remarks by Frank Long, Burton Rexford, and John Hopson in descriptions of the effects (not yet systematically studied) of "seismic" research in whaling grounds. These invasive activities by oil companies and other industrial interests which are, in theory, under regulatory "agreements" obviously concern AEWC members, as their references to the scarcity of bowheads and other sea mammals and fish reflects. More research targeted at the less spectacular aspects of hunting such as long-term monitoring of invasive practices seems needed.

Second, there appears to be a disjunction between Native and non-Native identification and description of subsistence issues, a disjunction revealed in the very different discourses employed to discuss whaling. For example, while it is fairly clear that both AEWC commissioners and anthropologists have an abiding interest in Native whaling traditions, anthropologists tend to be professionally concerned with reconstruction of past history and with the technology of hunting for its own sake, while AEWC commissioners speak mainly in terms of contemporary cultural survival and cultural identity. Thus, scholarly reviews may omit the significant emotional components of contemporary whaling traditions which are critical features of contemporary whaling practice.

Third, there has been relatively little discussion, at least in publication, of the degree to which quotas create a new and artificially imposed level of competition upon whaling captains and their crews. Thus, the social fabric of whaling systems is disrupted through the process of limiting the hunt. Questions concerning the methods and strategies which whaling villages have developed to cope with these new pressures might be formulated and researched.

Fourth, anthropologists ought to recognize the truly occupational character of whaling. While most cultural anthropologists attempt emic perspectives as a matter of course, there is a tendency to regard large segments of whaling work as somehow "ritual" or "ceremonial" in nature. This may actually hurt northern cultures and communities if these "rituals" (often regarded as "archaic" practices by outsiders) are described as such in publication.

---

3. In St. Lawrence Island Yupik, the word for "real food" is *nekpek*.

This perspective is apparent in the panel discussion where questioners distinguish between spiritual or ritual matters and the work of whaling or the technology of whaling. Commissioners themselves and Ms. Ahmaogak do not make that same distinction. In fact, discourse around so-called whaling ritual or spiritual matters during the panel discussion suggested what is often implied in the vast literature on traditional whaling practice. If questions about spiritual matters seemed to slide eerily past the mark, it was probably because the answers offered by the commissioners were so grounded in a practical understanding of whaling. Native whaling is an excellent representation of a theory of practice articulated through a integrated socioreligious framework which combines thought and practice. In other words, whaling as a life system embraces both men and women in its web of responsibility by providing a pragmatic association of deep meaning with taken for granted acts of living.

Fifth, the complementary nature of men's and women's contributions to whaling are reinforced both through the organization of the panel discussion itself, which emphasized the independent[4] but cooperative quality of men's and women's roles and by Maggie Ahmaogak's extremely informative presentation which also emphasized those qualities. What is particularly important, I think, is Ms. Ahmaogak's characterization of women's work. She outlines that work in a yearly chronology which envisions what women do not as "ritual" or "ceremonial" duties, carried out as a kind of adjunct of daily living responsibilities, but rather as work at the core of responsible womanhood. It is this elemental quality which is so often overlooked in discussions about socioreligious and socioeconomic activities which are intertwined so closely that they have long since ceased to be considered as one or the other.[5]

Ms. Ahmaogak's very detailed outline of a woman's hunting duties articulates with questions on the spiritual addressed to Frank Long, John Hopson, and Elijah Rock and with comments by Burton Rexford. For example, in answer to the question concerning what happens to the spirit of the whale after it has been killed, Burton Rexford offers an answer which addresses the spiritual element only indirectly. His answer is both a function of politeness and respect for those who might not have the same ideas as he and a way of maintaining respect for the spiritual aspects of the whaling itself. His answer suggests that prayer accompanies and is integral to the real work of bringing in the whale and should not be regarded as separate from the hard job of bringing in the whale. It is clearly not classed as a type of performance.

Both the commissioners and Ms. Ahmaogak expressed similar views of whaling which emphasized the traditional duties, including substantial community participation, high levels of individual responsibility, traditional roles of respect toward elders, toward other members of the community, and toward the whale itself, and the critical function of passing on knowledge of whaling through traditional means. At the same time, each of the commissioners spoke to the hard issues of sustaining this type of socioeconomic

---

4. Ann Fienup-Riordan's work on the concept of mainland Yupik personhood also stresses these conceptual qualities of autonomy and independence.
5. Even the very interesting discussions by Barbara Bodenhorn (1990) and Edith Turner (1990) do not give full recognition to this notion that socioreligious activity is an integral component of women's subsistence work and not a sometime appendage to it. This same failing can be found in my own description of women's work on St. Lawrence Island (Jolles 1991).

and socioreligious activity in the face of increased pressures from oil industry activity, pollution, and resource management requirements.

Finally, I think it is appropriate to thank once again the commissioners and the executive director of the AEWC who participated in the conference: Burton Rexford, chairman, from Barrow; Leonard Apangalook, vice-chairman, from Gambell; Maggie Ahmaogak, executive director, from Barrow; Frank Long, Jr., the representative from Nuiqsut; John Hopson, the representative from Wainwright; Raymond Hawley, the representative from Kivalina; Joseph Kaleak, the representative from Kaktovik (Barter Island); Elijah Rock, the representative from Point Hope; and Herb Anungusak, a participant from Wales. Each participant, through the panel discussion, made significant and extremely helpful statements which gave breadth and depth to our understanding of contemporary whaling in western Alaska's whaling villages.

REFERENCES

Bodenhorn, Barbara
   1990   "I'm Not the Great Hunter, My Wife Is": Inupiat and Anthropological Models of Gender. *Etudes/Inuit/Studies* 14(1-2):55-74.

Fienup-Riordan, Ann
   1986   The Real People: The Concept of Personhood Among the Yup'ik Eskimos of Western Alaska. *Etudes/Inuit/Studies* 10(1-2):261-70.

Jolles, Carol Zane, with Kaningok
   1991   *Qayuutat* and *Angyapiget*: Gender Relations and Subsistence Activities in Sivuqaq (Gambell, St. Lawrence Island, Alaska). *Etudes/Inuit/Studies* 15(2):23-53.

Lee, Richard B., and Irven DeVore (editors)
   1968   *Man the Hunter*. Aldine Publishing Company, Chicago.

Turner, Edith
   1990   Ethnographer's Shared Consciousness on the Ice. *Etudes/Inuit/Studies* 14(1-2):39-54.

Whaling: A Ritual of Life

Herbert O. Anungazuk
National Park Service, Alaska Regional Office
2525 Gambell Street
Anchorage, AK 99503-2892

Abstract. *Whaling is a sacred affair to the Inupiat hunter. It involves an infinite preparation process, and is a duty that begins at childhood and rarely ends with death. The cycle continues as the name of the deceased is given to a newborn child, following the death of a noble hunter. The belief is that a child will be a successful hunter when he is given the name of an esteemed elder. The name transforms the newborn into part of a new family, as relatives of the deceased see the babe as the hunter returned. Massive jawbones of bowhead whales adorning the grave sites of whale hunters in coastal whaling communities serve as monuments to the whalers after their passing. The belief that spirits are embodied in sea mammals is very strong among the northern Inuit, and is reflected in the active role of the woman who is responsible for preparing hides for boat covers and clothing necessary for warmth. An appropriate ceremony must be conducted when a whale is landed by the whaling crew to avoid insult to the spirit of the whale and other sea mammals.*

It takes years for a person to assume a leadership role of a whaling crew. Some may assume this position through inheritance, following the death of a whaling captain. The use of traditional whaling equipment has been affected greatly by modern technology. In some whaling communities, the *umiak*, rawhide lines, or sealskin floats are no longer used. The skills needed in the preparation of natural resources is not being taught to the younger generations that are continuing the intricate whaling tradition.

The plunge into the 20th century has been severe, and some whaling communities have had their share of moments which severely tested their will to continue as hunters. Massive death brought about by influenza epidemics has drastically affected many communities and whaling. Today, the contest is strong in nine whaling communities throughout Alaska where again, each year, the moon tells the hunter, "It is time." The boats are launched and the quest begins anew.

The contents of this paper are not intended to conflict with the information presented by the Commissioners of the Alaska Eskimo Whaling Commission, and this paper does not detail whaling as it would be seen or described by an anthropologist. It is my perspective as a whaler, and it attempts to familiarize you with the importance of whaling as we see it.

*****

In the beginning, the great whale was observed by many hunters for many seasons. Each sighting revealed a trait of the whale previously unknown. Its actions, the season it made its presence, and the waters and ice were intensely studied by the hunter. All relevant observations, however small, were discussed and carefully studied when the hunters met at the *kashgi*.[1] Hunting of small sea mammals involved an intricate art form that balanced safety and danger equally. Because of unknown dangers to the hunter, the hunt for great whales was planned over several seasons or years before the hunters felt confident they could accept the challenge of a great hunt.

The technology for a successful whale hunt was there, but they knew that the weapons must be suitably large for the task. Rawhide lines were made that would test the massive strength of the whale, and sealskin floats were adorned with ivory attachments that signified spirits that all whales possessed. One float would not slow a great whale, and others were made to insure proper drag. Dead whales found on the beaches were studied very closely. The contour of their bodies and their vital signs were studied; nothing missed the eye of the hunter. Images of wood, bone, and ivory were made, and marks were made in the likely areas where the great whale could be mortally wounded. In their art, people showed intense respect for sea mammals that provided for their well being.

In very recent times, 'To whale or not to whale' has been an issue that has harbored intense interest among a people whose lifeways depended on the harvest from the sea. The animals of the sea and land have spirits, and it is firmly believed that the whale offers itself to the hunter. No hunter can be expected to be successful without following that belief. Following a successful whale harvest, hunger or possible starvation is not known by many in the fierce winters that follow summers of plenty. The challenges to survive are many in the coastal plain, and many are not given a second chance to test unsound judgment, where even the smallest error can be fatal.

Recently, whaling has been jeopardized by outside factions, which has instilled anger among indigenous whalers. It has been alleged that bowhead populations are insufficiently large to sustain continued hunting. Such claims that bowheads are being hunted into extinction draw attention away from commercial whaling as the real cause in the dangerous reduction of their numbers. Testimonies of hunters were deemed unscientific, and large amounts of money were spent verifying the wisdom of Inuit[2] and Siberian Yupiq[3] whalers that they have held from lifelong studies of whales. No accurate numbers shall ever be known for the bowhead population, or for any species of sea mammals, because our seas are too vast for the whales to be counted. The land is vast. The land sustains man with plants and animals, but the sea is the "lifeblood" of many people. Disputes between factions have caused bitter arguments that can be rekindled if pressures to discontinue whaling are applied. Outside pressures to end the ancient whaling lifeways have tested the will of indigenous whalers, but they have not conceded.

---

1. In olden times, a community meeting house reserved especially for men.
2. English translation is "the people" among the people of Alaska, Canada, and Greenland who speak the Inupiaq language.
3. Literally meaning "the people" or "real people" among Siberian Yupiq language speakers of St. Lawrence Island in Alaska and eastern Chukotka. Other variances are Siberian Yuit or Siberian Yupiat, depending on how the word is used.

Today, we find whalers who have been fined, incarcerated, or prevented from whaling by courts of law. The hunters must run a gauntlet of rules and regulations, and face allegations in federal or state courts for alleged wrongdoing. The testimonies of indigenous people have continued to add light to the ancient whaling story, but the picture remains clouded. Untested ideas, borne in university circles across the globe, are repeated as academic gospel. Catastrophe follows as sudden, unforeseen change invades serene habitat and environment. Indigenous people are thrust into a new age far from the days of their ancestors, when change was far between. The scientist cannot see beyond the crucial moment of change, or appreciate that a shift from an age-old pattern such as whaling will have dire consequences.

Today, academics strive to change the natural world, and ask questions even they cannot answer. Populations of fish and game resources fluctuate because of industrial or agricultural poisons entering the food chain. The Eskimos, the Athapaskans, the Aleuts, and the southeastern Alaskan Indians provide for themselves from the land and the sea, and have done so since the dawn of creation. Each and every year, since time immemorial, they have anticipated the movement of mammals, birds, and fish. To them, there is no change, and they continue to harvest the gifts from the land and the sea. In the hearts and the minds of the people, the gift of sustenance provided by the Creator was never intended to be a subject of strife.

The boundaries separating indigenous people are there, and languages form the line between them. These boundaries were established ages ago. Now, the rules and regulations of a dominant culture may cause unforeseen changes to occur. A new culture woos the old, but the marriage of hunter and the judicial robe fails.

The land is the spirit of the people, and the seas, the rivers, and the forests are a part of this spirit. At death, our spirit passes into another dimension, of which we have no control. The memory of our ancestors drives the spirit within us. This spirit is angry, as it was never intended for man to put ownership on the animals placed here by the Creator.

Time waits for no one. From a seemingly sedentary land or seascape, the hunters' domain brims with animals on migration, twice a year. The hunter must anticipate this movement, as the animals disappear swiftly into the vastness of the land and the sea, until they must return to winter climes. The vigil heightens each time the moon says it is time. And, the southward return is never always fruitful for many communities. Man cannot contest the forces of nature as well as his prey.

An interesting argument would ensue if we were to review historical documentation of various court cases involving contested rules and regulation that lack cultural sensitivity. Although cooperation with the Native American sector is actively pursued through diplomatic procedures about sensitive issues, the final application points to rather brazen judiciary tactics that are applied to indigenous people, as a minority. The procedures show racial favoritism to members of nonindigenous groups. Court decisions must be diplomatic and within the limits of the law, but wisdom fails the very people entrusted with the law, as calls of injustice rise from the judicial bench. Many whalers would be branded criminals if national or international courts of law found measures to discontinue whaling.

To be a whaler is to be a part of an elite group in Inupiaq and Siberian Yupiq societies, and whaling serves important functions that are extremely necessary in their social structure. It begins at a very tender age and seemingly does not end with the death of a hunter. When death does occur, the name is given to a newborn, with the belief that he will acquire the knowledge and fine traits of a man skilled in the arts of survival. Death begets memory, and memories of the accomplishments and the words of caution of those who pass follow the hunters even onto the ice. The wisdom of survival that the 'ancient ones' followed guide novice hunters in their path. The intricate knowledge is passed on orally, as it has been done since the first hunters went onto the ice. The methods of sharing knowledge among hunters has changed little since its onset ages ago.

Little does a young child realize, when he is given his first bow, sling, or bola, that the intricate history and knowledge he must learn is just beginning. Soon, he is released from the safe confines of the female society and is thrust into a world that tests his will, stamina, and manhood. "Leave the sparrow, but take the sandpiper," he was instructed. His learning never ceases. Soon, the seal, the walrus, and the great whale shall become his prey.

The bowhead is majestic. Its awesome strength cannot be fully described, nor the feeling of majesty that it represents. The bowhead is caring. Its mothering ties are strong, and the great whale is swift to protect her young. At close encounters, you can imagine the mother telling the hunter, "This is my child," while she is really saying to her child, "See who they are," "Hear their noise," "Learn their smell, as they must be avoided." "It is only because you are with me that our meeting is different." The bowhead has spirit.

The preparation processes before the hunt are as numerous today as they were in ancient times. Each year, new clothing and weapons were used and the walrus hide covers of their *umiaks* were replaced. Each task was done carefully; the spirit of the great whale must not be angered. The person who will strike must be pure; marriage to a woman can affect his standing as the one who is chosen to thrust the harpoon into the whale. The woman must not be in her moon during this period, or the great whale would consider her husband unclean if he were to strike. The woman is a very essential part in the organization of a whale hunt. Her fine stitches prepare the boat for safe passage through the cold waters, and the clothing she sews adds safety from the cold, but her presence in the hunt is not allowed. It is believed her presence will insult the spirit of the whale, and all precaution is taken to avoid insult.

Certain rituals are followed after the landing of a whale. A portion of *muktuk* is removed at its navel and taken to the village.[4] This is done to remove doubt from some villagers that the whale hunt has been successful. There is always someone, everywhere, who does not believe until they have seen evidence of a successful harvest. One of the whale's jawbones is removed and sunk into the waters of the sea, with the belief that the whale spirit will return to its kind and say he was well treated by the people and that they would enjoy continued success.

The division of meat and *muktuk* is elaborate. The captain, his crew, and the other boat crews that rush to assist the victorious hunters all receive a certain share. All people at the butchering site also receive shares from certain portions of the whale. Even the youngest male child of the village receives a gift of baleen from the successful hunter. Much of the meat and *muktuk* appears to be taken as shares of the boat captain, and the phenomena of sharing among indigenous groups is often misunderstood. The task of the

---

4. *Muktuk* is the outer skin of all species of whales. It provides an added staple to the diet that is very nutritious.

captain has just begun, as it is now his duty to dispense his success as a hunter to the people. The elders receive special portions known for their delicacy, and choice portions are reserved for ceremonies that follow a successful season.

Today, as in the past, the tasks of the whaler are intricate. Each member of the crew has a specific function that holds no margin for error. The dangers of whaling are forever present, and will remain so, but the rewards are great and only fear for so great a mammal can prevent the crew from successfully landing a whale. "Do not be afraid," they tell us, "If you are, you shall never know success." Such is whaling. Once you have felt the excitement of the chase and once you have felt the blood surge through your veins, you are forever branded a whaler and no one can take that birthright from you.

The stories of the whale seem never to cease. Before the arrival of the whales, hunters always meet and recount the stories of past hunts. They tell of the booming sounds that reverberate beneath the sea and the high pitched squeals that are heard by holding paddles to the ears. They tell of the mother suckling her calf, and that the water surrounding them was stained by her rich milk. They tell of the whale which sounded, exposing its massive fluke. They know the whale tells them, "Be patient, there are more behind me."

The desire to whale was instilled into us by our forefathers. It continues today, as it has for generations, and it cannot be destroyed by regulation from the outside world. It has always been regulated by the wind, the ocean currents, or ice conditions that are, and always shall be, the environment of the great whale. The spirit of our ancestors watch over us as we ply the waters in search of the whale. We see nature as she is, and we shall always be here until the Creator decides otherwise.

Ancient graves show remarkable evidence that the Inupiat and Siberian Yupiq have pursued the whales for countless generations. At death, on the mountain side, are laid with them their elaborate tools and weapons needed in so great a quest. This ancient technology has changed little over time. Only we, the modern whalers, have changed with the times. Skulls and massive jawbones of whales signify mastery of their hunting expertise. The stature that whalers have in their community, before death, decides where on the mountain they shall lay. The graves do not tell us how long they have lain there since the passing of the whaler unto eternity, but some of the old graves do tell us they have been there since time immemorial.

The hunter, the whaler, and the elder; a person may be all these, and he enjoys a specific place among his society. The whaler's commitment is to be dutiful to the people. With duty comes responsibility, and with responsibility comes survival. Survival means commitment, however grave the circumstance may be, to the United States as our nation. The contribution and sacrifice to the United States is phenomenal, even though we ourselves are a nation. Our patriotism cannot and will not be surpassed by other groups within this nation. The descendants of the ancient hunter are inscribed forever in memorials of conflict and attest to our willingness to remain who we are. Although we are first the Inupiat, the Siberian Yupiq, or the Native American, we are American. We have stood with you in times of strife, and we ask that you stand with us as we pursue our traditions and our heritage.

It is not the intent of this paper to leave out anyone who is part of the whaling family. Your cries, too, are a part of our cries, as they come from the heart. The people of the plains, the mountains, the rivers, and the rain forests; you are a part of this plea, as we are all a part of creation and the earth is our home. We must endure all hardship as our ancestors have done, in one mind, body, and spirit.

Unity. Boundaries have changed with the times. Although we have pledged allegiance to different countries who have taken us in their fold, our unmarked borders still stand. We continue to separate ourselves one from another, and our language or dialect determines who we are. Although we have learned to live with legislation, policy, and government, the land and the sea continue to call us, as it beckoned our ancestors.

Remember that as we communicate today, information is shared between us that can never end. Other chapters adding to the story always arise and they blend together; they have blended very well together for centuries.

I acknowledge the elders of my community and the elders of all communities who have answered their call to duty, as their ancestors have for countless generations. Their mission remains the same today, as it began so long ago.

The culture and heritage of our people is now on the verge of extinction all across Alaska, and it is so among many people in all continents. The rich oral history is not being recorded, and much is lost with the passing of each elder. Language is the core of a people and it, too, is in danger. Once lost, can a people continue to call themselves who they are? Many outside our culture attempt to understand indigenous peoples, especially as their ancient cultures have changed dramatically in modern times. Many wish to learn about northern cultures, but few are they that actually come to us and see the bond we have created between ourselves and our environment. Our language is not with us for naught. Is it only to be lost because a domineering culture wishes it so? Our ceremonies compare with any found throughout the world. Is it being lost because of Western, Christian principles that harbored change with whip, abuse, and evil tongues so long ago? Our rituals and beliefs are real. Are they to fall victim with the changing of times? Or, is it my generation and yours that will sit together, employing all our efforts to gain back some of the knowledge that is being lost?

We council yesterday, today, and we shall council tomorrow. Much can be gained from exchange, and this information will continue to bear the fruit of understanding as we sit together. We can prove to everyone, everywhere that two distinct cultures can work well together in harmony and in peace. It is the gathering of many cultures, such as this, that determines the survival of a culture that is given to us by the Creator. Only He, through His wisdom, must decide who must stand and who must fall. Is it because of the foolishness of our generations that the last sunset awaits many cultures? Many have fallen to the grinding path of Western civilization, and untold numbers have vanished forever. My generation and yours can help decide whether those who have survived can continue. If we work together, it will be long before our task is complete. Let us do it before we complete the walk through this path of life.

Answers create many questions. Use them to understand our culture, even though it may be considered primitive. Learn how we have endured and why we have continued to endure in so harsh an environment, since time immemorial.

Life in itself is a ritual. The chapters we follow, between life and death, are rituals. It determines our existence. The distinguishing qualities of surviving an extremely harsh environment, passed on to us by our ancestors, have insured safe passage through the four seasons. The quiet voice of their teaching is extolled, and how we contribute our learning to the novice, determine our standing, among others. We must not erase that fact from our memory and from the memories of others. Remember, the story never ends. Always, another occurrence arises that beautifies the story, as it occurred with us. Learn from us. Keep the avenue of communication open, as it has opened between us today. Perhaps, someday we may see nature in the same way. Until that day comes, we shall always be in pursuit of the whale. To us, whaling is our keepsake, given to us by our ancestors; it is our birthright and our destiny.

The training we received from our parents, our grandfathers, and grandmothers prepared us to blend into the lifeline of the people. The memory of our youth is strong and it continues to reward us. You realize, as I do, that what we have been prepared for has no equal anywhere on this earth. The tasks that we have carried out as hunters, relying upon our learning, are based on enlightenment. From bird, to seal, and from walrus to whale, the picture remains vivid. Even that it may be the antelope or the bison and the tapir or the monkey, there is always the blessing in its memory. The land, the sea, or the plains and the rain forest have always provided.

I shall always hear the voice of the elder, even though some are no longer with us. I hear their laughter in story, and I see their smile, embedded forever in my memory. Memories endear the soul. And, they are rich memories.

Acknowledgments. My journey to distant lands as Native Liaison and Heritage Specialist with the National Park Service has been extremely rewarding. I am grateful for the opportunity to walk in the pathways of many others. I have met representatives of many indigenous nations at conferences across the country. Attendance at these conferences adds light and weight to the cause of preserving our dwindling cultures. Negative thoughts that arise about the continual loss of culture and heritage to Western influence are eased when one realizes that the burden to continue and survive rest on many shoulders. Many of you have added to the story, and we must continue to add to the picture before it fades from our view forever. *Quyanna.*